A History of Light and Colour Measurement
Science in the Shadows

A History of Light and Colour Measurement

Science in the Shadows

Sean F Johnston

University of Glasgow, Crichton Campus, UK

IoP

Institute of Physics Publishing
Bristol and Philadelphia

British Library Cataloguing-in-Publication Data

A catalogue record for this book is available from the British Library.

ISBN 0 7503 0754 4

Library of Congress Cataloging-in-Publication Data are available

Commissioning Editor: James Revill
Production Editor: Simon Laurenson
Production Control: Sarah Plenty
Cover Design: Frédérique Swist
Marketing Executive: Laura Serratrice

Published by Institute of Physics Publishing, wholly owned by The Institute of Physics, London

Institute of Physics Publishing, Dirac House, Temple Back, Bristol BS1 6BE, UK

US Office: Institute of Physics Publishing, The Public Ledger Building, Suite 1035, 150 South Independence Mall West, Philadelphia, PA 19106, USA

Typeset in the UK by Text 2 Text, Torquay, Devon
Printed in the UK by MPG Books Ltd, Bodmin, Cornwall

CONTENTS

PREFACE

This book is about how light was made to count. It explores a seemingly simple question: How was the brightness of light—casually judged by everyone but seldom considered a part of science before the 20th century—transformed into a measurable and trustworthy quantity? Why did the description of colour become meaningful to artists, dyers, industrialists and a handful of scientists? Seeking answers requires the exploration of territory in the history, sociology and philosophy of science. Light was made to *count* as a quantifiable entity at the same time as it came to *count* for something in human terms. Measuring the intensity of light was fraught with difficulties closely bound up with human physiology, contentious technologies and scientific sub-cultures.

Explorations often begin with meanderings, tentative forays and more prolonged expeditions. This one ranges over a period of 250 years, and pursues social interactions at every scale. As the title hints, the subject was long on the periphery of recognized science. The illustrations in the book reinforce the reality of social marginalization, too: depictions of light-measurers are rare. Certainly their shrouded and blackened apparatus made photography awkward; but the reliance on *human* observers to make *scientific* measurements came to be an embarrassment to practitioners. The practitioners remain shadowy, too, because of the low status of their occupation, commercial reticence and—somewhat later—military secrecy.

The measurement of brightness came to be invested with several purposes. It gained sporadic attention through the 18th century. Adopted alternately by astronomers and for the utilitarian needs of the gas lighting industry from the second half of the 19th century, it was appropriated by the nascent electric lighting industry to 'prove' the superiority of their technology. By the turn of the century the illuminating engineering movement was becoming an organized, if eclectic, community promoting research into the measurement of light intensity.

The early 20th century development of the subject was moulded by organization and institutionalization. During its first two decades, new national and industrial laboratories in Britain, America and Germany were crucial in stabilizing practices and raising confidence in them. Through the inter-war period, committees and international commissions sought to standardize light and colour measurement and to promote research. Such government- and industry-supported delegations, rather than academic institutions, were primarily responsible for the construction of the subject.

Along with this social organization came a new cognitive framework: practitioners increasingly came to interpret the three topics of *photometry* (visible light measurement), *colorimetry* (the measurement of colour) and *radiometry* (the measurement of invisible radiations) as aspects of a broader study.

This recategorization brought shifts of authority: shifts of the dominant social group determining the direction of the subject's evolution, and a shift of confidence away from the central element of detection, the eye. From the 1920s, the highly refined visual methods of observation were hurriedly replaced by physical means of light measurement, a process initially a matter of scientific fashion rather than demonstrated superiority. These non-human instruments embodied the new locus of light and colour, and the data they produced stabilized the definitions further.

The rise of automated, mechanized measurement of light and colour introduced new communities to the subject. New photoelectric techniques for measuring light intensity engendered new commercial instruments, a trend that accelerated in the 1930s when photometry was taken up with mixed success for a wide range of industrial problems. Seeds sown in those years—namely commercialization and industrial application, the transition from visual to physical methods and the search for fundamental limitations in light measurement—gave the subject the form it was to retain over the next half-century.

Nevertheless, changing usage mutated the subject. Light proved to be a valuable quantity for military purposes during and after the Second World War. A wholly new body of specialists—military contractors—transformed its measurement, creating new theory, new technology, new standards and new units of measurement.

Following this variety of players through their unfamiliar environments illuminates the often hidden territories of scientific change. And two themes run throughout this account of the measurement of light and colour from its first hesitant emergence to its gradual construction as a scientific subject. The first traces changing attitudes concerning quantification. The mathematization of light was a contentious process that hinged on finding an acceptable relationship between the mutable response of the human eye and the more readily stabilized, but less encompassing, techniques of physical measurement. The diffident acceptance of new techniques by different technical communities illuminates their value systems, interactions and socio-technical evolution.

A second theme is the exploration of light measurement as a science peripheral to the concerns of many contemporary scientists and the historians who later studied them, and yet arguably typical of the scientific enterprise. The lack of attention attracted by this marginal subject belies its wide influence throughout 20th century science and technology. Light measurement straddled the developing categories of 'academic science' and mere 'invention', and was influenced by such distinct elements as utilitarian requirements, technological innovation, human perception and networks of bureaucratization. Unlike more conventionally recognized 'successful' fields, the measurement of light did not evolve into an academic discipline or technical profession, although it did attract career specialists as guardians of a developing body of knowledge. By studying the range of interactions that shaped this seemingly diffuse subject, this book seeks to suggest the commonality of its evolutionary features with other subjects underpinning modern science. This richly connected region, belatedly gaining attention from historians and sociologists of science, has too long been in the shadows.

Perhaps unsurprisingly, the initial motivation for this study came from my own background as a physicist in industry and academe, and from doctoral work in the history of science. My acknowledgements are equally diverse. Charles Amick, Dick Fagan and William Hanley of the Illuminating Engineering Society of North America, Susan Farkas of the Edison Electric Institute, David MacAdam at the Institute of Optics in Rochester, Deborah Warner of the Smithsonian Institution, and the librarians of the Universities of Leeds and Glasgow helped in locating source material. Geoffrey Cantor, my doctoral supervisor during the time much of this work was gestated in the History of Science Division of the Philosophy Department at the University of Leeds, gave continual warm encouragement and advice, and Graeme Gooday, Colin Hempstead, Jeff Hughes and colleagues at the Universities of Leeds and Glasgow provided welcome suggestions, discussions and/or interest in my subject and draft at various stages. Some of the material in this book has appeared previously in the journals *Science in Context* and *History of Science*, and benefited from the comments of anonymous referees. Portions of this work presented at meetings also elicited supportive discussion, particularly those organized by the British Society for the History of Science (Edinburgh 1996), the CNRS Maison des Sciences de l'Homme (Paris 1997), the Society for the History of Technology (London 1996 and Baltimore 1998), the University of Gothenberg (Göteborg 1998) and the Katholieke Universiteit Leuven (Leuven 2000). Comments at those conferences from Jaap van Brakel, Bruno Latour, Barbara Saunders, Terry Shinn and John Staudenmaier were particularly helpful. I am no less grateful to Charles Thomas Whitmell, whose name appeared with surprising regularity as the collector of documents that attracted my attention at Leeds[1].

I dedicate this work to my family: to my parents, who planted the seeds of my interests; to my wife Libby, who nurtured them and supplied constant support and encouragement; and to my sons Daniel and Samuel.

Sean Johnston
Dumfries, April 2001

[1] C T Whitmell, born 1849 in Leeds; MA (Cambridge 1875); schoolmaster 1876–1878; Inspector of Schools 1879–1910; author, *Colour: an Elementary Treatise* (London 1888); died 1919, Headingley, Yorkshire.

CHAPTER 1

INTRODUCTION: MAKING LIGHT COUNT

On a cool Ides of March in 1858, a handful of people across central England stood outdoors and watched the sunlight fade. One peered at a newspaper; another carefully positioned a lit candle as he squinted at the sun; a third held up a thermometer. Near Oxford an enthusiast tried to cast shadows with an oil lamp, while in Northamptonshire another uncovered his last slip of photographic paper.

The inspiration behind these activities involving flames, newsprint, rulers, exposures and watery eyes was the Astronomer Royal, George Biddell Airy. In the previous month's number of the *Monthly Notices of the Royal Astronomical Society*, Airy had set out a programme to observe the forthcoming annular solar eclipse. Among other tasks, he urged his readers 'to obtain some notion or measure of the degree of darkness'. His suggestions included determining at what distance from the eye a book or paper, printed with type of different sizes, could be read during the eclipse, and holding up a lighted candle nearly between the sun and the eye to note at how many sun-breadths' distance from the sun the flame could be seen. Later in the article, under the heading 'meteorological observations', Airy advised that 'changes in the intensity of solar radiation be observed with the actinometer or the black-bulb thermometer'[1].

The observers' submissions covered the range from qualitative to quantitative observations. One noted that the change in intensity during the eclipse was 'not greater than occasionally happens before a heavy storm'[2]. Another held a footrule to the glass of a lantern, and found that, before the eclipse, 'at 12 inches distance the sunlight was still so strong that the lantern cast no circle of light on the paper held parallel to the glass. It was, however, perceptible at a distance of 9 inches. Whilst my pencil, held before it, cast a shadow at no greater distance than an inch.' During the eclipse, on the other hand, 'the lantern cast a very perceptible light, and the shadow was made at a distance of 8 inches from the paper'[3]. This observer had responded to Airy's exhortation for intensity data, but had made no attempt to manipulate the numbers obtained. By contrast, using an extension of Airy's text-reading technique, C Pritchard obtained a numerical estimate of the reduction in intensity during the eclipse. Cutting up 'a considerable number of exactly similar pieces ... of the leading articles of the Times newspaper', he affixed them to a vertical screen. He then noted the distance at which he could distinctly read the type as the sunlight faded, recording the distance to a tenth of a foot. Assuming 'that the distinctness with which a given piece of writing may be read varies inversely as the square of the distance and directly as the illumination of the writing; then the amount of light lost at the greatest obscuration of the sun was 2/5ths that of the unobscured illumination'.

James Glaisher, one of Airy's assistants at the Greenwich Observatory, employed the actinic method[4]. This involved exposing photographic paper at regular intervals during the eclipse. He noted both the times required to produce 'a slight tinge' of the paper, and to colour the paper to 'a certain tint'. This method, producing a seemingly objective record on paper, nevertheless relied on human judgement regarding the equality of tint. The observer cautioned, though, that 'since fixing the photographic impressions, it should be borne in mind that the deeper tints have become lighter in the process, whilst the feebler portions marking the occurrences of the greatest phase remain unaltered'[5]. None of the observers had much time; the sun was behind the entire disc of the moon for scarcely 15 seconds.

Airy was a strong supporter of 'automated' and quantifiable methods in astronomy, to permit large-scale and reliable data collection. He looked to photography as one means to achieve that end[6]. Another was via quantitative instruments—devices that could yield a numerical value from an observation instead of a qualitative impression. The most observer independent of the methods he proposed for the eclipse observations was measurement with the black-bulb thermometer. The temperature indicated by a blackened bulb thermometer, particularly 'when the bulb is inclosed in an exhausted glass sphere'[7], was related to the intensity of radiant heat (infrared radiation, in modern parlance) rather than to heat conduction from the ambient air. It was thus a direct measure of solar intensity. Glaisher and others monitored temperature to 0.1 °F, but did not attempt to analyse their data to infer changes in intensity.

The records of the 1858 eclipse suggest the ambivalence of these astronomical observers towards quantitative intensity data. There was no consensus about what methods were relevant, nor on what degree of 'quantification' was useful. Nowhere in Airy's article or his respondents' accounts was a clear *purpose* for intensity measurement expressed. The data were to be acquired for descriptive use rather than to test a mathematically expressed theory. As previously mentioned, most observers failed even to reduce their data to an estimate of the change in intensity during the eclipse: Pritchard's '2/5ths' estimate was the only one from over two dozen reports. The observers did not use their results to determine the obscuration of the solar disc, for example, nor to infer the relative intensity of the solar corona to that of the body of the sun. Instead, the estimates of brightness filled out an account having more in common with natural historians' methods than those of physical scientists. Despite astronomy's long history of accurate angular, temporal and spatial measurement, there was little attempt by these mid-19th century observers to bring such standards to the measurement of light intensity. The observers supplied Airy's request by obtaining merely *a notion* instead of *a measure* of the degree of darkness.

The case of the 1858 eclipse is noteworthy because it typifies attitudes current then and still circulating in some quarters for decades afterwards. Contrasting the inchoate observations of his respondants, the episode illustrates Airy's own desire to quantify the measurement of light, to make it more in accord with what he saw as the changing status of other scientific subjects[8]. Light measurement was increasingly being portrayed as a subject out of step with modern science. In 1911, the engineer Alexander Trotter observed:

> The study of light, its nature and laws, belongs to the science of optics, but we may look to optical treatises in vain for any useful information on [the distribution and measurement of light]. Illumination, if alluded to at all, is passed over in a few lines, and it has remained for engineers to study and to work out the subject for themselves.[9]

This perceived disjunction—jarring, at least, for engineers infused with the new fashion for quantification—was not restricted to practitioners of optics. Writing as late as 1926, the Astronomer Royal for Scotland, Ralph Allen Sampson (1866–1939), complained of the provisional character still maintained by astronomical photometry:

> One is apt to forget that the estimation of stellar magnitudes is coeval with our earliest measures of position.... The six magnitudes into which we divide the naked eye stars are a legacy from... sexagesimal arithmetic. The subsequent development of the two is in curious contrast. The edifice of positional astronomy is the most extensive and the best understood in all science, while light measurement is only beginning to emerge from a collection of meaningless schedules.[10]

Indeed, the quantitative measurement of light intensity was not commonplace until the 1930s. To modern observers, usually imbued with a strong faith in the merits of numbers, it may seem anomalous that scientists and engineers came routinely to measure such an ubiquitous attribute as the brightness of light so long after quantification had become central to other fields of science[11]. Why was it seen as being so decoupled from the observational criteria of other, seemingly similar, subjects? In the study of light alone, for example, 18th century investigators took great care in measuring refractive indices. They also cultivated theories of image formation, comparing their predictions with precise observation. In observational astronomy, the refinement of angular, positional and temporal measurement underwent continual development. Practitioners of these numerate subjects strove to improve the precision of their measurements. In astronomy, clocks were improved, angle-measuring instruments made more precise, and the vagaries of human observation reduced[12]. Even practitioners of the considerably less analytical subject of physiology conformed to evolving practice, readily adopting the routine quantitative measurement of variables such as respiration and pulse rate in the mid-19th century. By contrast, light measurement was characterized by a range of approaches and precisions through the 19th century[13]. Why did those interested in characterizing light resist a quantitative approach, and what were their motivations ultimately for adopting such methods? How fundamental or 'natural' was the resulting numerical system[14]? How, too, was the course of the subject determined by its segmentation between separate communities[15]?

This book explores the ideas and practice of light measurement from the 18th to the late 20th century, and discusses the factors influencing its development. I argue that the answers to these questions relate primarily to the particular *social* development of light measurement practices and, to a more limited extent, to the little appreciated technical difficulties of photometry. Underlying the cases examined is the question: Why was the subject mathematized at all? As Simon Schaffer has observed, 'Quantification is not a self-evident nor inevitable process in a science's history, but possesses a remarkable cultural history of its own'[16]. Moreover, quantification is not value free, and 'the values which experimenters measure are the result of value-laden choices'. Thus:

> Social technologies organize workers to make meaningful measurements; material technologies render specific phenomena measurable and exclude others from consideration; literary technologies are used to win the scientific community's assent to the significance of these actions.[17]

He suggests, however, that the spread of a quantifying spirit is linked ultimately with the formation of a single discipline of measurement, that is, a universally employed technique and interpretation of the results. By contrast, I argue that quantitative measurement can spread even in such culturally and technically fragmented subjects as light measurement, and support this view with an examination of the industries and scientific institutions emerging during the late 19th and early 20th centuries that became involved with the subject. The diffused distribution of light measurement between technical subcultures is important in itself. Svante Lindqvist has called the 'historiographical threshold' the level of fame that must be exceeded to attract the interest of historians. This book supports his argument that the 'middle' levels of science are worthy of attention, and that 'the network itself may be more important than its nodes'[18].

1.1. ORGANIZATION OF CHAPTERS

The book explores different levels and nodes of the network of light measurement in separate chapters. Chapter 2 traces early interest in the measurement of light intensity. Work in the 18th century by cautiously optimistic observers such as Pierre Bouguer, Johann Lambert and Benjamin Thompson was intermingled with more dismissive publications by their contemporaries. The subject was essentially re-invented to suit each successive investigator. What motivated this work, and how was it expressed? Bouguer's interest derived from a concern about the effect of the atmosphere on stellar magnitudes; Lambert's, from a desire to extend the analytical sciences to matters concerning the brightness of light; Thompson's, from a wish to select an efficient lamp and to design improved illumination for buildings. A second factor in contemporary responses was the deceptive simplicity of intensity measurement. In making their measurements, early practitioners commonly denied physiological relationships limiting the eye's perception of brightness. Their variable results consequently attributed a poor reputation to the subject. The more careful of the early investigators refined observing techniques to minimize the effects of the changes they noted in the sensitivity of the eye.

The 19th century witnessed profound changes in the manner in which science was practised. This was true also in the particular case of the practice, and attitudes towards the value, of light measurement. A survey of papers published on the general subject of light measurement shows an acceleration in publication towards the end of the century; its rate of increase was considerably greater than for more established subjects such as gravitational research or the standardization of weights and measures. What distinguished the work of this period from earlier investigations? Chapter 3 discusses the late 19th century as a crucial period in the gradual transition from qualitative to quantitative methods in the measurement of light. Despite the enthusiasm of a few proselytizers like William Abney, who published prolifically on every aspect and application of light measurement, general interest remained restrained. Part of the reason remained the difficulties imposed by vision itself. The human eye was increasingly identified as a very poor absolute detector of light intensity. The perception of brightness was found to vary with colour, the mental and physical condition of the observer and the brightness itself. By the first decade of the 20th century practitioners had evolved a thorough mistrust of 'subjective' visual methods of observation and inclined towards 'objective' physical methods that relied upon chemical or electrical interactions of light. This simplistic identification of 'physical' as 'trustworthy, unbiased and desirable' came to be a recurring theme in the subject. The rejection of visual methods for physical detectors was nevertheless a matter of scientific fashion having insecure roots in rational argument.

A major factor in the trend towards the acceptance of quantitative methods was the demonstration of the benefits of numerical expression. Among the first practical motivations for measuring the brightness of light were the utilitarian needs of the gas lighting industry. Photometers in use by gas inspectors outstripped those available in universities in the late 19th century. The nascent electric lighting industry began to seek a standard of illumination, too, by the early 1880s. The comparison of lamp brightnesses and efficiencies was an important factor in the marketing and commercial success of numerous firms. A major incentive for standards of brightness thus came from the electric lighting industry. So intimately did electric lighting and photometry become linked that practitioners of the art were as often drawn from the ranks of electrical engineering as from optical physics.

During the same period, independent researchers increasingly proposed systems of colour specification or measurement. Most had a practical interest in doing so. The principal goal of these early investigators was the development of empirical means of using colour for systematic applications[19]. The invention and use of such systems by artists, brewers, dye manufacturers and horticulturalists is evidence both of the creation of a strong practical need for metrics of light and colour measurement, and of lack of interest in academic circles. The utilitarian incentive for light and colour specification was thus a driving force in establishing a more organized practice of light measurement near the end of the century.

The benefits of light measurement were increasingly heralded and applied to industrial and scientific problems between 1900 and 1920. Professional scientists, engineers and technicians specializing in these subjects appeared during this time. Just as importantly, the 'illuminating engineering movement' became an influential community for the subject, with dedicated societies being organized in America and Europe. Here again, social questions are of central concern: How and why did such communities foster a culture of light measurement? The transition from gentlemen amateurs to lobbyists is discussed in chapter 4.

Sensitive to the growing needs of government and industry alike, the national laboratories founded in Germany, Britain and America between 1887 and 1901 were tasked with responsibility for setting standards of light intensity and colour. Broader cultural questions begin to emerge: Why did these institutions soon come to influence all aspects of photometry? How did the centre of control shift from the domain of individuals and engineering societies to state-supported investigation? Academic research was affected through the development of measurement techniques; government policy, by the recommendation and verification of illumination standards; and industry, by defining norms of efficiency and standards for quality control. This is a case of the pursuit of utilitarian advantages leading to fundamental research: the search for a photometric standard broadened to the study of radiation from hot bodies, and thence to Planck's theory of 'blackbody' radiation. Chapter 5 centres on the important influence of the national laboratories on the subject.

From the turn of the century, photometric measurements increasingly used photographic materials in place of the human eye. With two types of detector available—the human eye and photographic materials—investigators could now quantify light in two distinct ways. On the one hand, light could be measured in a 'physical' sense—that is, as a quantity of energy similar to electrical energy or heat energy. On the other hand, light could be measured by its effect on human perception. Disputes over the characterization of this *perceptual* sense as 'psychological', 'psychophysical' or 'physical' are discussed in chapter 7. The disparity between these two viewpoints, scarcely noticed in the preceding decades, was to introduce problems for both that remained unresolved for years.

The investigation of the photoelectric effect had been a convincing demonstration of the value of quantitative measurement in academic circles. From the 1920s, the development of new photoelectric means of measuring light intensity led to commercial instruments. This trend accelerated in the next decade, when engineers and chemists applied photometric measurement with limited success to a range of industrial problems. The successive transition between visual, photographic and photoelectric techniques was fraught with technical difficulties, however. As Bruno Latour has discussed, the 'black-boxing' of new technologies can be a complex and socially determined process. A central problem concerned the basing of standards of brightness on highly variable human observers, and on the complex mechanism of visual perception. Other problems revolved around the use of photographic and photoelectric techniques near the limits of their technology, and yet important to human perception of light or colour. While some of these difficulties submitted to technological solutions, others were evaded by setting more accessible goals and by recasting the subject. Chapter 6 centres on the rapid technological changes that transformed photometry in the inter-war period.

The technical evolution was frequently subservient to, and directed by, cultural influences. The inter-war period witnessed the dominance of technical delegations in constructing the subjects of photometry and, even more self-consciously, colorimetry. There was a profound conflict between a psychological approach based on human perception, and a physical approach based on energy detectors. The subject suffered from being of interest to intellectual groups having different motivations and points of view—so much so that the only resolution was by inharmonious compromise. Chapter 7 argues that the social and political climate between the world wars significantly influenced the elaboration and stabilization of these subjects.

Seeds sown in the 1920s were to be cultivated in the following decade. A 'fever of commercialized science' (as one physicist put it) was invading not only industry, but also academic and government institutions. Links between government laboratories and commercial instrument companies strengthened. Industrialists were imbued with the values of quantification by the commercial propaganda of large companies. The drive towards industrial applications faltered before the Second World War, however, owing to mistrust after the overoptimistic application of the principles of quantification. Plant managers and industrial chemists were to complain that their new photoelectric meters could not adequately quantify the many factors affecting the brightness or colour of a process or product. The previously simplistic and positive view of quantification was supplanted by a more cautious approach. These early efforts to commercialize light measurement are explored in chapter 8.

The closer identification of science with military technology was an outcome of the Second World War. Radiometry consequently was well funded in the post-war years, and carried innovations to the now 'cognate subjects' of photometry and colorimetry. Chapter 9 discusses the effects on technical practice and social organization.

Chapter 10 explores the general historical features of the subject of light measurement. The creation of a quantitative perspective, the development of measurement techniques, the organization of laboratories and committees and the design of commercial instruments can be discussed most profitably from a perspective that emphasizes the social and intellectual interactions[20]. This approach supports the view that dichotomies such as 'technology/science', 'internal/external technical history' and 'pure/applied science' are inadequate to understand this topic. Indeed, the history of light measurement provides evidence for the statement by Bijker, Hughes and Pinch that 'many engineers, inventors, managers and intellectuals in the 20th century, especially in the early decades, created syntheses, or seamless webs'[21]. Rather than discussing compartmentalized disciplines and well articulated motivations, these authors portray science as a complex interplay of cultural and technological forces. Engineers, scientists, committees, institutions, technical problems and economic factors combined in complex ways to shape the subject of light measurement. The subject can be related in these respects to quite different scientific endeavours. A quotation from a paper on the regulation of medical drugs illustrates the commonality found also in the subject of light measurement:

> The stabilization of technological artifacts is bound up with their adoption by relevant social groups as an acceptable solution to their problems. Such groups... may be dispersed over social networks. [This] involves complex processes of social management of trust. People must agree on the translation of their troubles into more or less well delineated problems, and a proposed solution must be accepted as workable and satisfactory by its potential users and must be incorporated into actual practice in their social networks.[22]

The importance of traditions of device design, important in the present study, have recently been analysed in a different context. Peter Galison has written extensively on the history of microphysics, and has argued persuasively that instrumentation has been a central factor in the emergence of distinct scientific subcultures[23]. The growing experimental complexity of all these instruments created an almost impenetrable wall between experimental traditions. Researchers could no longer cross over from one methodology to the other, or even fully understand each other. Those scientific workers at the boundaries between sub-cultures of measurement, or between theory and experiment, military and civilian science, had to develop local languages—pidgins and creoles—to translate between them. This fertile analogy works very well for what Galison to some extent disparages but acknowledges to be a seductive and ubiquitous idea in science studies: the notion of science as 'island empires, each under the rule of its own system of validation'[24]. The present book explores the emergence, coalescence and decay of subcultures closer to the borders of recognized science.

The subject of light measurement is a particular case of a more general socially mediated process. But in addition to this, as previously mentioned, the subject has skirted the periphery of science and evades easy definition. Light measurement can be interpreted as a case of an 'orphan' or 'peripheral' science neglected both by engineers and academic scientists. Although not typical of the cases studied by historians of science, it is nevertheless representative of a wide and flourishing body of activities that attained importance in the 20th century.

My operational definition of peripheral science includes the following characteristics:

- a lack of 'ownership' of, and authority over, the subject by any one group of practitioners;
- a persistent straddling of disciplinary boundaries;
- absence of professionalization by practitioners of the subject;
- a shifting interplay between technology, applied science and fundamental research that resists reconciliation into a coherent discipline.

Peripheral sciences are not merely the applied science and technology that have dominated the 20th century, but a particular class of such subjects. Focusing on French and German developments, Terry Shinn has discussed a class of similar subjects under the name 'research technologies'. Lacking easy definition, these have hitherto been little studied by either historians of science or historians of technology. Nevertheless, many subjects in modern science and technology are demonstrably of this class and would profitably be treated in these terms. I shall return to these ideas in chapter 10 to explore the value of this designation as an explanatory idea in the history of modern science and technology.

1.2. TERMS

The terminology employed in this subject is frequently opaque. Researchers concerned with light measurement have fallen into three distinct camps, each measuring intensity for its own reasons, using methods developed at least partially in isolation from the other two distinct groups of practitioners. These three camps were (and are) *photometry*, *colorimetry* and *radiometry*. The precise definitions of these terms have varied over the decades, but can be approximated as follows: photometry deals with the measurement of the intensity of visible light; colorimetry involves the measurement or specification of colour or coloured light and radiometry refers to the measurement of non-visible radiation such as infrared and ultraviolet 'light'. The grouping together of these subjects is a modern construct, because the practitioners have generally mixed them only peripherally, and only in a concerted way since the 1930s. The interaction and eventual merging of these subjects is, however, one of the threads traced in this work. For convenience, I will generally use these terms and *light measurement* interchangeably whether the measurement of visible, coloured or invisible 'light' intensity is concerned, except where I refer to a specific topic.

A more central terminological problem relates to discussion of the amount of light itself. Since standards of light measurement were first discussed in the last decades of the 19th century, a detailed terminology has evolved to differentiate between, for example, the measurement of light emitted by a source, falling on a surface, radiated into a given solid angle or perceptible to an average human eye. The respective terms and definitions have changed as national standards and languages clashed. Some of the historical confusion surrounding the definition of these quantities is discussed in chapter 7. For the purposes of this work, though, all of these are aspects of the central problems of determining *how much* light is present at some location or *how concentrated* it is, i.e. of quantity and intensity, respectively. Early practitioners often used the term *luminosity* and the unit *candle-power* for the intrinsic brightness of a light source. Following the lead of one of the first writers on photometry, Pierre Bouguer, I employ two general ideas. First, I use the term *quantity of light* to refer to the light reaching either the human eye or the variety of physical detectors that have come into use since 1870. This idea, called by convention *flux* in modern terminology, represents the total amount of light reaching the detector by integrating over the field of view of the detector, or over the range of wavelengths to which it is sensitive, or over the area that the light illuminates in unit time[25]. Second, I use the terms *intensity* or *brightness* to refer to the concept of variations in perceived brightness. Intensity is a measure of the *concentration* or *density* of light in some sense. A lens can focus a given quantity of light to a more intense spot of smaller area, making it brighter. Intensity can thus be represented as a quantity of light per unit area, or per unit solid angle, or per wavelength range. In modern terminology these are distinguished by the names *illuminance*, *radiance* or *spectral flux*. While these distinctions are not crucial to the content of this book, the non-intuitive basis of these terms encapsulates some of the complexities faced by practitioners of the subject.

NOTES

1 'Suggestions for observation of annular eclipse of the sun, 1858, March 14–15', *Mon. Not. Roy. Astron. Soc.* **18** No 4 129; 'Observations of the annular solar eclipse', *Mon. Not. Roy. Astron. Soc.* **18** No 5 184.

2 *Ibid.*, p 188.

3 *Ibid.*, p 184.

4 Glaisher, appointed in 1833 as Airy's second assistant, was an early advocate of meteorology and an innovator in photography.

5 *Mon. Not. Roy. Astron. Soc.* **18** No 5 196–7.

6 For an account centring on transits of Venus, see Rothermel H 1993 'Images of the sun: Warren De la Rue, George Biddell Airy and celestial photography', *BJHS* **26** 137–69.

7 *Mon. Not. Roy. Astron. Soc.* **18** No 4 131.

8 Indeed, even in other aspects of optics such as the angular measurement of diffraction fringes.

9 Trotter A P 1911 *Illumination: Its Distribution and Measurement* (London) p 1.

10 Sampson R A 1926, 'The next task in astronomy', *Proc. Opt. Convention 2* 576–83; quotation p 576.

11 For 17th and 18th century roots of 'l'esprit géométrique', see Frängsmyr T, Heilbron T J L and Rider R E (eds) 1990 *The Quantifying Spirit in the Eighteenth Century* (Berkeley).

12 Differences in the 'personal equation', relating an observer's muscular reflex to aural and visual cues, were minimized by various observational techniques and instrumental refinements. See, for example, Schaffer S 1988 'Astronomers mark time: discipline and the personal equation', *Sci. Context* **2** 115–45.

13 See, for example, Olesko K M and Holmes F L 1993 'Experiment, quantification and discovery: Helmholtz's early physiological researches, 1843–50', in D Cahan (ed) 1993, *Hermann von Helmholtz and the Foundations of Nineteenth-Century Science* (Berkeley) pp 50–108.

14 Philip Mirowski, for example, has concluded that measurement standards and seemingly 'natural' schemes derived by dimensional analysis are tainted by anthropomorphism: 'measurement conventions—the assignment of fixed numbers to phenomenal attributes—themselves are radically underdetermined and require active and persistent intervention in order to stabilize and enforce standards of practice' [Mirowski P 1992 'Looking for those natural numbers: dimensionless constants and the idea of natural measurement', *Sci. Context* **5** 165–88; quotation p 166].

15 Thomas Kuhn defined a *community* as a group that shares adherence to a particular scientific 'paradigm' [Kuhn T 1970 *The Structure of Scientific Revolutions* (Chicago, 2nd edn) p 6]. I have used the term to label a loosely knit group that, while sharing common goals, methods or vocational backgrounds, is not as firmly centred on a core-set of knowledge and self-policing activities as is a *discipline*. This distinction is discussed further in chapter 10.

16 Schaffer *op. cit.* note 12, 115.

17 *Ibid.*, p 118.

18 Lindqvist S 1993 'Harry Martinson and the periphery of the atom' in S Lindqvist (ed) 1993 *Center on the Periphery: Historical Aspects of 20th-Century Physics* (Canton) pp ix–lv.

19 Ames A Jr 1921 'Systems of color standards', *JOSA* **5** 160–70.

20 For an overview of the 'first wave' of sociological studies, see Merton R K and Gaston J (eds) 1977 *The Sociology of Science in Europe* (Carbondale). For more recent introductions, see Collins H M 1982, *Sociology of Scientific Knowledge: A Source Book* (Bath) and Barnes B and Edge D 1982 *Science in Context* (Milton Keynes).

21 Bijker W E, Hughes T P and Pinch T J (eds) 1987 *The Social Construction of Technological Systems* (Cambridge, MA: MIT Press) p 9.

22 Bodewitz H J, Buurma H and de Vries G H, 'Regulatory science and the social management of trust in medicine', in *op. cit.* note 21, 217.

23 Galison P L 1997 *Image and Logic: A Material Culture of Microphysics* (Chicago).

24 *Ibid.*, p 12.

25 The term *quantity of light* is sometimes used to mean the total amount in a given time period, i.e. the *time integral* of flux. The difference between these two meanings will be clear from the context.

CHAPTER 2

LIGHT AS A LAW-ABIDING QUANTITY

The measurement of light and colour began in darkened rooms. But it also started on mountain tops and on sea voyages. And at the centre were individual observers, idiosyncratic techniques and personal beliefs.

The measurement of light intensity cannot be traced backward to a distinct lineage, or forward to a coherent discipline or purpose. It had many independent and *repeated* origins; the early development was more akin to the seasonal variations of a field of scrub grass than to the growth of a branching tree. These disparate activities (and more) nevertheless came to be described by a single term.

During this period, characterized by a lack of social cohesion and interaction between investigators, a collection of practices developed that came to value the brightness of light as a quantity. Their motivations and methods were particular, seldom involving social interactions tied to organized applications of light measurement or the sharing of research results by like-minded individuals. Indeed, an investigator during this period who became aware of another's work was as likely to discount it as to build upon it. The period lacks much coherency in theory or practice and reveals little cumulative intellectual evolution. This handful of isolated investigations of light measurement, while devoid of a unifying impetus, nevertheless evinces three general areas of interest: the study of brightness, of radiant heat and of colour description.

2.1. BEGINNINGS

Given this rejection of a clear evolutionary line, we can merely sketch the emergence of a 'subject' by discussing the incoherent variety of co-existing ideas. The range of early attitudes, methods and uses of light measurement can be illustrated with a number of loosely connected examples.

The few 17th and 18th century publications referring to the intensity of light usually took the form of untested proposals for its measurement or unsubstantiated assertions regarding its dependence on distance from the light source[1]. Thus the Capucin cleric R P François-Marie, in a book on the measurement of light intensity published in 1700, proposed the construction of a scale of intensity by passing light through cascaded pieces of glass, or reflecting light repeatedly from mirrors, to diminish the light in equal steps corresponding to an arithmetic progression. He was careful to 'convince his conscience and his superiors that it is not impious to try to measure light, the gift of God'[2]. Others, usually assuming a geometric rather than arithmetic progression of intensity diminution, attempted to study the naturally available sources of light. Christian Huyghens reported that he compared the light of the sun with that of Sirius, looking at the sun through a long tube with a hole at the top, and making the two lights equally bright[3]. The observations were criticized by his near contemporary, Pierre Bouguer, because they were not made at the same moment with the external conditions and the state of the eye itself the same.

Bouguer (1698–1758) first wrote critically about questions of illumination in an essay published in 1729[4]. In the preface, he describes that he took up the subject after reading a memoir by J J d'Ortous de Mairan[5]. Mairan had attempted to show (without success) how, with a knowledge of the amount of light from the sun reaching the earth from two altitudes, the amount from other altitudes could be calculated. In a note in 1726, Bouguer initially tried to solve this specific problem, and published his successful results using the moon as subject and a candle as a comparison. From this, he developed means of attenuating light in measurable ratios. His *Essai* discusses how the brightness of light varies with distance from the light source, and discussed the means of determining it. He assumed an inverse-square law of illumination, which appears to have been appreciated by at least some writers at least a century earlier, although enunciated in various forms[6]. Bouguer concluded that the eye was unreliable in measuring absolute brightness, and should instead be employed only to *match* two light sources[7]. To make such a comparison, he devised a 'lucimètre' consisting of two tubes to be directed at the two light sources, and converging at a paper screen viewed by the eye. To use the device, the observer pointed the two tubes towards the two sources. The light through one tube could be attenuated partially by masking its aperture with an adjustable sector to make the two patches of light appear equal. From the reduction in aperture area, the ratio of the two intensities could be judged. In an alternate version, one tube could be lengthened, so that the light reaching the screen was reduced according to the inverse-square law (figure 2.1).

This first foray into the 'gradation of light', published at the age of 31, was separated from his second work on the subject by 28 years. Bouguer spent 11 years on a voyage to Peru to measure an arc of the meridian for the Académie Royale des Sciences de Paris; he was later appointed Royal Professor of Hydrography at the Hague[8]. Besides writing up the results of the expedition, Bouguer afterwards published treatises on navigation and ships. His practical experiences had considerable relevance to his formulation of photometric questions. During his travels he climbed several mountains to measure the dependence of barometric pressure on height, noting at the same time the visual range, and became interested in further developing his early ideas on the transparency of the atmosphere:

> I did not foresee that one day I should climb the highest mountains

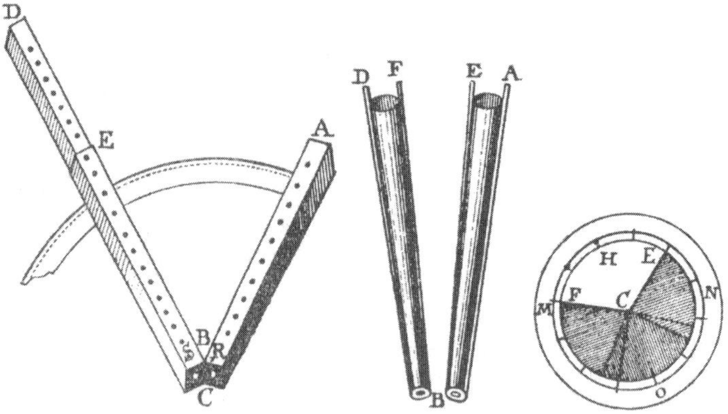

Figure 2.1. Comparing and grading lights: Pierre Bouguer's light-measuring apparatus. Left: the lucimètre. Centre: a telescopic version consisting of two equal-length tubes some 2 meters long, one having an adjustable sector aperture (right). The ends of the tubes B, covered with fine white paper, are viewed through a tube to reduce stray light. From Bouguer P 1760 *Optical Treatise on the Gradation of Light* (transl. by W E K Middleton).

> of the earth, and make a very large number of observations which would make it possible for me to make a better determination of the logarithmic curve whose ordinates express the various densities of the atmosphere.[9]

Similarly, on board ship he made observations of the visibility of the sea floor and related it to variations in the transparency of sea water, to scattering of light through the water, and to surface reflections. In the last five years of his life, Bouguer returned to the subject of photometry. The resulting book detailing his researches was published shortly after his death[10].

This second, and more extensive, work was not merely a revision of Bouguer's *Essai*. The first of its three parts dealt with 'means of finding the ratio between the intensities of two different lights'. He used his experimental techniques to evaluate, for example, how the brightness varied across the sky, and by how much 'the parts of the sun near its centre are more luminous than those which are near the edges of this body'. The second part was entirely new, and dealt with reflection from rough and polished surfaces. Bouguer examined, too, the scattering of light by the atmosphere, developing a theory of visual range to explain his South American observations. With his lucimètre he measured, and provided data for, most of the quantities he dealt with theoretically.

The 18th century polymath Johann Lambert (1728–77) made his own study of illumination in 1760 at the age of 32. In a treatise on the subject, Lambert coined the term *photometry* and discussed the need for a light-measuring device, observing that the eye is not an instrument analogous to a thermometer[11]. Lambert was familiar with at least two previous works: Bouguer's 1729 *Essai*, and the German translation of a text on optics by the Englishman Robert Smith[12].

According to Lambert, he had heard of, but not read, Bouguer's *Traité*, but refers to the *Essai* about a dozen times in his own book. The two investigators, however, employed very different approaches. Where Bouguer had favoured geometrical arguments and extensive experiments to confirm his ideas about nature, Lambert's work started from a foundation in analytical mathematics. W E K Middleton, translator of Bouguer's *Traité*, observes that, to Lambert, 'it was entirely fitting that all phenomena should at once be subjected to mathematical analysis. His instinct was to develop theory as far as possible, often on the basis of little experiment.'[13] Lambert's treatise covered an impressive array of topics, ranging through the intensity of direct, reflected and absorbed light; the photometry of the atmosphere; the illumination of planets; and an investigation of colour and shadows.

The measurement of light provoked occasional interest in the second half of the 18th century as sources of artificial lighting were improved, partly to meet the demand for street lighting and production by the new industries. Manufacture often now continued beyond the hours of daylight. Particularly in France, the study of light and lighting came to be recognized as a worthy scientific activity. Antoine-Laurent Lavoisier was awarded a gold medal by the Académie Royale des Sciences for an essay on the best method of lighting city streets[14]. Better oil burners and lamp chimneys date from this period, examples being Argand's centre-draught oil burner (1786), which replaced the solid wick, and the cylindrical lamp chimney (Quinquet 1765), both touted as major achievements[15]. There is nevertheless little evidence that the writings of Bouguer and Lambert were applied during this time. Indeed, in a subject that each investigator seemed eager to reinvent, Bouguer's contributions were slighted not only in the 18th, but also in the 19th and 20th centuries. One commentator wrote, 'there is very little evidence of any mathematical treatment of problems, or satisfactory definitions of the conceptions in Bouguer's work', but 'Lambert developed a system of conceptions... the principle of which is still in use unchanged today'[16]. Bouguer's approach, however, had much in common with opinions of the late 19th century, e.g. in arguing the limitations of the eye as a detector of 'absolute' intensity, and in limiting his experiments and discussions to those relating to a *ratio* of intensities.

A third extensive investigator of light intensity during the 18th century—but employing distinct methods and for different reasons—was the American Benjamin Thompson or Count Rumford (1753–1814)[17]. In 1794, Thompson devised a visual photometer for measuring light intensity, with which he measured the transmission of glass, the reflectance of mirrors and the relative efficiency of candles, lamps and oil burners. Thompson's work is notable for its breadth, attention to experimental detail and pervasively quantitative nature.

Where Bouguer had aimed at scientific answers to natural phenomena and Lambert sought mathematical justification, Thompson's work was grounded in meticulous experiment. His photometer consisted of a sheet of white paper and a cylinder of wood fixed vertically a few inches from it (figure 2.2). The two light sources to be compared were placed on moveable stands some 6 to 8 feet from the paper and from each other. The observer compared the shadows of the cylinder

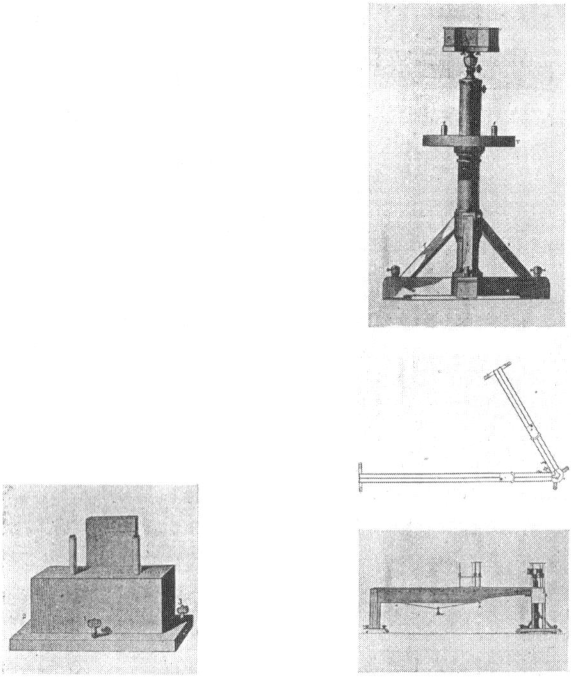

Figure 2.2. Bringing precision to measurement. Rumford's photometers. Left: portable photometer. Right, top to bottom: Rumford's laboratory photometer, in perspective, plan and elevation views. From Buckley H 1944 *Trans. Illum. Eng. Soc.* **9** 73–88.

cast by the two lights, and moved one or the other light further away until the densities of the shadows appeared to be exactly equal. Thompson concluded that the 'real intensities of the lights in question at their sources' were then 'to each other as the squares of the distances of the lights from the centre of the paper'.

Thompson used his devices in a series of carefully organized experiments covering a broad programme of research. Much concerned with *efficiency*, he measured the illumination produced by various lamp fuels. He calculated their relative expense, observing the light emitted by an Argand lamp and by a wick lamp of common construction and finding that the Argand lamp used 15% less oil for the same illumination. Thompson's general concern for practice and efficiency is also indicated by his development of the Rumford stove and work on the nature of heat. In studying the fluctuations of the light emitted by candles, he discovered a variation 'from 100 to 60' for a good quality candle, and as much as 100:16 for 'an ordinary tallow candle, of rather an inferior quality'. His observations guided the further development of his experimental method. He cautioned that 'in all cases it is absolutely necessary to take the greatest care that the lights compared be properly trimmed, and that they burn clear, and equally, otherwise the results of the experiments will be extremely irregular and inconclusive'.

Thompson's experiments investigated not only the brightness of light sources, but also the effect of common materials. He measured the loss of light through plates of different kinds of glass, providing a suggestion for commercial use:

> With a very thin clean pane of clear, white, or colourless window-glass, not ground, the loss of light, in 4 experiments, was .1321; .1218; .1213; and .1297; the mean .1263. When the experiment was made with this same pane of glass, a very little dirty, the loss of light was more than doubled.—Might not this apparatus be very usefully employed by the optician, to determine the degree of transparency of the glass he employs, and direct his choice in the provision of that important article in his trade?[18]

Mirrors, too, came under his scrutiny. Thompson noted that 'the mean of 5 experiments, made with an excellent mirror, gave for the loss of light .394; and hence it appears, that more than 1/3 part of the light, which falls on the best glass mirror that can be constructed, is lost in reflection'. Besides measuring the reflectance of various mirrors, he studied the effect of angle ('the difference of the angles of incidence at the surface of the mirror, within the limits employed, namely 45° to 85°, did not appear to affect, in any sensible degree, the results of the experiments').

Other experiments dealt with more fundamental questions. The first described in Thompson's paper concerned 'the resistance of the air to light'. He measured this 'transparency of air' by verifying the inverse-square law over the 20-foot length of the photometer room. Thompson investigated the transparency of flame by comparing candles alternately in a line parallel and perpendicular to the screen (finding little difference, he concluded that flame was transparent). Six years later Thompson used what he had learned in planning the lighting of the Royal Institution.

Thompson makes no mention of previous work, although his apparatus was similar to that described by Lambert some 34 years earlier. Nor does he make any reference, apart from the inverse-square law, to theoretical relationships; his photometry was strictly empirical and directed towards answering immediate questions of illumination.

Thompson's unique and potentially fruitful approach, like those of Bouguer and Lambert, excited little interest. There appears to be no citation by his contemporaries either of his methods or results. Indeed, commenting on their work and the state of photometry as late as 1868, a French observer lamented:

> Nothing is more delicate, more difficult than the measurement of luminous intensities. In spite of all the progress achieved in the science of optics, we do not yet possess instruments which give this measurement with a precision comparable to those of other physical elements... we are struck that modern physicists have not thought at all about the subject.[19]

These 18th century examples of photometric study, although sparse, reveal qualities of the subject that characterized it into the 20th century:

- First, differing *perceptions* of its feasibility and value are evident. On the one hand, characterized by Huyghens, Mairan and François-Marie, the measurement of light intensity was interpreted as a straightforward task susceptible to trivially simple methods and analysis. The eye was considered to be an unproblematic and reliable detector of brightness—indeed, 'brightness' had no meaning independent from 'seeing'. On the other, epitomized by Bouguer, Lambert and Thompson, photometry was portrayed as a potentially misleading subject requiring careful experiment and analysis (there was, of course, a third, implicitly held, majority view: that photometry did not constitute a 'subject' worthy of 'study' at all). These contradictory perceptions, by practitioners seeking a quick answer to solve a larger problem, on the one hand, and investigators concerned with the foundations of the subject on the other, introduced confusion, dissatisfaction and lack of consensus.
- Second, the techniques of measurement were diverse, relying as they did upon glass-stacking, extendable tubes or shadow-casting.
- Third, the style of engagement was highly variable. From the analytical approach of Lambert to the utilitarian fact-finding of Thompson, the motivations and methods of photometry were redefined by each investigator.

2.2. A LAWLESS FRONTIER

A view of light as an entity that could or should be quantified was slow to become established. As discussed earlier, quantitative intensity relationships were proposed sporadically during the 18th century and earlier. Bouguer, Lambert and (later, in 1852) August Beer described eponymous intensity relationships. These state that the logarithm of the quantity of light received is inversely proportional to the thickness ('Bouguer's law') and to chemical concentration ('Beer's law') of an absorbing material, and the quantity of light to the cosine of the angle of incidence on the illuminated surface ('Lambert's law'). Several of their predecessors had proposed their own laws but with various unverified formulas.

The rather casual exposition of empirical intensity relationships without experimental confirmation was not an unusual mode of scientific discourse during the early 19th century. For example, in an 1809 paper Étienne Malus, discoverer of polarization by reflection, inferred a law of intensity as a function of polarizer angle by a dubious method[20]. Malus' law relates the amount of light transmitted and reflected by two polarizers in series to the angle between polarization axes. Knowing no means of accurately determining intensity, he never experimentally confirmed the relationship. Henry Fox Talbot later devised such a means and, in the process, raised some of the issues that were to become central to light measurement. Prompted by an 'article in a foreign journal', and seeking a method 'to determine experimentally the intensity of a polarized ray' he published in 1834 the investigations of photometry he had made nine years earlier:

> Photometry, or the measurement of the intensity of light, has been supposed to be liable to peculiar uncertainty. At least no instrument that has been proposed has met with general approval and adoption. I am persuaded, nevertheless, that light is capable of accurate measurement, and in various ways; and that the difficulties which stand in the way of obtaining a convenient and accurate instrument for photometrical purposes will ultimately be overcome.[21]

Talbot's claim that 'light is capable of accurate measurement' was to be repeatedly challenged until the end of the century. As he noted, there was no general agreement on the adequacy of photometry for any purpose. Talbot's method, related to persistence of vision, sought to redress the difficulties. Recalling that a glowing coal whirling around appears as a continuous circular ring (an observation made by Isaac Newton, if not earlier), he reasoned 'that *time* may be employed to measure the intensity of *light*' (emphasis in original). To do so, a light source would repeatedly be eclipsed by a rapidly rotating wheel having one or more sectors cut away. An observer viewing the light would see an interrupted beam, but flickering too quickly to perceive. Talbot postulated that the apparent brightness should be proportional to the fraction of the cut-out diameter of the wheel. Thus, to avoid one of the problems he saw with photometry—that of obtaining a quantifiable reference intensity—Talbot appropriated a new physical effect. He saw this principle as being generally applicable not only to photometry, but indeed to many other forms of sensation:

> it offers a method (and perhaps the only possible one) of subjecting to numerical comparison some qualities of bodies which have never, I believe, been even attempted to be measured, such as the intensity of odours, &c; for this principle seems to have a general application. We may always find means of dividing the experiment into minute intervals of time, and we may cause that quality of the body which we wish to estimate the intensity of to act upon our senses or upon our instruments, only during a certain number of those intervals, but regularly and rapidly recurring in a stated order.[22]

Talbot thus broached another theme that was to dog the subject: that of relating human perception to physical effect. His 'simple and natural' law was generally accepted by his successors and used as a reliable means of altering the intensity of light for photometric researches[23]. Talbot also extended his technique to colour research by painting his rotating wheels with various proportions and tints. His methods failed to alter contemporary attitudes concerning the usefulness or applicability of photometry itself, though. Talbot's colour research with rotating discs attracted little interest for a half century[24].

Talbot and a handful of predecessors concluded, then, that the brightness of light *could* be quantified to provide answers to both scientific and practical questions. The subject nevertheless failed to gain the direct attention of their scientific and engineering contemporaries. Yet these technical sub-cultures had good reasons for their attitudes. The clearest examples of subjects that might be expected, from a naïve modern perspective, to have embraced photometry, but did not, are photography and astronomy.

2.2.1. Photography: juggling variables

Developed from the 1830s, photography is seemingly tied closely to issues of light intensity. Ostensibly obvious questions—all quantitative—could be posed: How much light is needed to darken a photographic plate? How much are plates of different compositions darkened by the same amount of light? How much do different colours of light affect the results? How much does an optical filter reduce the intensity of transmission? But questions such as these reveal the gulf between the contexts of the mid-19th and 20th centuries. Such questions were quite irrelevant to the concerns of the first practitioners; they were not, in fact, posed. Talbot himself, a seminal British innovator in photography and a photometric investigator, never combined the two studies.

Early photographers were concerned with the *effect* of light on the photographic plate rather than with the intensity itself. The two were not synonymous. A correctly exposed plate was the goal of the photographic method, and light intensity was merely one of the factors that could affect the result. Instead of a fundamental interest in light, the photographer had an interest merely in its control as an exposing agent. The control of light was straightforward, in principle, for most photographic work: the intensity could be varied over wide limits simply by altering the aperture of the camera lens. But early cameras had little need for adjustable apertures: there was always too little light available. Light intensity was largely an uncontrollable factor in photography, as artificial lighting was generally too weak for exposure. Photographic processes of the period were sensitive mainly to ultraviolet and blue light, which was weakly emitted by flame and later incandescent lamp sources—and strongly absorbed and scattered by smoke-filled Victorian skies. Intensity control was confined largely to designing photographic studios with skylights, large windows and adjustable mirrors to make best use of natural light.

Another factor of more practical concern than light intensity was the sensitivity to light of various photographic processes. Great gains in sensitivity could be obtained by devoting attention to photochemistry. The first decades of photographic technology were thus dominated by the investigation of new light-sensitive materials, methods of development and 'fixing' processes[25].

Of greater importance to the photographer was exposure *time*, which was precisely controllable simply by shielding the plate from the scene to be photographed. Within very broad limits, photographers discovered, exposure time and light intensity could be traded off[26]. Moreover, neither was critical in its effect on photographic density: a factor of two either way (typically amounting to a latitude of a minute or so) did not seriously influence picture quality. Thus exposure time, readily controllable to a few seconds for an exposure lasting several minutes, could be regulated easily to the necessary precision.

Even when a gross error in exposure did occur, the later methods of plate development could compensate. Common practice with the relatively 'slow' materials of the period was to hold the plate up to a dim lamp periodically during development and wash it free of chemicals when it was judged to be sufficiently dark. Writing in 1883, C Ray Woods noted:

> in studio work...there is a certain amount of uniformity; but in landscape photography the question becomes more complex. Quantity and quality of light, nature of subject and colour, atmospheric effects &c.—all these and more have to be considered. Arm yourselves with a photometer if you will, it is simply a matter of impossibility to correctly time the exposure, to give it, say, the theoretically exact quantity of light to produce the desired effect with a certain strength of developer.[27]

Wood's rough solution was to abandon any attempt to measure a 'theoretically exact quantity of light' and instead to expose the plate by about 'half as much again as the estimated exposure time' and then to develop very slowly in a bromide developer while observing the plate's density. One of his contemporaries noted that exposure was seldom a problem because both under- and over-exposed plates could be developed correctly by using 'strengthening' and 'restraining' developers, respectively[28].

So the use of an instrument to measure light intensity seemed pointless to the practical and adept Victorian photographer, because there were simply too many extraneous factors influencing the exposure that could *not* be quantified. Light intensity was by no means the crucial factor in obtaining a good photograph.

The occasional forays into light measurement by photographers were seldom appreciated by their contemporaries. As an evaluator of the 'Simonoff photometer' noted, 'the actinic or photographic energy is by no means always proportionate to its intensity', citing as an example the 'trebled' exposure required on days when the sky had a faint yellow caste. The second drawback, he noted, was that 'the eye of the observer may not always be in the same condition of sensitiveness to light; the iris being more or less expanded according to the brilliancy of the general illumination'[29].

For early photographers, then, photometry was a solution in search of a problem. Photography until the late 19th century relied upon exposure time and processing conditions more than on control of light intensity to influence results. The problem of quantitative measurement of light was successfully avoided or recast in terms of other variables.

2.2.2. Astronomy: isolated forays

Nineteenth century astronomers weighed up the measurement of light as diffidently as did photographers. While there were potentially a number of applications—determining stellar magnitudes, the brightness of variable stars, and eclipse phenomena, for example—none of these practices was central to the main concerns of astronomy at that time and only isolated cases of interest can be found.

William Herschel, who brought a quantitative point of view to astronomy as he was later to bring to the study of radiant heat, was one such case[30]. His interest was provoked by reading a paper by John Michell in 1767 proposing to measure the distance of stars by their brightness[31]. Michell knew of Bouguer's earlier work in light measurement, and had devised a crude photometric method: enquiring how far away the sun would have to be to appear as bright as a typical star, he used Saturn as a reference. Saturn's brightness depended on the sun, and in opposition (i.e. illuminated face-on as seen from the Earth) was as bright as a first-magnitude star. Its intermediate brightness, directly linked to the dazzling light of the sun, made it a convenient photometric 'stepping stone' to relate solar and stellar brightness. By estimating a factor for the amount of sunlight Saturn received, he made a reasonable estimate of the distance of Sirius[32]. Theoretical calculations of planetary brightnesses had been published by Lambert, based on their distances, size and probable composition. Herschel carried this idea further over a period of years, by 1813 publishing a list of a series of reference stars for a range of magnitudes. To do so, he observed pairs of stars through his telescope and reduced the intensity of the brighter one; from estimates of the amount of reduction needed to equalize the intensities, he inferred their relative brightness. Herschel related his scale of apparent intensity to one of actual distance. His procedure was poorly received, however. The simplistic relation between brightness and distance was attacked by several contemporaries, undoubtedly colouring their perceptions about the usefulness of photometric methods in astronomy.

2.3. TECHNIQUES OF VISUAL PHOTOMETRY

The cases cited earlier, and the accounts of the 1858 eclipse described in chapter 1, illustrate the range of methods used to gauge or report light intensity through the 19th century. These techniques were frequently re-invented or recast into seemingly new forms. From a modern perspective the methods used fall into three categories of observation.

2.3.1. Qualitative methods

Intensity was related to a familiar value such as the brightness prevailing during various weather conditions. The report served simply to draw a familiar impression or to paint a 'mind picture'.

2.3.2. Comparative methods

Bouguer had observed that the human eye adapts to a large range of ambient lighting and so is intrinsically unsuitable for determining intensity. It can, however, be sensitive to temporal or spatial differences in intensity. Bouguer had recommended that brightnesses be evaluated by direct comparison of an unknown intensity with some known reference. The methods can be classified as either

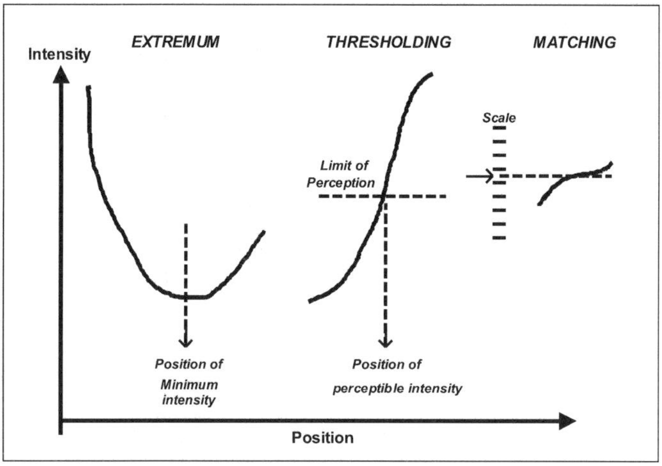

Figure 2.3. Methods of visual photometry.

extremum detection, *thresholding* or *matching*. Each of these related methods needs a reference or standard of comparison (figure 2.3).

- In an *extremum* technique, the observer notes the point of maximum or minimum intensity by comparing the light with itself at a prior time or different position. This technique located the extrema of intensity. Augustin Fresnel, author of the first quantitative theory of diffraction which predicted particular angular positions for intensity minima, verified his predictions in the 1820s by an extremum technique. He reasoned that while the eye can determine the brightest point of a pattern with relative accuracy, it can judge the dimmest even more surely (the eye, once dark adapted with the iris fully dilated, cannot 'accommodate' any further to weak lighting).
- In a *thresholding* or *extinction* technique, the observer compares the intensity to a minimum detectable level. The intensity is reduced by some means until it is below the threshold of visual detection. The amount of reduction required is then a measure of the relative brightness. Airy's 'candle versus sun' technique for determining the intensity of the eclipsed sun adjusted the apparent intensity of the candle flame (the reference) by changing its distance relative to the disc of the sun until the flame disappeared. The text-reading method employed by Pritchard for the eclipse also had used thresholding as the comparison: he noted the distance at which text could be read to a certain standard of clarity. The reference in his case was therefore a definition of visual distinctness[33]. His method appears to have been shunned by serious investigators, however. Some of them argued that visual thresholding is limited by eye accommodation, and depends on background lighting, the rate of change of intensity, and the characteristics of the observer. One attempt to obviate the effect of eye

accommodation was to employ an aperture smaller than the smallest pupil diameter[34].

- *Matching* or *nulling* compares the intensity directly with a standard. The observer either adjusts the standard intensity until its difference from the unknown is 'nulled' or cancelled, or else uses several fixed standards for comparison. Bouguer, Lambert and Thompson all matched their subject to another known source such as a star, planet or standard candle.

2.3.3. Physical methods

Unlike visual methods, physical techniques relate intensity to some other physical effect. The actinic method used by Airy's assistant, James Glaisher, relied on a photochemical effect: light intensity was determined by the amount of darkening it produced on a photosensitive material. Similarly, the blackened-bulb thermometer indicated the intensity of irradiation by the length of its mercury column.

These techniques were adequate to give a good estimate of the brightness of light sources or surfaces. Indeed, the capabilities of visual photometry exceeded what was demanded of it. There was little evolution of technique through the period; instead, old ideas were recycled in new combinations and for new purposes.

Observers thus had an assortment of methods at their disposal, ranging from the descriptive to the numerical. Until a consensus regarding the *value* of such observations was established, however, the methods remained diverse and unfocused. Scientific culture as much as material technology controlled the subject. The dual importance of these influences is revealed by two concurrent subjects related to intensity measurement which contrast sharply with the case of photometry. Researchers of *radiant heat* (a subject later to be linked strongly to the theoretical framework of energy physics) had long been performing careful quantitative experiments, while a number of pragmatic investigators were attempting to describe and measure *colour* by quite different techniques.

2.4. STUDIES OF RADIANT HEAT

The heat produced by the sun, fires and lamps has a distinct phenomenology to that of the light generated by those sources. Unsurprisingly, the investigation of the intensity of radiant heat had an early history distinct from that of the brightness of light, and an equally distinct historiography[35]. Seventeenth-century investigators had observed the reflection and transmission of 'heat rays' using their skin or thermometers as sensors, frequently making quantitative estimates. The French investigator Mariotte, for example, in 1682 noted that covering a concave mirror with a glass pane reduced the heating effect on a thermometer at the mirror focus by about one-fifth. A flurry of activity in the late 18th century, using better thermometers, culminated in a series of experiments made by William Herschel in 1800. Herschel, too, used thermometers as quantitative instruments, mapping the relative heat intensity provided by different colours. By equating the heat intensity to the change in scale reading of the thermometer upon illumination, Herschel was able to report, for example, that a sample of red glass stopped 692/1000 of the heat rays in the red part of the spectrum[36]. Others quickly extended his work, seeking to verify or disprove his claim that most heating occurred beyond the red end of the spectrum. In the process of investigating a plethora of discordant results, researchers studied the emissivity, absorptivity and transfer of heat between bodies[37].

Unsurprisingly for the study of invisible radiations, research was centred on non-physiological detectors. While Herschel's 'radiant heat' was detectable by the skin, the radiation detector he used from 1800 was a sensitive thermometer[38]. And from the beginning there was no question but that it was quantifiable: his first experiments recorded not the *presence* of this radiation, but the temperature change it produced in his thermometers.

In the following decades, Herschel's sensitive thermometers were joined by detectors exploiting electrical phenomena dependent on heat. Seebeck reported a new 'thermoelectric effect' in 1821 and then demonstrated the first *thermocouple*, consisting of junctions of two metals which produced a potential difference (voltage) when at different temperatures. In 1829 Nobili constructed the first *thermopile* by connecting thermocouples in series[39]. Macedonio Melloni, a Professor and Director of the Institute of Physics at the University of Parma, helped to modify the design in 1833 to adapt it for radiant heat measurements rather than for temperature differences produced by contact and conduction[40]. In 1880, Samuel P Langley announced the *bolometer*, a temperature-sensitive electrical resistance designed to detect weak sources of radiant heat[41]. And in 1883 Willoughby Smith discovered the photoconductive effect, the equivalent phenomenon using visible light. Despite some cross-fertilization of photometry and radiometry during this period[42], physical detectors of visible light were largely rejected for reasons discussed below.

Radiant heat remained a study distinct from photometry through the 1830s and 1840s, even though it was by then increasingly interpreted as a form of light[43]. By the 1850s, radiometry was linked to questions of heat transfer and energy, both 'hot' topics at the time[44]. Light and radiant heat remained separately categorized in the scientific mind. The effects of 'actinic', 'luminous' and 'thermal' radiation were seen as distinct[45]. As the three types of radiation acted preferentially on different types of detector (photographic materials, the eye and temperature-sensitive instruments, respectively), it was natural to employ the most sensitive for each, and to construct the subjects along observational lines (figure 2.4).

By the late 19th century, two principal varieties of invisible radiation were broadly accepted by men of science. Their characteristics, however, were distinguished initially by how they related to visible light. One variety lay beyond the deepest violet portion of the spectrum and was denoted, from the early 19th century, *ultra violet*; the other lay beyond the red, and was called *infra red* (being written 'infra-red' by 1880 in Britain and, by 1920, 'infrared' in America). Experiments demonstrating the interference of light, particularly from the late 19th century, convinced many investigators that infrared, and possibly ultraviolet,

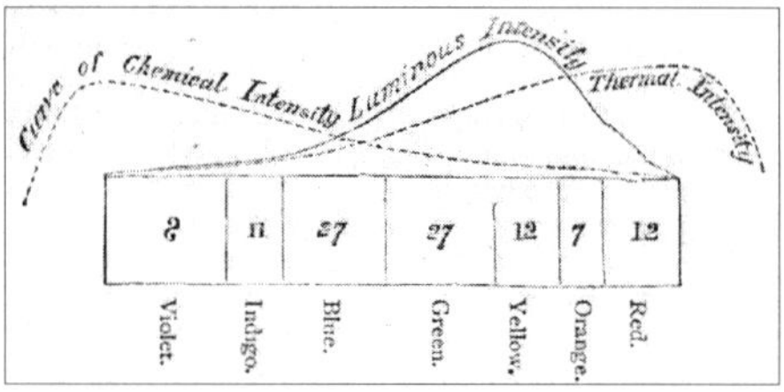

Figure 2.4. Categorizing light: radiation as tripartite. Buckmaster J C 1875 *The Elements of Acoustics, Light and Heat* (London) p 83. The numbers indicate the proportionate parts of the colours in the solar spectrum.

rays were 'waves' having wavelengths longer and shorter, respectively, than those of visible light. James Clerk Maxwell's theory of electromagnetism of 1862 led others to predict the existence of electromagnetic waves. Heinrich Hertz, in 1887, reported the discovery of such emissions from electric sparks[46]. Yet the acceptance of a spectrum of radiation that incorporated visible light, invisible light and radio waves took hold only in the early 20th century, and had little currency as a unifying principle for light measurement in Victorian times[47]. Far more sensible was a division of subjects along observational lines: into what could be *seen* and what could be *detected*.

2.5. DESCRIBING COLOUR

Just as the study of radiant heat was constituted as a distinct subject, colour description was conceived as independent of photometry by most 19th century investigators. A brief sketch of the period's categorization of the subject of colour measurement will illustrate its separate and considerably later origins from the measurement of light intensity and radiant heat. During its rise in the 19th century, the subject was dominated by utilitarian need and pragmatic solutions. It was, moreover, of interest to distinctly separate communities comprising a schismatic collection of parties speaking mutually incomprehensible languages. Artists, industrialists and scientists had distinct ideas of colour measurement.

The 19th century preoccupation with colour measurement began with empirical means of using colour for systematic applications[48]. Mid-century efforts to characterize colour were frequently focused on the qualitative. Artists, having more practical experience with the subject than most men of science, were the instigators of several systems. David Ramsay Hay (1798–1866), for example, wrote on 'the numerical powers and proportions of colours and hues' in 1846. His rather arbitrary numerical descriptions intermingled with the flowery language of the artist: 'Blue... belongs more to the principle of darkness or shade... and is consequently the most retiring of the three. It is also of these elements the most cool and pleasing to the eye, associating, as it does, with the groundwork of the retina itself'[49]. Hay's method of quantifying colour was to assign rather arbitrarily proportions of 'light and darkness' with little reference to either experiment or theory. In this scheme, 'the phenomenon of colour seems to arise by a different mode of action', with yellow, for example, being embodied in 45 parts light and 15 parts darkness. Attempts to develop a 'notation' for colour generally centred upon expressing it as a combination of quantifiable characteristics. Besides the 'brightness' that was central to photometry, such attempts factored colour into the separable characteristics of 'hue' (or tint) and 'saturation' (or colour purity)[50]. By treating these properties as coordinates, colours could be 'mapped' onto three-dimensional spaces.

The Boston artist Albert Munsell, in his turn, devised a colour 'tree' to express all possible colours, intending it as a tool for industry and teaching[51]. The director of a French dye works developed another of the first such systems to characterize his colours. His motive for developing a system of colour specification had initially been to investigate complaints from a customer about the fading of the colours of dyed fabrics[52]. Such systems proliferated by the turn of the century and fulfilled a practical need. For example, Robert Ridgway, Curator of Birds at the US National Museum, published his own *Nomenclature of Colors for Naturalists* in 1886. La Societé Française des Chrysanthémistes published its *Repertoire des couleurs* in 1905 to describe flowers, but the catalogue found widespread use in other domains. Numerical languages for colour met the requirements of commercial specification. Such systems were characterized by a certain rigidity of definition coupled with empirical details. The number of hues might be 10 (Munsell) or 36 (Ridgway) values; the number of grey levels, 6, 9 or 15; the number of colours defined, typically several hundred to a few thousand.

Besides matching fabrics, paints and flower colour, early efforts to characterize colour emphasized quantitative uses. Chemists began using the term *colorimetry* in the 1860s to refer to the determination of the quantity or concentration of a substance by the colour it imparted to a solution[53]. Although more complex than in the case of photometry, *matching* proved the most successful strategy, and various methods of colour matching were developed. One of the most successful of these was the 'Tintometer' invented by Joseph Lovibond (1833–1918), a former English brewer[54]. Based on the comparison of the coloured sample to a graded set of glass filters, the Tintometer found use in industries as diverse as steel production, water quality measurement and the valuing of flour. Such early applications had a strongly empirical basis. Although Lovibond spent several years investigating schemes of colour matching, he had no time for theorizing. He confined himself to empirical experiment, which 'enabled the author to devote much of his time and energy to actual work, which would otherwise have been employed in profitless controversy'[55].

Despite the efforts to render colour into numerical form, 19th century colorimetry made little attempt to *measure*; instead, it compared samples to arbitrarily defined colour standards. Such an activity was in no way quantitative. As a philosopher–photometrist was to argue early in the next century, 'the assignment of numerals to represent telephones or the articles of a salesman's catalogue is not measurement; nor—and here is a more definite representation of properties—the assignment of numerals to colours in a dyer's list'[56].

Through the first half of the 19th century, then, a few isolated approaches tried to make sense of the brightness and colour of light and the nature of radiant heat. These three subjects, evaluated with distinctly different motives and techniques, were constructed along individualistic lines by a small number of investigators improbably convinced of the value and feasibility of intensity measurement. Only studies of radiant heat—a subject perceived as being more akin to thermal physics than to optics—adopted early the quantitative approach that was a more thoroughly integrated part of its sub-culture. Colour seemed more amenable to a cataloguing or taxonomic strategy, a pragmatic solution to problems for which utilitarian considerations were paramount. Physical scientists for the most part ignored the measurement of visible intensity, or deferred it until other, more fruitful avenues for research had been explored. Neither early photographers nor astronomers—later to become proponents of a quantitative approach—made photometry an important component of their technical repertoire. Each had ample new phenomena to explore qualitatively before the more mundane work of quantitative measurement was felt necessary to yield new results.

Light measurement was thus weakly impelled from two directions, simultaneously encouraging and discouraging its investigation. A handful of investigators developed *reasons* to measure light, and means to do so. But several factors limited their interest. The uncertain nature of the visual process, inherent complexities in visual photometry, dearth of theories to drive experimental verifications, and abundant problems amenable to non-quantitative methods, all kept photometry in the background until the second half of the 19th century. Indeed, Airy's 1858 eclipse—occurring mid-day, in mid-month, mid-century and in the middle of England—was not merely a transitory spectacle; it marked a threshold for the emerging self-realization of the subject.

NOTES

1 Walsh J W T 1958 'Was Pierre Bouguer the "father of photometry"?' *Am. J. Phys.* **26** 405–6.

2 François-Marie R P 1700 *Nouvelle Découverte sur la Lumière pour la Mésurer et en Compter les Degrés* (Paris), discussed in W E K Middleton's translation of Bouguer P 1729 *Traité d'Optique sur la Gradation de la Lumière* (Paris; translation Toronto, 1961).

3 Huyghens C 1698 *Cosmotheoros Sive de Terris Coelestibus Earumque Ornatu Conjecturae* (The Hague).

4 Middleton *op. cit.* note 2. See also Perrin F H 1948 'Whose absorption law' *JOSA* **38** 72–4.

5 d'Ortous de Mairan J J 1721 *Mém. Acad. R. Sci. Paris* 8–17.

6 See Ariotti P E and Marcolongo F J 1976 'The law of illumination before Bouguer (1729): statement, restatement and demonstration' *Ann. Sci.* **33** 331–40.

7 Middleton *op. cit.* note 2. Criticizing the observations of Huyghens (p 46), Bouguer wrote: 'apart from the fact that this clever mathematician may not have made all the necessary distinctions between the total quantity of light and its intensity, it is only too certain that we can only judge directly the strength of two sensations when they

affect us at the same instant. How can we assure ourselves otherwise that an organ as delicate as the eye is always precisely in the same state, that it is not more sensitive to a slight impression at one time than at another? And how can one remember the intensity of the first sensation when one is actually affected by the second and when an interval of several hours or even days has gone by between the two? To succeed in this determination he would have had to have recourse to an auxiliary light which he could make use of in the two observations, and which would serve as a common term of the comparison.' Deriding the methods of François-Marie (*op. cit.* note 2 p 47): 'His results must depend more or less on the transparency of his pieces of glass, and not only this, but on the differing state of his eyes, which would be more or less sensitive at one time than another. When his sight was a little fatigued all lights would ordinarily appear to him stronger. He would then need a greater number of pieces of glass to weaken them to the same extent. Each observer would in this way attribute a different degree of the scale to the light which he was measuring. People would not be able to agree when observing at different times or in different countries, and the measurements would never give exact ratios.'

8 Bouguer P 1749 *La Figure de la Terre... Avec une Relation Abrégée de ce Voyage* (Paris).

9 *Ibid.*, p 209.

10 *Ibid.* Bouguer's biographical details are from the translator's introduction and from *DSB* vol 2, 343–4.

11 Lambert J H 1760 *Photometria Sive Mensura et Gradibus Luminis, Colorum et Umbrae* (Augsburg). Abridged German transl. Anding E 1892 in *Ostwald's Klassiker der exakten Wissenschaften*, nos 31, 32 and 33 (Leipzig).

12 See Bouguer *op. cit.* note 8 vol III p 57. R Smith's 1738 *A Compleat System of Optiks in Four Books* (Cambridge) was translated into German in 1755.

13 *Ibid.*, p ix. Middleton quotes a passage illustrating Lambert's preference for analysis rather than physical observation in his study of the hygrometer [from de Saussure H B 1783 *Essais sur l'Hygrométrie* (Neuchâtel) p ix]: 'Le célèbre Lambert... ce grand géometre, considérant ces objets sous son point de vue favori, semble s'être occupé du soin de tracer géometriquement la marche de l'hygromètre... plutôt que de l'hygromètre proprement dite.'

14 Buckley H 1944 'Some eighteenth-century contributions to photometry and illuminating engineering' *Trans. Illum. Eng. Soc.* **9** 73–88.

15 Schrøder M 1969, transl. H Shepherd *The Argand Burner: its Origin and Development in France and England, 1780–1800* (Odense).

16 Keitz H A E 1955 *Light Calculations and Measurements* (Eindhoven) p 8.

17 Brown G I 1999 *Scientist, Soldier, Statesman, Spy: Count Rumford* (Sutton).

18 Thompson B 1794 'A method of measuring the comparative intensities of the light emitted by luminous bodies' *Phil. Trans. Roy. Soc.* **84** 67–82.

19 Guillemin A 1868 *Les Phénomènes de la Physique* (Paris) p 272 (my translation).

20 Buchwald J Z 1985 *The Rise of the Wave Theory of Light* (Chicago) pp 45–8. Malus observed qualitatively that the brightness of light refracted through a crystal of Iceland spar varied in a complementary way to that of the reflected component as the crystal was rotated. Assuming the total intensity to be conserved, he deduced that the reflected component was proportional to the cosine of the angle squared and that the refracted component was proportional to the sine of the angle squared.

21 Talbot H F 1834 'Experiments on light' *Phil. Mag.* **5** 321–34; quotation pp 327–8.

22 Talbot *ibid.* 333–4.

23 However, 'Talbot's law' failed when used to alter the exposure of photosensitive materials, especially when the flicker frequency was slow. See, for example, Baker E A 1926 'On the validity of Talbot's law for the photographic plate' *Proc. Opt. Convention* **1** (London) pp 238–44.

24 By William Abney, whose contributions to the subject are treated at greater length in chapter 4.

25 This is illustrated by the great diversity of processes available by 1860. The earliest reported process of Niépce had relied upon the effect of light on the solubility to oil of a preparation of asphalt; the later *daguerreotype* employed a surface of silver, sensitized with iodine vapour, developed after exposure by mercury vapour, and 'fixed' by immersion in hot brine; the *calotype* process, by contrast, used paper soaked in silver salts, and was fixed by sodium iodide. Each successive process required less exposure time and preparation than did its predecessor. See, for example, Fabre C 1890 *Traité Encyclopédique de Photographie* (Paris).

26 A photosensitive medium integrates light, changing its optical density in proportion to both the exposure time and intensity. In such a detector, either time or intensity can be used to control results. This relationship breaks down (the subsequently termed *reciprocity failure*) for extremes of intensity, exposure time or wavelength.

27 Woods C R 1883 'On latitude of exposure' *Photog. News* **27** 67–8.

28 Anon. 1883 'Latitude of exposure' *Photog. News* **27** 113–14.

29 Anon. 1884 'The Simonoff photometer' *Photog. News* **28** 610. This was a device in the form of a telescope incorporating an adjustable aperture wheel and graticule with scribed letters. The appropriate aperture, calibrated in terms of intensity, was selected to make the smaller letters illegible while the telescope was pointed at the light source of interest.

30 On Herschel's novel astronomical style, see Schafer S 1981 'Uranus and the establishment of Herschel's astronomy' *J. Hist. Astron.* **12** 11–26.

31 Michell J 1767 'An inquiry into the probable parallax, and magnitude of the fixed stars, from the quantity of light which they afford us, and the particular circumstances of their situation' *Phil. Trans. Roy. Soc.* **57** 234–45

32 Hoskin M A 1963 *William Herschel and the Construction of the Heavens* (London).

33 Bouguer *op. cit.* note 2, reported that the Swedish astronomer Celsius had used a similar method based on printed slips or black and white patterns. Geminiano Montanari, of the University of Bologna, published a comparable method in 1676; see Ariotti *op. cit.* note 6 332, 338. The idea of reading text as a means of determining a threshold of intensity was current until at least the turn of the 20th century. Such 'acuity' devices, based on the faculty for discriminating small details in patterns, were a class of photometers unique in that they did not rely on an observation of intensity.

34 Heyde's Aktinophotometer of 1905; see Thomas D B 1969 *The Science Museum Photography Collection* (London) p 37 catalogue no 267.

35 See, in particular, the work of Cornell E S 'The radiant heat spectrum from Herschel to Melloni.—I. The work of Herschel and his contemporaries' *Ann. Sci.* **3** (1938) 119–37 and 'The radiant heat spectrum from Herschel to Melloni.—II. The work of Melloni and his contemporaries' *Ann. Sci.* **3** (1938) 402–16; see also Barr E S 'The infrared pioneers—I. Sir William Herschel' *Infr. Phys.* **1** (1961) 1–4 and 'The infrared pioneers—II. Macedonio Melloni' *Infr. Phys.* **2** (1962) 67–73; Arnquist W H 1959 'Survey of early infrared developments' *Proc. Inst. Radio Engrs* **47** 1420; Lovell D J 1968 'Herschel's dilemma in the interpretation of thermal radiation' *Isis* **59** 46–60.

36 Herschel W 1800 'Experiments on the refrangibility of the invisible rays of the sun'

Phil. Trans. R. Soc. **90** 293.

37 Olson R E 1969 'A note on Leslies' cube in the study of radiant heat' *Ann. Sci.* **25** 203.

38 Herschel W 1800 'Experiments on the solar, and on the terrestrial rays that occasion heat' *Phil. Trans. R. Soc. London* **90** 90, 293, 437; Herschel 1800 'Experiments on the refrangibility of the invisible rays of the sun' *Phil. Trans. R. Soc. London* **90** 284; Herschel 1800 'Investigation of the powers of the prismatic colours to heat and illuminate objects: with remarks, that prove the different refrangibility of radiant heat' *Phil. Trans. R. Soc. London* **90** 255.

39 For later variants, see Cartwright C H and Strong J 1938 'Vacuum thermopiles and the measurement of radiant energy' in Strong J 1938 *Procedures in Experimental Physics* (New Jersey).

40 Schettino E 1989 'A new instrument for infrared radiation measurements: the thermopile of Macedonio Melloni' *Ann. Sci.* **46** 511–17. Melloni also invented a device to display the temperature change caused by radiant heat as a coloured surface: see Melloni M 1850 *La Thermochrose ou la Coloration Calorifique* (Naples; reprinted in facsimile edition, Bologna, 1954).

41 Hudson R D Jr and Hudson J W (eds) 1975 *Infrared Detectors* (Stroudsburg).

42 Particularly via W de W Abney, who dabbled in all aspects of light measurement. See Abney 1882 'On the influence of the molecular grouping in organic bodies on their absorption in the infra-red region of the spectrum' *Proc. R. Soc. London* **31** 416.

43 Cornell E S 1938 'The radiant heat spectrum from Herschel to Melloni. II. The work of Melloni and his contemporaries' *Ann. Sci.* **3** 402–16.

44 Brush S G 1970 'The wave theory of heat: a forgotten stage in the transition from the caloric theory to thermodynamics' *BJHS* **5** 135–67.

45 For a discussion of the effects of these radiations on selenium, see Hempstead C 1977 *Semiconductors 1833–1919: an Historical Study of Selenium and some Related Materials* (unpublished PhD thesis, Durham University) pp 34–5.

46 Hendry J 1986 *James Clerk Maxwell and the Theory of the Electromagnetic Field* (Bristol); Buchwald J Z 1994 *The Creation of Scientific Effects: Heinrich Hertz and Electric Waves* (Chicago).

47 From the turn of the century through the First World War, one line of research was to seek the existence of radiation beyond the visible, and to explore its properties [e.g. Rubens H and Hollnagel H 1910 'Measurements in the extreme infra-red spectrum' *Phil. Mag.* **19** 764; Nichols E F and Tear J D 1923 'Short electric waves' *Phys. Rev.* **21** 587]. These investigations explored radiations of similar wavelength using distinctly different sources, detectors and methodologies. This programme was largely detector centred, seeking to show the connections—indeed, to bridge the perceived gap—between infrared 'optical' radiation and electrically related 'radio' waves. Thus infrared spectroscopy emerged as a subject of study, attracting a small but active band of physicists who developed an analogue of visible-light spectroscopy, using infrared-transmitting lenses and prisms. See Johnston S F 1991 *Fourier Transform Infrared: A Constantly Evolving Technology* (Chichester) chapters 5 and 6.

48 Ames A Jr 1921 'Systems of color standards' *JOSA* **5** 160–70.

49 Hay D R 1846 *A Nomenclature of Colour* (London) pp 20–6.

50 Luckiesh M 1915 *Color and its Applications* (London).

51 Munsell A H 1907 *A Color Notation* (Boston). Munsell (1858–1918) lectured on colour harmony at the Massachusetts Normal Art School from 1890 to 1915. His colour system was influenced by the idea of a colour 'sphere' proposed by Nicholas Ogden Rood in *Modern Chromatics* (1879).

52 Chevreul M E 1858 *The Laws of Contrast and Colour* (London).

53 The use of indicator solutions to infer content from colour change dates back at least to Gabriel Fallopius in 1564, and to Robert Boyle a century later. See Debus A 1962 'Solution analyses prior to Robert Boyle', *Chymia* **8** 41–61 and 'Sir Thomas Browne and the study of colour indicators' *Ambix* **10** 30.

54 Lovibond J W 1897 *Measurement of Light and Colour Sensations* (London).

55 Lovibond J W 1915 *Light and Colour Theories* (London) p 3.

56 Campbell N R 1928 *An Account of the Principles of Measurement and Calculation* (London) p 1. See also chapter 3.

CHAPTER 3

SEEING THINGS

Late Victorian photometry was shaped by new players wielding new devices and seeking new goals: gas inspectors and astronomers with practiced eyes or sensitive emulsions. With observers closeted in darkened rooms, the measurement of light remained an intensely individualistic affair, based on a personal judgement of a light source by a single pair of eyes. But a variety of processes—social, technological and scientific—transformed the brightness of light in the late 19th century from the passing concern of a few disparate individuals to a subject employed and studied by groups.

It was shaped by new perceptions and expectations, too. This cultural transformation was accompanied by the growing identification of the subject as a part of physical science, steering it towards an increasingly quantitative expression. Despite this recategorization, photometry remained, by the end of the century, an undisciplined and fragmented study. This chapter discusses the changing perception of photometry among emerging communities of engineers and scientists, isolated by distinct backgrounds and goals. The disjointed status of the emerging subject is reflected in the heterogeneous case studies and issues discussed in this chapter.

But to discuss quantitative measurement we must adopt definitions. Among the clearest analyses of quantification were those devised by the physicist and philosopher of science Norman Campbell (1880–1949). Having a strong personal stake in light measurement, Campbell in 1928 cited photometry as a study still suffering from inadequate foundations, an evaluation common to his generation[1]. Setting aside his judgements for the time being, we can nevertheless profit from his categorizations of quantification. Campbell defined measurement as 'the assignment of numerals to present properties in accordance with scientific laws'. He described quantification as being of three possible classes (table 3.1). In his first class, Campbell categorized values that are simply ordered or ranked according to a lesser-than, greater-than criterion. A scale of hardness is of this type. Values on such a scale can be compared and even equated, but it is not possible to quantify by *how much* various values differ.

In a second class of measurement, values may be ordered on a scale that has regular increments; the temperature scale is such a case. This scale still is

Table 3.1. Classes of measurement as defined by N R Campbell.

Class	Characteristics	Example
1	Ranking, ordering	Rock hardness scale
2	Ordering with uniform scale	Temperature
3	Arithmetic operations	Mass, length

not completely quantitative, because it does not support arithmetic operations. Temperatures, for example, cannot be added or subtracted.

'Countability' is the defining characteristic of the third, fully quantitative class of measurement[2]. In this type, the quantity has a direct relationship with the order of natural numbers. Campbell used the example of illumination to illustrate this class[3].

Photometry, as employed by various practitioners through the 19th century, could fall into any one of these classes, although the first and second were the most common. The mere *ranking* provided by class 1 measurement was a characteristic of stellar magnitudes in the first half of the century and earlier. Class 2 *ordering* of intensities typified usages such as early gas photometry. Class 3, involving wholly quantitative *measurement*, became popular only in the last decade of the century, and then only with limited precision. Campbell himself noted that light intensity is a difficult case of his 'laws of measurement', because it is additive only for isolated wavelengths: if two colours are mixed, they do not in general add to a unique sum, because the results depend on how the detector responds to different colours. Thus the hesitancy of researchers to adopt quantitative methods in late Victorian photometry can be attributed in part to the lack of assurance in the validity of this approach—in short, it did not appear to work well and had dubious relevance. Comprising an inchoate collection of techniques and usages in the mid 19th century, photometric practice was, a few decades later, striving for numerical expression.

3.1. RECURRING THEMES

Interest in the quantitative measurement of light intensity increased in the second half of the 19th century owing to the creation of new research problems, especially in the areas of astronomical and lighting photometry. Chronicling the tentative evolution of light measurement by practitioners struggling to make sense of its perceived complications, this chapter discusses the scientific, social and technological factors responsible for the growth of a quantitative perspective up to the first years of the 20th century. The subject was approached in different fashions by different communities of practitioners, and remained a discordant collection of techniques, apparatus and applications at the end of the century. Throughout the precarious establishment of the subject, however, certain recurring themes can be distinguished.

With the increasing employment of photometry, practitioners discovered the limitations imposed by the human eye. Its reliance on visual observation proved a serious hindrance to the application of photometry because agreement between investigators was poor and because considerable labour was required for precise observations. Successive practitioners repeatedly faced the same questions. Was the eye reliable, and to what extent? Could apparatus be designed to improve its accuracy? Could another means of measuring light replace the eye entirely?

The 'human factors' in photometry were to crop up repeatedly. Intensity measurements could be perturbed not only by the vagaries of the eye, but also by those of the brain. Careful practitioners concluded that they could be misled by inadvertent prejudice, and that the matching of two lights by eye was prone to psychological bias. Probably the first investigator to voice this concern was Benjamin Thompson who, in 1794, had employed a double-blind method to avoid the problem. He adjusted the positions of light sources on his photometric bench by a hand-winch, giving notice to his assistant to

> observe, and silently write down, the distance of the lamp or candle, so that I did not even know what that distance was till the experiment was ended, and till it was too late to attempt to correct any supposed errors of my eyes by my wishes or expectations, had I been weak enough to have had a wish in a matter of this kind. I do not know that any predilection I might have had for any favourite theory would have been able to have operated so strongly upon my mind, ... but this I know, that I was very glad to find means to avoid being *led into temptation*.[4]

Most practitioners ignored such niceties, and either accepted what they recognized as an imprecise measurement or carried on unaware of the potential systematic errors.

A second characteristic of the subject was its growth in popularity quite divorced from scientific and technological evolution. Growth—as evidenced by the number of papers published, number of practitioners, or number of photometric laboratories—was high in the latter decades of the century. This burgeoning popularity resulted from an increased perception of the utility of photometry. The elaboration of techniques and the evolution of a scientific basis, however, evinced no such trend: the practice of photometry, in relation to other sciences and technologies during the period, changed slowly. One reason for its slow development was the identification, oft repeated, of practical difficulties in what appeared superficially to be a straightforward measurement technique. Among the several hundred photometric investigations published during the 19th century, few were directly concerned with such limitations[5]. With little serious exploration of their complexities, photometric methods were consequently abandoned as often as they were refined. Owing to the unexpected subtleties of visual observation, photometry was to gain a reputation as an imprecise or even impossible technique. Most practitioners by the end of the century were engineers

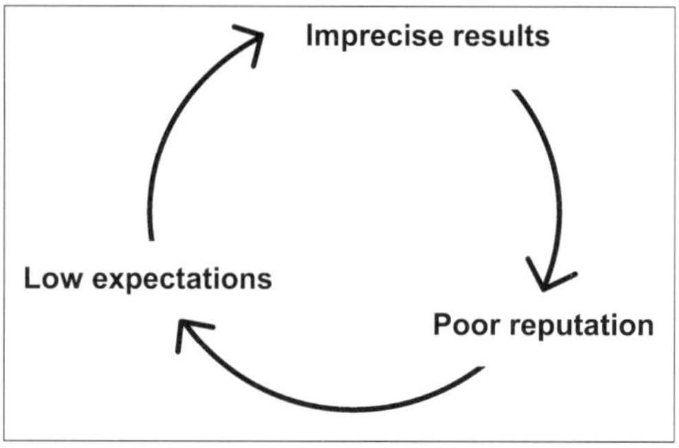

Figure 3.1. Circle of development for photometry.

rather than scientists, and they relegated photometry to routine verifications rather than to continued development.

As was to be demonstrated repeatedly through the century, the reputed imprecision of photometry restricted the usages to which it was applied; in turn, the undemanding usages placed little pressure on practitioners to improve their technique. This circle of low expectations → imprecise results → poor reputation → low expectations thus relegated light measurement to the depths of the scientific toolbox (figure 3.1).

A final recurrent theme in 19th century photometric practice is the scarcity of collaborative development. The *value* and *credibility* of photometry were to be repeatedly questioned and re-evaluated between communities, times and locales. Consigned to mundane applications, its reputation as a straightforward if inaccurate technique promoted its unenthusiastic usage by independent groups having little contact. This 'balkanization' of the subject inhibited change at the end of the century and relegated light measurement to a peripheral science.

3.2. ALTERED PERCEPTIONS

Chapter 2 described a period of independent investigation of light measurement, during which few connections existed between individual investigators. This situation began to change in the period 1850–80, however, when technological and cultural innovations combined to increase the influence and applicability of photometry. While the cause-and-effect relationships between these agents are difficult to map, their combination transformed the measurement of light intensity into a useful—if highly specialized—tool for diverse groups of scientists and engineers. The new networks grew first around newly valued *uses* of light measurement; that is, they had *cultural* nuclei. But the groups of practitioners remained disconnected. What had been studied by isolated individuals came to be studied by independent communities.

3.2.1. Astrophysics and the scientific measurement of light

From the 1850s onward, a handful of astronomers nurtured the first durable interest in photometry, increasingly interested in extending their domain from mere astronomical time and position measurement. Among the new phenomena gaining attention, the brightnesses of stars and planets were identified as being amenable to systematic observation and classification. There had already been a number of published catalogues that included visual estimates of magnitude as an adjunct to positional coordinates[6]. In 1851, though, W R Dawes criticized what he saw as weaknesses of previous estimates:

> The differences among observers of great experience and celebrity are much greater than would probably be imagined by those who have not been led to examine the subject, and clearly show that widely different *scales* of magnitude have been adopted...[7]

According to Campbell's classification, stellar magnitudes at this time were of the first class, merely ranking values along an unreliable scale. To illustrate the poor precision of magnitude estimation, Dawes listed stars for which the magnitudes had been reported as anything from 5.3 to 8.5, discrepancies corresponding to differences of about eightfold in estimated intensity[8].

Some practitioners sought to improve the precision of their visual techniques and to trace the experimental factors that limited it. More commonly, however, scientists intrigued by the possibilities of photometry applied the technique unaware of its difficulties. In 1878, Charles Zenger reported a method of measuring the relative intensity of planetary discs and satellites: he noted the time of disappearance of planetary features near twilight[9]. Zenger based his work on that of Bunsen (of prior fame in spectrum analysis) who had used a photographic technique to measure the background intensity of the sky versus the zenith distance of the sun, this serving as the reference for the threshold technique. Zenger reported no particular precautions concerning the sensitivity of the eye to differing levels of light nor indeed any reference at all to the uncertainties of observation.

Surveys of the *Monthly Notices of the Royal Astronomical Society* for the latter half of the 19th century show that intensity measurement came to be adopted increasingly for special studies, and evolved towards a more quantitative and accepted technique in astronomical practice. In the same year as Zenger's work, for example, W H M Christie made visual measurements of the disc of Venus, attempting to fit them to a theory of specular reflectance and diffusion by the planetary atmosphere[10]. Christie, appointed Chief Assistant at Greenwich in 1870 at the age of 25, was later to succeed Airy as Astronomer Royal. His interest in relating theory and experiment was new to late 19th century photometry. The emerging quantitative attitude was shared by the American Samuel Langley in the description of his new bolometer:

> I therefore tried to invent something more sensitive than the thermopile, which should be at the same time equally accurate,— which should, I mean, be essentially a 'meter' and not merely an

> *indicator* of the presence of feeble radiation. The distinction is a radical one. It is not difficult to make an instrument far more sensitive to radiation than the present, if it is for use as an indicator only, but what the physicist wants, and what I have consumed nearly a year of experiment in trying to supply, is something more than an indicator,—a *measurer* of radiant energy (emphasis added).[11]

Practitioners now labelled the obtaining of an *indication* of light intensity as inferior to a *measurement*, in contrast to Airy's *notion/measure* equivalence of a quarter-century earlier. Measurement to Langley and his contemporaries was more than the mere ranking of magnitudes. Inherent in the idea was the ability to reproduce observations and to relate them in a precise, repeatable way to other physical quantities—a strategy to extract more from observations. This linking with other forms of measurement was a key to promoting the quantification of light. The change in emphasis was reflected in the birth of a new subject of study: astronomy was joined by 'astrophysics'[12]. A typical article of the newly renamed journal *Astronomy and Astrophysics* in 1892 (the year of Airy's death) was on the 'Distribution of energy in stellar spectra'[13]. This work paralleled similar studies of the sun made by Herschel nearly a century earlier, but now appropriated it for the use of astronomers. The new community of astrophysicists saw clear reasons for measuring the intensity of starlight:

> The problems of stellar photometry are closely connected with many cosmic questions, primarily with the light changes of variable stars; but they have an equally important bearing on the questions of stellar distribution and evolution. It has been said by good authorities that it is of more importance to measure the light than the place of a star, and if one considers merely the astonishing number of variable stars now being discovered, it will be admitted that the importance of stellar photometry can scarcely be overestimated.[14]

Having created a *need* to measure light, then, what strategies did these practitioners use to tame this difficult subject? One of the 'good authorities' mentioned by Parkhurst was probably the astronomer Edward C Pickering (1846–1919), who provided Parkhurst with his instruments. Professor of physics at the Massachusetts Institute of Technology and director of the Harvard College Observatory, Pickering was then at the centre of developments in astronomical photometry and spectroscopy and important in influencing the acceptance of these subjects by astronomers[15]. He was not, though, solely responsible for the growth of this research area. Stellar photometry, the first concerted usage of light measurement for scientific applications, had begun at Harvard with its first director, William C Bond (1789–1859). In 1850, Bond applied photographic methods to the making of photometric measurements of stars[16]. His work attracted other astronomers to photometric observations soon afterwards. N R Pogson, in 1856, employed a visual photometer to evaluate starlight, and found that Hipparchus's scale of magnitude gave approximately a factor of 100 between the intensity of first and sixth magnitude stars. To create a scale of uniform increments (in effect moving stellar photometry from Campbell's 'class 1' to 'class 2' measurement), he therefore proposed the definition of a magnitude change of 1 as a change in intensity of $100^{1/5}$ (approximately 2.5-fold). The definition was probably the first numerical interval to be applied to light measurement. It proved even more useful than technical developments because it promoted the sharing of observations between subsequent astronomers. At Oxford, Charles Pritchard (1808–1893) used a wedge photometer to measure the magnitudes of stars visible to the naked eye at up to 100° from the north pole[17]. His catalogue, the *Uranometria nova Oxoniensis* published in 1866, agreed 'quite well' with Bond's work, 'providing a generally acceptable magnitude sequence for the brighter stars'[18]. An assistant at Harvard, Charles S Peirce (1839–1914), published the work he carried out between 1872 and 1875 as *Photometric Researches*[19]. Such comparisons and collaborations signalled the beginning of the social phase of astronomical photometry. Indeed, these photometric atlases promoted networks of individuals and institutions just as they created relationships between stellar objects.

Sharing Bond's conviction of the usefulness of such observations, and building upon the work already done at Harvard College Observatory, his successor Edward Pickering initiated an extensive programme of stellar photometry at Harvard College Observatory when he became director in 1877. Pickering introduced several innovations to convert photometry from a volatile to a sound subject. The first of these was in promulgating a standard. By adopting Pogson's scale of magnitude, and choosing Polaris as the reference star against which all others would be compared, he defined a photometric scale that other workers found straightforward to accept. Second, Pickering established a reliable technique. Working with the firm of Alvan Clark & Sons, he devised new types of visual photometer adapted for telescopic use. By means of adjustable mirrors, his 'meridian photometers' combined an image of Polaris with the target star as it crossed the meridian[20].

Pickering's third tool of persuasion was sheer volume of data. To command attention, the new photometric systems had to map a representative number of stars. The first *Harvard Photometry*, published in 1884, catalogued some 4000 stars. On its completion, Pickering immediately promoted a more extensive stellar survey. Between 1889 and 1891, Solon I Bailey took the equipment to South America to catalogue the stars of the southern hemisphere. By 1908, Pickering and his co-workers had extended the work tenfold, cataloguing 45 000 stars in their *Revised Harvard Photometry*[21], Pickering alone recording some 1.4 million observations[22]. John Parkhurst, the final recipient and user of Pickering's instruments from the opening of Yerkes Observatory in Chicago in 1897, carried on through the 1920s, having by then switched to photographic photometry[23]. By defining an observational method, publicizing his data, and training and supporting energetic acolytes, Pickering thereby legitimated astronomical photometry and enlisted the support of the astronomical community.

Apart from this American concentration of photometric research, most 19th century astronomical photometry took place in Germany. As in America, an observing community spread from an observatory where the practice of photometry was stabilized. Johann Zöllner (1834–82) became interested in stellar photometry as a student, and defended perhaps the first PhD dissertation on photometric research in 1859[24]. Zöllner marshalled technique and training to extend the influence of stellar photometry as Pickering was later to do. His 'astrophotometer', which incorporated a petroleum-burning reference lamp, was adopted by other German observers[25]. Established in 1877, the Potsdam Observatory became a centre for photometric observations and produced a line of researchers[26]. Zöllner's student, Hermann Carl Vogel (1834–98) while working at observatories in Kiel and Potsdam from 1870 undertook an extensive programme of stellar classification using spectroscopic and photographic techniques. Gustav Müller, in his turn, gained an interest in photometry while working as an assistant to Vogel at Potsdam. Between 1886 and 1906, he planned and carried out an extensive programme of stellar photometry. Adopting Pogson's scale of magnitude as Pickering had done, Müller's *Photometrische Durchmusterung des nördlichen Himmels* catalogued over 14 000 stars[27]. The measurement precision of this generation of catalogues was considerably better than that of their predecessors[28].

The isolated but extensive and respected work of the Harvard College and Potsdam observing communities influenced the following generation of astronomers. Ralph Sampson, for example, (1866–1939), later Astronomer Royal of Scotland, was to specialize in photoelectric photometric studies through the inter-war period because of their influence. According to one chronicler, the 'advent of Harvard photometric eclipse observations of satellites of Jupiter stimulated him to re-examine previous observations' and instigated his interest[29].

The success of photometric and photographic methods in astronomy led the astrophysicists to more complex but vastly more fruitful techniques. By the turn of the century, *spectro*photometric observations were being made. As early as 1899, Karl Schwarzchild (1873–1916), then an observatory assistant in Vienna, developed techniques for combining spectroscopy with photographic photometry. These allowed the relative intensity of a star to be mapped as a function of wavelength, by applying the photometric method successively to narrow bands of wavelengths[30]. From this colour information, experimentalists could classify stars by type, and theorists were able to estimate temperature[31]. Stellar classification, based on spectral lines and photometrically determined temperatures, became a major activity in astrophysics[32].

The isolation of the observing communities diminished as the number of practitioners grew. Hans Rosenberg (1879–1940), for example, began working with Schwarzchild around 1907, where he analysed spectrograms using a Hartmann microphotometer[33]. In the following decade Rosenberg worked at Yerkes Observatory, where Parkhurst had started a photometry programme in 1897 with the help of Pickering. Starting from a handful of centres in the second half of the 19th century, astronomical photometry had become a cooperative international network before the Second World War[34].

By the beginning of the 20th century, then, astronomical photometry was an established technique employed by a growing community of astrophysicists. Their *motivations* had been transformed during this period, however. Where Herschel's enthusiasm for photometry was unshared by his contemporaries, and Bond's interest in the 1850s had been provoked by a desire to catalogue the heavens more fully, the growth of stellar photometry was due in large part to successful lobbying by a few individuals. The demonstration of the feasibility of the technique and the supply of voluminous data from the Harvard and Potsdam observatories, owing to the energetic programmes of Pickering, Zöllner and their followers, served to render the measurements trustworthy. From the 1880s, however, the additional information provided by spectroscopy became a major incentive in astronomers' adoption of photometric techniques.

3.2.2. Spectroscopy

While serving eventually as an impetus to astrophysics, the study of spectroscopy was at first only peripherally concerned with light intensity[35]. Quantitative measurement became increasingly attractive to its practitioners, however. Following Bunsen's and Kirchoff's investigations in the late 1850s, investigators began to use spectrum analysis to infer chemical composition. The presence or absence of particular spectral lines was originally the sole criterion of analysis. Spectral lines were initially classified by their relative positions in the spectrum (e.g. Fraunhofer's alphabetic ordering of prominent solar lines), followed somewhat later by wavelength values. Towards the end of the 19th century, astronomical spectroscopists began to describe certain spectral lines by their appearance. They noted, for example, that particular lines always appeared sharp, or diffuse, and that certain lines were always characteristic of a substance. Semi-quantitative descriptions such as *sharp*, *principal*, *fine* and *diffuse* gained currency[36].

Initial interest centred upon the *identification* of small quantities of material rather than on determining its quantity. In popular lectures given in 1869, J Norman Lockyer (1836–1920) emphasized spectroscopy's potential for detection and discovery, a role seemingly divorced from quantification:

> not only are we able to differentiate between different bodies, but the most minute quantities of substances can be determined by this method of research... for instance, Kirchoff and Bunsen have calculated that the 18-millionth part of a grain can be determined by the spectroscope in the case of sodium.[37]

The example of ubiquitous sodium, and the discovery of new elements, was to reappear in many popular accounts of spectroscopy[38].

For laboratory spectrum analysis, the neglect of intensity measurements by experimenters was in part a consequence of the instability of the light source: the flames commonly used to heat specimens varied in intensity and temperature, and thus were far from stable subjects. Also, the intensities of different spectral lines from a single source could differ by 1000:1 or even 10^6:1, making photographic methods ill suited owing to their limited dynamic range[39].

Interest in this minor subject grew as new spectroscopic phenomena emerged[40]. Technology and organization also shared significant responsibility for a growth in popularity. From 1870, the availability of dry gelatine photographic plates made photographic spectroscopy more practical. Units of wavelength had been standardized by 1890, promoting the comparison of results and strengthening the links of the social network. The new techniques had an immense scientific pay-off. Spectroscopy (both visual and photographic) was being used to infer the velocity, temperature and composition of stars and planets, and to probe new phenomena[41]. The potential of the new research programmes convinced practising spectroscopists of the need for further development of intensity measurement.

3.2.3. Shifting standards: gas and electrotechnical photometry

Photometry had hitherto been an intensely personal affair. The apparatus had to be designed and calibrated by each investigator, the observations were performed in a light-tight room or at a telescope eyepiece, and the results relied solely on the evidence of his eyes. Communication of results demanded, however, that intensity calibrations be regularized. The socialization of the subject relied upon standards.

Such intensity standards were not trivial to generate. The astronomer John Parkhurst, for instance, calibrated his graduated wedge for stellar photometry using two methods: first, by making measurements 'of standard stars whose magnitudes have been well fixed'; and second, 'by measurements of an artificial star whose light can be reduced by a known amount either by (a) polarization, (b) a revolving wheel or (c) reduced apertures by stationary diaphragms'[42]. Because each estimate of intensity was imprecise, averaging was necessary. The comparison of individual instruments was tedious: Parkhurst reported making 2700 measurements on standard Pleiades stars, 3000 readings for a comparison with a Zöllner photometer and 500 readings for comparison with a 'wheel' (Talbot) photometer. Even with such careful photometric methods, though, astronomers felt compelled to emphasize that they still 'found it by no means easy to get good concordant observations'[43]. The brightness of fluctuating light sources such as twinkling stars was difficult to measure by relatively slow visual or photographic observations. Measurements were further hampered by changing sky conditions.

The use of 'standard stars' 'well fixed' by other observers can be seen as Parkhurst's attempt to enrol an ill defined community to support his measurements. Stellar catalogues served a social role in forming that community. But the difficulty in obtaining 'good concordant observations' illustrates the fragility of this grouping of practitioners at the mercy of their technology. While such time-consuming methods of characterization were practical for some scientific work, they were wholly unacceptable for industrial problems. If photometry was to be accepted widely, reasoned some practitioners, generally available standards of light measurement and intensity were required.

3.2.4. *Utilitarian connections*

Light standards were impelled by utilitarian requirements, and photometry gained new supporters through its connection with questions of illumination. Intensity standards in commerce and industry became widely sought and employed during the second half of the 19th century, when the regulation of gas lighting provided an incentive for development. The quest for a standard, in its turn, supported the growth of new communities recruited to maintain and employ it.

Until the late 18th century, open oil lamps and candles had undergone little active development. The Argand lamp of 1786 demonstrated the value of thoughtful design, and promised a more stable light standard. The Carcel, developed in France in 1800, was another successful oil lamp containing a clockwork pump for supplying oil to the wick[44]. In 1860, its burner and chimney dimensions were standardized for use as a reference for testing the illuminating power of Paris gas. The English standard, the *Parliamentary candle*, was similarly defined for the same reason. Gas testing, the first routine use of photometry, gave the technique a legal and economic dimension.

The illuminating gas industry, originating in England in the early decades of the century, provided the dominant source of domestic and public lighting in most cities within two decades[45]. The first company in London was set up in 1810, and the number of companies supplying gas in the capital reached 13 before falling back to three in the 1880s as a result of mergers. The Metropolitan Board of Works (MBW) was given extensive powers to supervise the industry in the early 1860s when the number of companies proliferated. Following public concern about the accuracy of gas metering and the purity of gas, Parliament passed legislation to give supervisory powers to magistrates. When this measure proved ineffective, the Metropolitan Board of Works was given responsibility[46]. The first gas examiner was appointed in 1869, followed by four more a year later. A unified department concerned with the legislation and regulation of the gas supply grew out of the MBW[47].

The gas standards to be verified centred on illuminating power and purity[48]. Groups of gas examiners were responsible for particular areas of London, with an inspector responsible for one metering house. By 1889 some 22 locations were specified[49]. The legal requirements created a new community of photometrists. These first salaried light-measurers were highly trained with respect to the other administrative staff: half had studied at a university or equivalent, compared with 6% of the other departments of the MBW, and all employed photometric and chemical analysis in their work[50]. The major users and adapters of photometric equipment, and the most numerous photometrists, were the gas examiners of London and other gas-supplied cities between at least 1860 and 1880.

The scientific practices of the staff, and physical standards of illumination, were set by a body of experts known as the Metropolitan Gas Referees. The Superintending Gas Examiner, William Joseph Dibdin (1850–1925), Chemist to the MBW in the late 1880s, thoroughly investigated the available photometric methods and published one of the first widely available books summarizing the subject[51]. Observing that 'the present chaotic condition of the Photometer

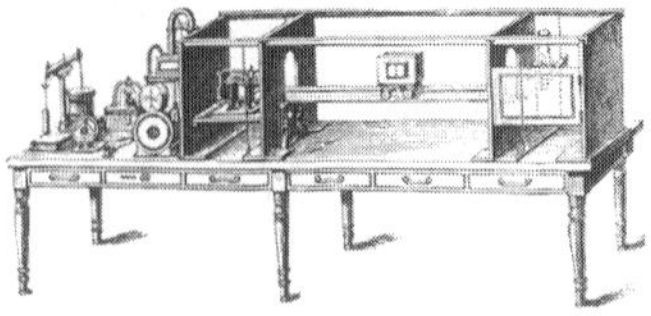

Fig. 2.—Letheby Photometer.

Fig. 14.—Sugg's Travelling Photometer.

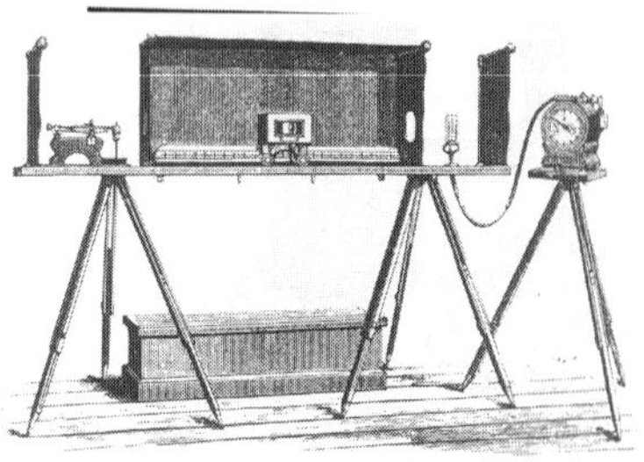

Fig. 8.—The Portable Photometer.

Figure 3.2. Late Victorian commercial photometers by William Sugg & Co. Despite the apparent variety of form, one practitioner noted that 'the traditional way to make a "new" Photometer is to alter the wooden casing as much as possible; and then to call this outcome of the cabinet maker's art a new Photometer' [Dibdin W J 1889 *Practical Photometry: a Guide to the Study of the Measurement of Light* (London) pp 29, 34, 68]. The meter in the lower figure is a gas flow gauge.

itself is a fruitful source of much uncertainty', and attempting to reassure the 'newly-appointed and possibly somewhat nervously constituted Gas Examiner', he sought to give 'a full narration of the various systems now before the public' (figure 3.2)[52]. Not only did Dibdin strive to provide practical answers to utilitarian problems of gas testing; he also prescribed procedures for measuring electric lights, and made an examination of stellar photometry. By providing a comprehensive text, recommending standardized methods and training scientific staff, the Metropolitan Gas Referees thus became the *de facto* arbiters of photometric standards in England[53].

One of the first tasks of the Referees was to seek improved intensity standards. The accuracy of the Parliamentary Candle, the first standard defined by the Referees, was poor: although intended to burn 120 grains of spermaceti per hour, initially only the candle weight (one-sixth of a pound) was specified. By 1871 the specification had been elaborated to provide permissible limits (114–126 grams/hour or ±5%) and a description for the manufacture that included wick and wax characteristics[54]. Yet standards based on candles were, according to one observer, 'not more scientific, and hardly more accurate, than the barley-corn, of which three went to the inch, as a standard of length'[55].

The prevailing wax candle standards were widely recognized to be imperfect. The material burnt was of indefinite composition, prompting some writers to claim that the spermaceti available had changed from that in the originally defined candles. By the end of the century wax candles had been extensively investigated and universally condemned. The subject of intensity standards had become of pressing concern to a range of parties[56]. Electric lighting, increasingly promoted from the late 1870s, was a primary motivation. Intense competition between the gas industry and the nascent electric lighting companies was a consequence of the new lighting technology. Within months of the commercial availability of electric lighting systems, the streets and squares of some towns were converted. Among the important technical factors in the competition were the relative cost and quality of gas and electric illumination. For meaningful comparison of the technologies, accurate intensity standards were needed.

Having an immediate financial incentive, photometric investigations proliferated. A committee on the Standard of Light for the British Gas Institute investigated the precision of intensity standards in 1883, finding variations of between 1% and 16% in the standard candle. A committee for the British Board of Trade found similar variations, and the American Institute of Electrical Engineers set up its own panel. Improved standards were proposed, investigators usually settling on refining the composition of the combustible agent as the best strategy. The German Association of Gas and Water Engineers had defined the *Vereinskerze*, or 'Association Candle', in 1868, which it also manufactured and sold. A paraffin candle having 2% stearine added, it was defined by weight, with 10 candles weighing 0.5 kg. They, too, found their wax candle to be unsatisfactory, rejecting it for the 'Hefner' lamp less than two decades later.

The Hefner proved a more long-lived standard. This unit represented the intensity radiated horizontally by a standard light source consisting of an oil lamp burning amyl acetate. Its inventor, Jacob von Hefner Alteneck (1845–1904), a senior engineer at the Berlin electrical firm of Siemens & Halske, chose a simple hydrocarbon of known composition as the fuel to remove one source of variability from the problem of standardization. Similarly, the British chemist and inventor A G Vernon Harcourt (1834–1919) developed, over the last two decades of the century, standard lamps based on pentane (figure 3.3). These were adopted by British industry, and eventually by the national laboratory.

The setters of standards recognized early on that, like other flame-based standards, the Harcourt and Hefner lamp intensities varied with humidity, air pressure and carbon dioxide concentration. This variability was not seen initially

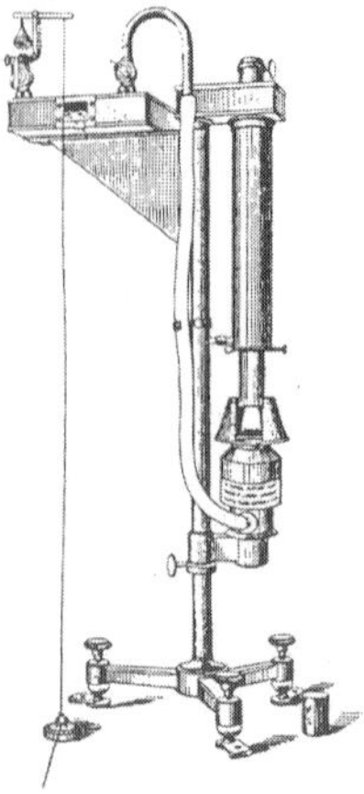

Figure 3.3. Britain's answer to a Victorian standard of intensity: the Vernon–Harcourt pentane lamp. Dibdin W J 1889 *Practical Photometry: a Guide to the Study of the Measurement of Light* (London).

as a disadvantage. On the contrary, gas industry representatives argued that, since the flame standards were to be used to evaluate the quality of illuminating gas, both would be similarly affected by atmospheric conditions, and so less variable measurements would be obtained. For those interested in the comparison of electric lamps and the more difficult inter-comparison of gas and electric sources, however, this argument seemed specious; in their view, a photometric standard had to be stable and represent a known value of illuminating power. The judgement of the appropriateness of a standard was consequently far from objective; flavoured by industrial allegiances, it favoured the then-dominant illuminant, gas.

Other practical difficulties with flame standards included controlling the size of the flame and (in the case of the Hefner lamp) its yellow-orange caste. 'Our German friends may bask in the ruddy rays of their 0.9 candle Hefner lamp, or our French neighbours enjoy their 10-candle Carcel', wrote the first president of the Illuminating Engineering Society of London, extolling the virtues of inter-comparable, if nationally distinct, intensity standards[57]. The perturbing factors were carefully detailed in texts on illuminating engineering by the turn of the century. Laboratories were beginning to employ incandescent filament lamps as working standards, and a controlled flame as the best available primary standard. The testing of gas lamps necessitated peripheral equipment such as a consumption meter, pressure regulator, pressure gauge and calorimeter to monitor the gas supply and its quality, and apparatus for determining atmospheric pressure, temperature and humidity. To promote stability, each room was ventilated only between measurements to replenish the oxygen and reduce carbon dioxide levels. Even then, atmospheric changes were a sometimes serious problem. One annual report stated that 'a further mild winter has made it impossible to secure very low values of atmospheric humidity in connection with the realization of the pentane unit in terms of the values of electric sub-standard lamps... the second successive winter this has been impossible'[58].

An indication of the difficulty of using flame standards is given by the Assistant in the Photometry Section of the National Physical Laboratory[59]. To make a photometric comparison of the Harcourt pentane lamp with an incandescent lamp, the experimenter first lit the pentane lamp, carefully adjusted the flame height, then 'threw open the doors and windows of the room' to allow the flame to stabilize for a half hour (the purity of pentane was critical, too, having to be prepared by a procedure specified by the London Gas Referees)[60]. He then gradually increased the voltage of the incandescent lamp to avoid thermal shock to its filament. Once the lamps were ready, the doors and windows were closed, whereupon the visual photometric comparisons could be carried out for 10 or 15 minutes. During the photometric measurements, hygrometer and temperature readings were taken by other observers at several points around the Harcourt lamp—moving slowly and quietly to avoid perturbing the flame. The readings were later averaged and used to compensate for the known humidity and temperature dependence of the flame. When the pentane lamp began to diminish in intensity, the experimenters had to repeat the ventilating process. The photometry room was necessarily large, both to accommodate the long optical benches needed to match different lamp intensities, and also to provide enough oxygen for the lengthy comparison of flame-based standards. On the other hand, only one photometric measurement could be made at a time, so multiple rooms were required to avoid lost time.

Partly owing to difficulties such as these when maintaining flame standards, the working standards in use in Britain, America and France, based on various designs of incandescent lamp, were rationalized into an international photometric unit in 1909[61]. The German-speaking countries retained the Hefner lamp as the primary standard, although it was calibrated with respect to the international standard[62]. Here again, different communities disputed the qualities that were essential to an intensity standard. Supporters of electric lamp standards contended that the Hefner demanded critical measurement of, and correction for, humidity and temperature, rendering the measurement both time-consuming and unreliable. By contrast, supporters of the Hefner argued that its environmental influences were well characterized, and that the lamp itself was straightforward to fabricate by any laboratory. On the other hand, they pointed out, the characteristics of incandescent lamps depended greatly on the materials employed and the method of manufacture, and could not be standardized. Any particular lamp would have to be individually calibrated with respect to a known primary standard. Electric lamps were also critically dependent on the power supply. The use of such electric 'glow-lamps' as at least interim standards of intensity required standards laboratories to make very exact measurements of electric current: a photometric measurement of electric lamps required the supply voltage to be measured to an accuracy from 0.1% to 0.02%. This demanded a large storage battery to be used and maintained, usually in a specially constructed room[63]. Generators were kept as far as possible from the measurement room to avoid interference with the sensitive galvanometers required for precise electrical measurements. Most seriously of all, the illuminating power of an incandescent lamp changed unpredictably with age. The only means of minimizing this problem were to operate the lamp at reduced power, to limit the time it was on, to scrupulously avoid mechanical shock and to compare it periodically with another type of standard.

Thus intensity standards, whether based on candles, oil lamps or electric filament bulbs, were disturbingly precarious and contentious. Their combination of physical and social instability rendered them ineffectual; the lack of consensus in these standards, as in other aspects of light measurement, restricted the development of photometry during the following decades. The discord existed at all levels, extending down to groups of investigators in different industries, towns or laboratories.

Despite this lack of consensus, engineers at the local scale employed photometry unproblematically to provide routine information for specific tasks[64]. The Edison company, for example, used a permanent photometric installation as part of the control system for electrical power in one of its generating stations. The photometer, mounted on a graduated iron bar, verified the luminous intensity of the lamps, and a galvanometer monitored the strength of the supply current. The reference source was a 'standard gas mantle, perfectly adjusted to normal luminous intensity'[65]. The town's electricity supply was thus in the incongruous position of being regulated in terms of the locally available illuminating gas. Again, the dominant commercial light source was shaping the practice of photometry.

Gas photometry was the principal usage of light measurement. Consider, for example, an 1870 book in which W M Williams proposed an explanation for the continued prodigious heat and light emission from the sun[66]. His explanation relied upon the assumption that light would pass unattenuated through successive layers of flame, and thus could build up to the level of brightness observed from the solar surface, even if the temperature of the flame was modest. Seeking measurements of flame intensity and transparency to confirm his theory, the author consulted not the optical scientists of the day, but the local gas examiner in Sheffield[67]. This official employed his 'photometer of the best construction' in a series of practical experiments. In a period when the majority of the adepts were to be found in the gas industry, most photometric measurements had this pragmatic and utilitarian flavour.

The dominance of gas photometry began to falter as electric incandescent lamps increasingly were seen to be feasible. By the 1880s, the emphasis in industrial photometry was rapidly shifting away from gas testing to the evaluation of electric lamps[68]. The commercial availability of filament lamps dates from 1879 in America, and a few months later in England and other European countries[69]. An indication of the rapid trend towards 'electrotechnical photometry' is given by the laboratories set up for the judging by Committee

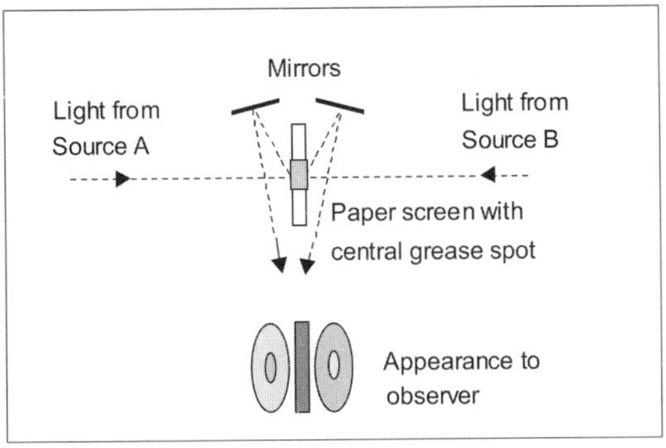

Figure 3.4. Principle of the Bunsen grease-spot photometer head.

of Experiments at successive Electrical Exhibitions. In the 1882 exhibition at Munich, the photometric laboratory used numerous intermediate gas-burner standards. The following year, the Exhibition at Vienna did away with these in favour of electric lamps. The organizers justified the change in terms of the ease of use and stability, at least over short terms, of the latter[70]. In common with the previous examples, the choice of intensity standard in this case had other than a purely technical motive—but now the electric lamp, not gas, was in control.

3.3. THE 19TH-CENTURY PHOTOMETER

As photometry was increasingly employed, its technology stabilized. Photometers came to exemplify the goals of precision and reliability increasingly sought of their users, but paradoxically revealed the unavoidable weakness of human observers in the process.

All standards work, and the majority of scientific applications, employed visual photometers. Devices for light measurement had been designed sporadically through the century for specific researches. By the end of the century, these had evolved into impressively refined products which nevertheless employed the observational principles established by previous generations. Typical instruments often included prisms, polarizers, viewing telescope, translucent or reflective screens (prepared with great care to yield particular viewing characteristics), graduated goniometers or scales. But of the dozens of elaborated versions, serious practitioners used only a few in their work[71]. The principal technical innovation was improvement in the 'photometric heads' used to combine and observe the illumination produced by two light sources. Visual photometry relied upon comparing two sources of light, one the sample and the other a known reference. Comparison proved more accurate when the two intensities were in proximity.

The most enduring photometer design was Bunsen's 'grease-spot' photometer, invented in 1843 for an investigation of the chemical action of light (figure 3.4)[72]. It relied on the fact that a spot of grease or wax on paper

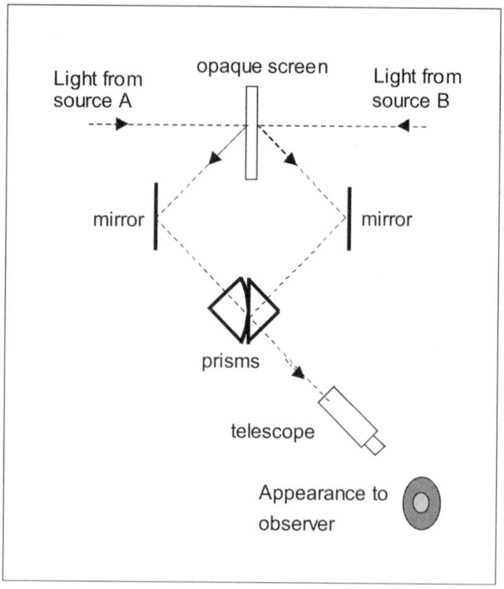

Figure 3.5. Principle of the Lummer–Brodhun photometer head.

appears bright when illuminated from behind, and dark when lighted from the front. By placing the two lamps to be compared on either side of such a screen, the intensities could be adjusted to equality by noting when the grease spot disappeared[73]. The design, employing readily available materials, embodied the majority view that light measurement could be made an everyday task. Experimenters nevertheless invented numerous variants of Bunsen's apparatus. Mirrors were added to allow both sides of the screen to be viewed simultaneously or to alternate the side of the screen illuminated; the simple greased paper was replaced by materials having more optimal transmission and reflection characteristics, or more stable properties. By the end of the century, practitioners of photometry had evaluated the ease of use and repeatability of many types of visual instrument and generally favoured the new head invented by Otto Lummer and Eugen Brodhun in Germany in 1889. This scheme, designed to counteract the perturbing factors by then identified, provided a 'visual field' consisting of two or more immediately adjacent regions from the two light sources (figure 3.5). The screen, instead of being a combination of reflecting and translucent areas, was simply a diffuse reflector and thus easier to fabricate. The precision-manufactured prisms caused the images of the two sides of the screen to be combined when viewed through an eyepiece, yielding a central spot for one side and an outer ring for the image from the opposite side of the screen. As in the grease-spot head, the balance of the two sources was indicated when the division disappeared or had minimum contrast. Its inventors claimed their photometer to be some eight times more precise than the grease-spot photometer. The Lummer–Brodhun

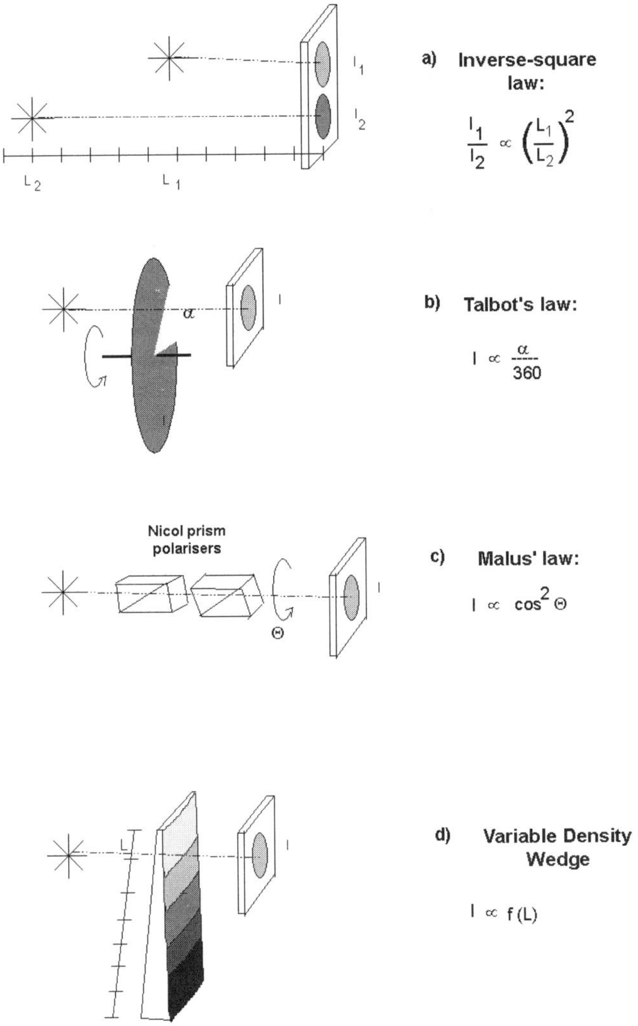

Figure 3.6. Methods used to adjust the reference intensity in visual photometry.

version became the standard for the German gas and electric lighting industries following its commercial manufacture beginning in 1893. This photometer head and its variants, incorporating the values of 'precision' and 'reliability', served routinely in photometric laboratories for the following 40 years. There were, nevertheless, detractors. A dissatisfied British user, for example, complaining that 'the telescope or microscope is considered to be an indispensable adjunct to any instrument in Germany', concluded that the need for one-eyed observation was fatiguing and that the photometric measurement depended too sensitively on the quality of focus[74].

While it comprised the instrumental heart, the photometric head was not the entire photometer. To match the sample intensity to that of the reference light source, the reference intensity had to be adjusted by some convenient means (figure 3.6). Most of the preferred methods related the adjustment of intensity to a simple mathematical relationship. A laboratory-based photometer had few constraints on physical space or on the duration of a measurement, unlike an instrument designed for astronomical use, and so the adjustment of the reference intensity used in the photometric comparison usually relied on moving the lamp away from the screen so that the brightness decreased according to the inverse-square law. The photometer 'bench' contained one or more 'carriages' to move either the photometer head or one of the light sources. To measure light sources of very different intensity, long photometer benches were necessary. One constructed at the National Physical Laboratory in 1905 was 90 feet long, running the length of a specially constructed building[75]. With such apparatus, rapid adjustment of the reference intensity proved cumbersome. Operators increasingly became aware that practical factors such as speed, ease of adjustment and comfort were critical to the measurement accuracy obtained. One practitioner described his technique for equating two lights:

> The secret is this. First you oscillate the photometer until you get the best balance you can, then you oscillate one of the standards, one person oscillating it while the second person is getting a final adjustment of the photometer.[76]

Application of the inverse-square law was ill suited to astronomical usage, however, where apparatus was necessarily mounted on the telescope. In the rotating sector method devised by Talbot, the experimenter exposed the reference screen to light from an opaque disc having a cut-out sector. In later versions devised by William Abney, the sector angle could be adjusted as the disc rotated, allowing continual and rapid matching of its intensity to that of the unknown.

For laboratories having less space or fewer assistants, other methods of intensity adjustment found application. The second most popular adjustment method was based on Malus's law of polarization. The rotation of one polarizer by up to 90° relative to another provided a precise method of varying intensity by 100%. Other, less reliable, methods relied on tilting a reference surface (which provided an analytically known variation in reflectance only for 'ideal' materials) or on estimates of visual acuity that were based on viewing text. These latter were employed mainly by enthusiasts or inventors unfamiliar with the practicalities, and were avoided by serious practitioners.

Optical density wedges found frequent application in astronomy and photography. As standards they were, however, less fundamental than the preceding methods. A wedge was usually formed by a thin prism of grey or 'neutral' glass. Other alternatives included wedges of gelatine and fine lampblack, or coloured liquids[77]. If the glass was homogeneous, its thickness was proportional to the logarithm of its transparency. In practice, no such mathematical relationship was used; instead of relying on the theoretical relationship, the experimenter measured the transparency of the wedge at known positions along its length using one of the previously described techniques.

But, besides the increasingly sophisticated equipment, there was the central importance of the observer himself to the measurement[78]. Each careful photometric observer developed his own method for avoiding errors. William Abney wrote in 1891:

> This operation of equalizing luminosities must be carried out quickly and without concentrated thought, for if an observer stops to *think*, a fancied equality of brightness may exist, which other properly carried out *observations* show to be inexact.[79]

Abney's method of differentiating between 'fancied equality' and 'properly carried out observations' was thus simply to dissociate the mind from the eye. Far from being deemed intrinsically problematic, the reliance upon a mental technique was interpreted by practitioners as a mark of expertise. By the following decade, such unproblematic separation of psychological and physical effects no longer seemed practicable to most scientists.

3.4. PREJUDICE AND TEMPTATION: THE PROBLEMS IN JUDGING INTENSITY

Good photometric practice was arduous. Itemizing the precautions he took to ensure good visual comparisons in stellar photometry, John Parkhurst listed essential precautions in 1906:

(1) The two stars to be compared were made parallel to the line of the eyes. To the writer this precaution was of the utmost importance, for if two equal stars were placed in a vertical line the lower would appear more than half a magnitude the brighter.
(2) Two or three comparison stars were used at each observation if they could be found in proper distances and magnitudes, though this rule often conflicted with the two following.
(3) The stars to be compared should be in the same field, and
(4) The interval in brightness should be less than half a magnitude. If this limit was exceeded the comparisons were weighted in the reductions, inversely as the interval.
(5) Prejudice which would arise from anticipating the star's expected changes, was avoided by postponing the reduction till the maximum or minimum was completed. The observing list was long enough so that the previous observations were usually forgotten at the time of a comparison.
(6) The comparison of too bright stars was avoided by reducing the aperture when necessary.
(7) Light in the eyes was avoided by using for recording a one-candlepower incandescent lamp, so shielded as to illuminate faintly a circle one or two inches in diameter on the record book[80].

Figure 3.7. Gas inspector at work at a photometry table. From Dibdin W J 1908 'Gas photometry in London' *Illum. Eng.* **1**.

Parkhurst's item (5) stresses the measures necessary to avoid involuntary bias by the observer, and echoes the fears of Benjamin Thompson a century earlier in being 'led into temptation'. Parkhurst's other precautions indicate the physiological limitations of visual observation. His list emphasizes the sheer difficulty of obtaining meaningful results. For Parkhurst, the measurement of intensity was highly problematic.

The photographic photometry of small light sources such as stars entrained its own unique problems, the most serious of which was that it did not agree with visual determinations. The very scale of gradation was unstable. Instead of Pogson's ratio of about 2.5 for the difference between magnitudes, a value closer to 3 was usually found, depending on the particular type of star in question and the type of photographic plate used. The problem, astronomers concluded, was due to the different colour sensitivities of the eye and photographic materials. To settle the issue, the Permanent Committee of the Astrographic Congress meeting in Paris in 1909 resolved to equate photographic and visual magnitudes for white type Ao stars[81]. As the visual photometric scale had been defined previously by Pickering and was more firmly established due to the publication of extensive catalogues, this required an adjustment of the photographic photometric scale, also set by Pickering[82]. This *ad hoc* decision thus linked two techniques of light measurement according to a rather arbitrary criterion, namely the particular emission spectrum (and apparent colour) of a common type of star. Quantification in terms of visual and photographic magnitudes already relied on the arbitrary definition of magnitude. That astronomers accepted such a chain of definitions indicates their beliefs concerning the overriding utility of *some* numerical measure for relating and recording stellar intensities.

The increasing usages of photometry by the turn of the century were accompanied by criticism from their users and cautions from experts (figure 3.7). Hermann von Helmholtz had written of intensity measurement that

> the whole region is closely entangled with physiological problems of the utmost difficulty, and moreover the investigators who can make advances are necessarily limited, because they must have long practice in the observation of subjective phenomena before they are qualified to do more than see what others have seen before them.[83]

Even careful attention to technique by meticulous observers resulted in measurements that were of doubtful accuracy. Measurements were affected by several subtle considerations that could be easily missed by a novice investigator. 'Photometry is not a simple and well-defined subject', wrote the author of another book,

> Bare directions will not suffice, but the practitioner must bring to the task a judgement trained for instrumental manipulation and an appreciation for the many modifying influences that the measurements which he obtains may possess in value.[84]

Indeed, the *modifying influences* could seriously affect the accuracy of the measurement. Until these influences could be identified and themselves quantified, implied the author, photometry would yield imprecise and unreliable results.

Foremost among the modifying influences was the basic problem of estimating the brightness of light by eye. As early as 1729, Bouguer, criticizing his contemporaries' ideas of light intensity, had objected that the sensitivity of the human eye varied from time to time, and that too much variation would be found among different observers to allow precise and consistent results. Bouguer's Victorian successors, usually seeing photometry as a 'simple and well-defined subject', frequently started afresh only to rediscover the problems.

Another physiological factor frequently overlooked was the limited range of brightness over which the eye could precisely match two lights. One practitioner, studying photometry for various colours of light, noted:

> If the intensity is too strong, the tired eye partially loses its ability to recognize small differences of intensity; if the light is too weak, on the contrary, the eye no longer easily grasps the difference of intensity... and the measurements are similarly less precise.[85]

As noted earlier, too little or too much mental concentration also was undesirable. Similarly, the observing time and state of health of the observer were relevant to the results obtained. Writing 36 years later, another commentator seemed mired in subjectivity when he wrote:

> Looking at the photometer screen for too short a time reduces the precision, but this happens also if the period is made too long... the

> accuracy, or rather the precision, obtainable in photometric work depends largely on the individual.... As in everything, experience tells also in this class of work. Even the condition of the observer is of importance, and it will be quite obvious that a person out of health will be less reliable—under otherwise equal conditions—than a healthy individual.[86]

For accurate work, he admonished, no more than a dozen measurements could be taken before resting the eyes.

An ill defined range of acceptability seemed to pertain for each of these variables. Even the mental state and expectations of the observer were an important factor. 'The unconscious mental bias' that could result if an observer became aware of any progressive tendency in his readings was avoided in some laboratories by arranging that 'the observers shall work in pairs, each one noting down the readings obtained by the other'[87]. Taking into account these various factors, an unfatigued observer, using convenient apparatus and matching light sources that were neither too bright nor too dim, could obtain accuracies better than 1%; in poor conditions, accuracy might be an order of magnitude worse.

Ominously for the subject, it seemed difficult to countenance a fundamental relationship between the observations of the human eye and of any physical measurement. Alexander Trotter observed:

> Photometry is not the measurement of an external or objective dimension or force, but of a sensation. It is difficult to make a quantitative measurement of our sensations. Two pigs under a gate make more noise than one pig, and while it is possible to measure the amplitude of the vibrations of air which produce sounds, and to estimate those which correspond to the faintest audible sound and those which cause the roar of a large organ, we know little of the quantitative measurement of sound. The attempt to apply measurement to sensations of smell has not met with success, and in spite of the delicacy with which different sensations of taste may be discriminated, it not only seems impossible to measure taste, but there appear to be physiological reasons for a rapid approach to a saturated condition of the sensation. A similar difficulty arises in the action of light on the eye.[88]

For this author, photometry was synonymous with visual observation, being not *a measurement of an external dimension* but rather *a sensation*. He saw no natural connection between light intensity and a physical quantity such as energy. Such a view precluded replacing the eye by a physical detector, because such a replacement would somehow have to mimic the response of the eye, faults and all. At the turn of the century, in any case, practitioners saw few serious alternatives to human observation in the measurement of light. For engineers, there was no physical detector of light available that had the necessary attributes, namely ease of use, reliable properties and a spectral response similar to that of the eye.

By the end of the century, investigators were usually aware of physiological factors, and employed photometers that allowed the eye to make immediate, side-by-side comparative measurements as just described. Measurement again became problematic, though, when the light sources being compared were of different colours. If the flame (or star or light transmitted through a coloured medium) differed in colour from the standard used for comparison, the observer frequently found it difficult to determine a unique relationship between them. The subject could be matched by various combinations of coloured lights, and the match would differ for observers having different colour vision. As different light sources were composed of different distributions of colour, this situation posed severe problems: not only did the result depend on the observer, but on the type of light as well. Colour equality was a subjective attribute that could not be reified. Only when light sources could be compared colour by colour could an 'additive', unique mathematical relationship (Campbell's 'class 3 measurement') linking them be found. And, as astronomers had found, the relationship might hold only for particular varieties of detector or measuring conditions. But while this pessimistic conclusion was pointed out by other writers on the subject, it was by no means universally accepted. William Abney, for example, reported an extensive body of work on colour photometry, claiming to have no difficulty in matching different coloured lights precisely[89].

Beyond the measuring technique itself, the units used in the measurement and description of light could cause considerable confusion, even among engineers. What, exactly, was being measured? One authority related his experience with an American associate:

> An expert, called in to interpret a clause in an electric-lighting contract between a town near New York and the local electrical company, with regard to some 2000 nominal candle-power arcs, expressed his opinion as follows: 'The arc lamps are suspended at the cross roads, and each one, therefore, sends its light in four directions; one cannot, therefore, expect to get 2000 candles in each direction. The 2000-candle arc arranged for in the agreement was one sending 500 candles down each road'. We do not wish to make fun of this expert, for in truth he is a very sensible man.[90]

The arc lamps, explained this authority, produced the equivalent of the light of 2000 candles in every direction. The quoted expert had confused a unit of intensity (candle-power) with a unit of total quantity. With practitioners self-trained and originating from a variety of technical backgrounds, photometry had little prospect of advancement. As late as 1914 photometric concepts and the practice of photometry were perceived as difficult, non-intuitive and a serious hindrance to progress. In a preface to a book on illuminating engineering, Arthur Blok wrote:

> Prominence is given to the 'flux of light' conception, as this seems in great measure to remove a sense of intangibility which the problems

> of illumination so often present to those who approach them for the first time.[91]

Even the inverse-square law, accepted since the time of Bouguer, was disputed by some engineers:

> as far as the evidence goes... photometry is on a fundamentally wrong basis, and... it is absolutely impossible to compare and to express as the function of one and the same unit, the luminous intensity of a *source* of light reduced theoretically to a mathematical point, and that of a luminous beam of which the rays are parallel or sensibly so.[92]

The author was complaining about the theory of lighthouses[93]. British lighthouse lantern sizes had long been designated as 'first order', 'second order' etc. It was now (1893) proposed to replace these by candlepower ratings. The author concluded that 'the values of the luminous intensities attributed to lighthouses and to projectors have not any physical meaning'. In his mind, the quantitative measurement of light was simply not feasible. Many others agreed that the concepts of intensity were flawed. Hospitalier proposed relating light *intensity* to a magnetic field, and *candle power* to a magnetic pole, as analogies. The appropriate physical analogy to apply to light was far from obvious. By the end of the century, however, most engineers favoured the system of photometric units introduced in 1894 by André Eugène Blondel (1863–1938) based on the concept of 'luminous flux', and which defined illumination according to the flux received by a unit surface. His system was adopted in 1896 by the International Electrical Congress at Geneva, and subsequently by the International Illumination Commission and the International Conference on Weights and Measures in following decades. While still unintuitive, Blondel's system was self-consistent and presented a close similitude to other physical units.

Perhaps even worse than being contentious, the practice of photometry was more often ignored. Allied closely, as they were, to standards in the gas industry, developments in photometer design were largely unremarked among scientists. In accepting an award for his design at the 1893 Chicago Exposition, Lummer chided his academic colleagues for having treated photometry 'rather slightingly'. He claimed that they had neglected the subject until the needs of the illumination industry and the public had shown them its importance[94].

3.5. QUANTIFYING LIGHT: N-RAYS VERSUS BLACKBODY RADIATION

The scientific and engineering communities that were beginning to crystallize around the subject at the end of the 19th century followed parallel but independent courses in light measurement. A transition was occurring, among physicists at least, from acceptance of visual methods of observation to a preference for physical methods. The 20th century opened with some notable scientific applications of intensity measurement. Two contrasting and important cases illustrate this trend: n-rays and blackbody radiation.

The case of n-rays has popularly been cited as an example of 'unscientific' methods and 'anomalous physics'[95]. In the context of photometry, however, and perhaps less Whiggishly, it highlights the profound difficulties of visual observation when applied to subtle intensity differences. And for scientists of the day, the n-ray case came to represent more: it illustrated the dangers and undesirability of attempting to *measure* using the human senses.

On 23 March 1903, in the heady decade following the discovery of x-rays, α-rays and β-rays, the French scientist René-Prosper Blondlot (1849–1930) announced his discovery of what he termed 'n-rays'[96]. He reported that these rays were first produced from a heated filament in an iron tube, and emitted through a thick aluminium window. The primary demonstration of the rays was to increase apparent brightness. There were recent antecedents for such observations; indeed, Blondlot's method was current in electromagnetic research from the early 1880s, when Heinrich Hertz explored the characteristics of radio waves by noting the effect of ultraviolet light on the intensity of electric sparks, to the early 1900s, when Lee de Forest observed that a gas flame brightened when a spark gap was operating nearby, inspiring his invention of the triode valve. In the same way, Blondlot found that if a white card was illuminated with extremely dim light—just above the threshold of visibility—his n-ray source would make the card much easier to see. The same effect was produced on other objects illuminated by weak light sources such as fluorescent screens or electric sparks. He and several other investigators used this intensity variation to study the properties of n-rays.

Blondlot himself published ten papers on the phenomenon in 1903, and a dozen in 1904 in the *Comptes Rendus* alone. Over a 16 month period, British, German and American researchers tried with little success to replicate Blondlot's results. But at least 14 French scientists, most of them initiated by Blondlot himself, seemed to have the knack[97]. The observations required not only dark adaptation but also a progressive sensitization to extremely feeble light sources. Said Blondlot, 'to observe n-rays or similar agents, a special exercise of the vision is necessary...we must adapt our organs to a function completely different from that which we normally demand of them'[98]. Indeed, training in meticulous photometric observation was an important part of Blondlot's experimental protocol. He wrote:

> It is *indispensable* in these experiments to avoid all strain on the eye, all effort, whether visual or for eye accommodation, and in no way to try to *fix* the eye upon the luminous source, whose variations in glow one wishes to ascertain. On the contrary, one must, so to say, see the source without looking at it, and even direct one's gaze vaguely in a neighbouring direction. The observer must play an absolutely passive part, under penalty of seeing nothing. Silence should be observed as much as possible. Any smoke, and especially tobacco smoke, must be carefully avoided, as being liable to perturb or even entirely to mask the effect of the 'N' rays. When viewing the screen or luminous object, no attempt at eye-accomodation should be made. In fact,

> the observer should accustom himself to look at the screen just as a painter, and in particular an 'impressionist' painter, would look at a landscape. To attain this requires some practice, and is not an easy task. Some people, in fact, never succeed.[99]

While such visual training had been preached as standard practice in photometry, through 1904 several physicists raised objections about Blondlot's methods. Typical among them was a review of Blondlot's book, *'N' Rays*. Echoing the words in Helmholtz's *Physiological Optics*, the reviewer's central criticism dealt with the *subjectivity* of visual observations:

> the so-called proof of their existence depends, not on objective phenomena that can be critically examined, but on a subjective impression on the mind of the experimenter, who sees, or imagines he sees, or imagines he does not see, a slight change in the degree of luminosity of a phosphorescing screen.

And, in closing:

> these observers have been the subjects either of an illusion of the senses or a delusion of the mind.[100]

In response to his critics, Blondlot supplemented his visual detection method by a seemingly conclusive *physical* method of determining brightness: he exposed half a photographic plate to the light from a spark illuminated by n-rays, and the other half while the spark was shielded from the rays. For each exposure, Blondlot moved the plate manually back and forth a number of times between positions having these conditions to minimize the effect of any external perturbations such as a gradual change in the intensity of the source. The photographic results, like his previous visual observations, showed remarkable statistics (figure 3.8). Of 40 such experiments, just 'one was unsuccessful' in showing a 'notably more intense' impression under n-ray illumination. He concluded that the 'constancy of the results is an absolute guarantee of their worth', and that he had 'succeeded in recording their action on the spark by an objective method'[101].

For Blondlot, this physical technique was a direct analogue of his visual methods, and necessary only to convince experimenters not having the requisite observational skills. He made no attempt to exploit this physical technique nor to suggest that others develop it further. It was merely a gambit to silence his vocal critics. His writings suggest that Blondlot's aim was to discover new phenomena, not to restrict himself to the mere establishment of the exact mathematical relationship between intensity and n-rays. Quantification had a distinctly secondary role in such an agenda.

The *Revue Scientifique* carried out its own investigation in late 1904, and concluded that Blondlot and his followers were all victims of autosuggestion, that no accentuation of light intensity in fact occurred, and that n-rays did not exist. While *The Electrician* reported at the end of the year that 'this extraordinary controversy goes merrily on', Blondlot published no papers in the *Comptes Rendus* after 1904[102].

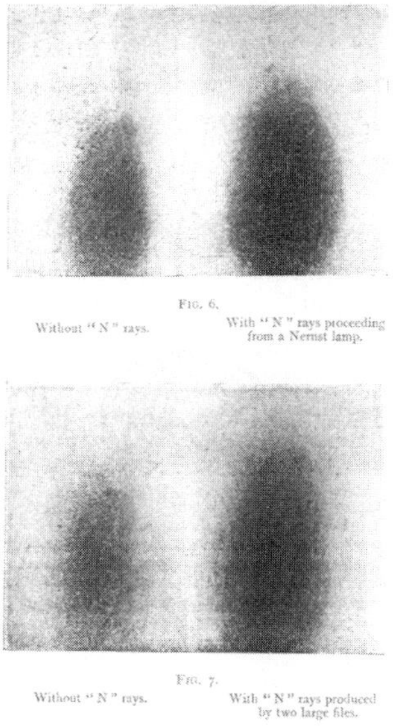

Figure 3.8. Physical proof of n-rays, from Blondlot R 1905 *'N' Rays* (London) facing p 66. Blondlot wrote that he employed a metronome to time the exposure of the photographic plate with and without n-ray illumination, but qualified this by noting that the method did not yield good photographs for publication. The reproduced figures did not use such timing.

This new scepticism over visual methods parallels and contrasts nicely with another case of the measurement of light from hot bodies. This second case was widely perceived as a notable success for 'physical' measurement by contemporary scientists. Radiometry, the close cousin of physical photometry, was mapping the blackbody spectrum between the 1880s and 1920s. Among the experimentalists were some like Heinrich Rubens (1865–1922) who were to seek Blondlot's n-rays without success. Indeed, Blondlot later corresponded with Rubens and attempted to publicly ally his own work with Rubens' researches[103]. Rubens refined the measurements of the emission from heated bodies and extended them from the visible to the far infrared spectrum. By the closing decade of the century, the experimental work had been sufficiently refined to permit some important laws to be postulated[104]. Between 1887 and 1906, this close interaction between experimental work and theoretical derivations culminated in the work of Max Planck (1858–1947). The results were later taken as the first evidence for the quantization of energy[105].

What did these radiometric studies have that n-ray research lacked? Why was their reliability almost unquestioned, and quickly accepted by theorists? The novelty of n-rays cannot be invoked: the period was swamped by novel phenomena that were unanticipated by either theory or prior experiments.

Yet in the eyes of contemporary scientists there were some key differences. First, the blackbody results were repeatable: measurements tended to agree between observers. Although Blondlot claimed that he had achieved excellent repeatability, his results could be reproduced only with great difficulty, if at all, by others. This was a disturbing characteristic of what appeared, on the face of it, to be a straightforward experiment. By contrast, the blackbody measurements, which involved meticulous experimental arrangements using *physical* rather than *physiological* detectors, could be understood by all interested physicists, and verified in at least a qualitative way. In contrast to Blondlot's 'threshold' method of observation, the blackbody measurements were intrinsically numerical; as such they could roughly be approximated by crude observations and then increasingly refined. The statistical calculation of the uncertainty of such measurements instilled more confidence than did the mere detection achieved by Blondlot.

So the blackbody experimental evidence was not an 'all or nothing' affair. Expressed in another way, the blackbody research was founded on what Campbell was to call 'class 3' measurement, i.e. fully quantitative determinations. The n-ray results, by contrast, never sought to go beyond demonstrating the *presence* or *absence* of an intensity change, even when Blondlot claimed to have produced excellent statistics for such detection. They constituted Campbell's crudest 'class 1' observation, in which intensity measurement is limited to a 'greater than' or 'less than' decision. What appears to have disturbed contemporary physicists was that Blondlot restricted his observations to this lowest common denominator and made no serious effort to use available and, in their view, superior techniques. His methods, in short, appeared perversely and persistently old fashioned[106].

A second difference between n-ray observations and blackbody measurements was that the latter were perceived as being 'objective'. The observer merely 'recorded the instrument reading' and played no part in judging the result. Even with Blondlot's photographic technique, his critics pointed out, he had to *judge* how long to leave his plate in the exposed and unexposed positions. Even so, such physical evidence could have been much more easily confirmed than the visual threshold technique Blondlot used almost exclusively; the photograph was capable of providing 'class 3' information if the grey scale were calibrated. There are few records of other investigators attempting to detect n-rays by physical methods, however[107]. This illustrates that scientists were concerned not just by the need to use the eye, but by the sum of Blondlot's experimental methodology. By the time Blondlot published his photographic evidence it was too late; the scientific community had already dismissed his results[108].

The putative differences of quality between visual judgements and radiometric measurements do not appear marked in retrospect. Both were vulnerable to numerous sources of systematic error, but, significantly, radiometric methods confined their systematic errors to physically determinable causes. Errors might be caused by stray light, drifts of readings caused by air fluctuations of the galvanometer, electrical interference of the detector caused by external sources and so on. Each such contribution, though, was seen by the physicist practitioners as potentially identifiable and avoidable. With visual observations, on the other hand, there seemed to be hidden contributions to error that could not easily be evaluated—at least by physicists: a judgement of brightness might be influenced by the observer's alertness, visual characteristics or unwitting bias. Indeed, this crisis for visual photometry between about 1890 and 1910 centred on its reliance upon tacit knowledge and a dominant technical sub-culture. At the root of the comparison was an unsubstantiated faith in physical measurement and a distrust of physiologically based perception.

To physical scientists by the early 20th century, the need to consider explicitly the condition of the observer along with the experiment itself had become distasteful. According to the physicists Richtmeyer and Crittenden:

> the question of the precision of photometric measurements is of peculiar importance in that in this field, more than any other, the precision obtainable is limited by other than physical factors; namely, by the ability of the eye to decide when two adjacent areas appear equally bright.[109]

This sentiment was echoed in a practical context: an engineer wrote, 'The existence of these phenomena [glare, etc] affords one reason why illuminating engineering differs radically from most other fields of engineering. The ultimate judgement... must be based on an appeal to the senses'[110].

These 'other than physical factors' and 'appeals to the senses' had to be avoided. Practitioners such as Richtmeyer sought something better than visual photometry. The solution, they believed, lay in physical methods. Early summarizers of the photometric state-of-the-art noted the trend away from visual measurement and towards 'physical' methods, even if they were pessimistic about the current success:

> As a department of physical science the subject does not seem to have been very attractive, probably because it is one of the least accurate kinds of measurement. Many attempts have been made to banish visual photometry altogether from the physical laboratory. At one time it was thought that the radiometer would supplant it, but it was soon found that the rotation of the 'light mill' depended on thermal rather than on luminous rays. The thermopile and the bolometer have been used to measure the whole radiant energy by means of electrical apparatus, and the dark rays or the luminous rays have been filtered out by selective absorption. Considerable accuracy is possible with such methods, but even if by great precautions changes of temperature have been avoided, and unsuspected radiation of heat guarded against, the proportion of luminous energy to thermal energy is so small that it is hopeless to arrive at any precise measurement of light alone.[111]

The practicalities of using a radiometric detector to measure visible light were indeed onerous. The 'great precautions' needed to avoid swamping the small visible contribution to radiant heating proved impracticable.

Addressing a meeting of the Illuminating Engineering Society of New York, it was left to an engineer to express their growing desire for a *quantitative* subject:

> All the natural sciences aim, then, at becoming exact sciences and become exact through the making, correlation and reduction of measurements. Any branch of natural science without measurements is not above the qualitative stage. The number and degree of precision of the measurements in a branch of science is a gage of the extent to which that branch has become exact.[112]

The latter half of the 19th century thus saw photometry reconceived as a useful tool, particularly by astronomers and engineers. The stimulus for this revised perception was, in each case, *utility*. Astronomers and spectroscopists saw photometry as a means of extending their grasp and of uniting their studies with those of an increasingly mathematized physical science. Gas and electric lighting engineers exploited it as a tool to regularize production and to gain commercial control. Standards of stellar magnitude and luminous intensity conferred legitimacy on the subject and promoted its expansion. With its rising application, however, the practitioners of photometry became increasingly aware of the technical weaknesses of visual methods; their enthusiasm to use photometry was tempered by dissatisfaction with its practical difficulties. The scientists developed increasingly elaborate strategies to minimize the effect of the observer, experimenting with photographic methods while the engineers employed visual techniques, which alone could provide a direct measure of the sensation of illumination at a speed adequate for routine work. The development of the subject over the following decades, though, relied more upon its perceived utility for the emerging communities than on improvements in its foundations or practice.

NOTES

1 Campbell's work spanned the philosophical and applied physics dimensions of light measurement, based on his experience successively at the Universities of Cambridge and Leeds, the National Physical Laboratory and the General Electric Company [*DSB* **3** 31–5]. See Campbell N R 1922 'The measurement of light' *Phil. Mag.* **44** 577–90, written when his research at GEC into photoelectric tubes was getting underway, and Campbell N R 1928 *An Account of the Principles of Measurement and Calculation* (London), written as commercial GEC phototubes were entering the market. In the latter (pp 45–6), he writes: 'Photometry lies outside the range of most physicists, but it offers very interesting problems in measurement. I have an especial interest in it, because I was wholly ignorant of it when I studied the principles of measurement, but have been led since to a close acquaintance with it. Accordingly it has provided a means of testing the principles to which the study of other fields has led.'

2 More precisely, the units follow the associative and distributive laws of arithmetic.

3 He noted, however, that while 'the luminous flux from a lamp is a very important theoretical magnitude', in practice 'the fluxes from two lamps can never be added accurately because one lamp always absorbs some of the light from the other'. See Campbell N R 1928 *An Account of the Principles of Measurement and Calculation* (London) p 44.

4 Thompson B 1794 *Phil. Trans. Roy. Soc.* **84** 362; author's italics.

5 Of 564 publications on light measurement listed in the *Royal Society Catalogue of Scientific Papers 1800–1900*, 41% deal with uses of light measurement, 36% with photometer designs, 15% with units of light, and 8% with spectrophotometry, according to the Royal Society subject divisions.

6 Stellar catalogues that included magnitude estimates appeared increasingly from the 16th century. In the 17th century, at least seven such catalogues were published. Fewer astronomers held an interest in stellar magnitudes in the 18th and early 19th century, however. See Lundmark K 1932 'Luminosities, colours, diameters, densities, masses of the stars', in Eberhard G, Kohlschütter A and Ludendorff H (eds) *Handbuch der Astrophysik* **1** (Berlin) pp 210–573, especially pp 224–73.

7 Dawes W R 1851 'On a photometrical method of determining the magnitude of telescopic stars' *Mon. Not. Roy. Astron. Soc.* **11** 187–90.

8 Applying Pogson's scale of magnitude. To improve the accuracy, he suggested using a threshold technique: a star would, he reasoned, be invisible to a telescope of a certain minimum aperture because the light collected would be insufficient to excite the retina of the observer. This is an example of the extinction method. So, by 'stopping down' the objective lens, one could estimate the stellar magnitude. Dawes pointed out that this sort of photometry merely *ordered* intensities, and did not give them fixed numerical identities that could be added and subtracted. This was the very point reiterated by Campbell 75 years later.

9 Zenger C V 1878 'On a new astrophotometrical method' *Mon. Not. Roy. Astron. Soc.* **38** 65–8.

10 Christie W H M 1878, 'Notes on the specular reflexion of Venus', *Mon. Not. Roy. Astron. Soc.* **38** 108–9.

11 Langley S P 1881 'Researches on solar heat' *Proc. Am. Acad. Arts Sci.* **16** (1881) 432–6 and 'The bolometer' *Nature* **25** (1881) 14–6. For biographical details, see Walcott C D 1912 'Samuel Pierpont Langley' *Biog. Mem. Nat. Acad. Sci.* **7** 245–68. The bolometer, which measures the change in temperature caused by incident radiation, is more sensitive than the thermocouple, which generates a voltage related to temperature difference, and the thermopile, consisting of thermocouples in series.

12 Plotkin H 1978 'Edward C. Pickering, the Henry Draper Memorial, and the beginnings of astrophysics in America' *Ann. Sci.* **35** 365–77.

13 Pickering E C *Astron. & Astrophys.* **11** 22–5.

14 Parkhurst J A 1906 *Researches in Stellar Photometry* (Washington, DC) p 1.

15 Bailey S I 1934 'Edward Charles Pickering' *Biog. Mem. Nat. Acad. Sci.* **15** 169–92.

16 Bond W C 1850 *Ann. Harvard Coll. Observ.* **1** 149.

17 Langley S P, Young C A and Pickering E C 1886 'Pritchard's wedge photometer' *Mem. Am. Acad. Arts Sci.* **11**. As with many photometric innovations, the origins of wedges of graded transparency are unknown. The use of a wedge was certainly described by L A J Quetelet in 1833, and by R Sabine for photographic use in 1882.

18 *DSB* **11** 155–6. The term 'uranometry' refers to the measurement of celestial objects, deriving from the Greek *ouranos* (heavens). Catalogues based on photographic photometry sometimes were entitled 'actinometries'.

19 Pickering's brother William Henry (1858–1938), also at Harvard, published a work with the same title in 1880.

20 Polaris, the north star, was useful in that it was relatively bright and maintained a fixed position in the sky, thereby making possible its observation during an entire night. As the two stars had different elevations, Pickering found it necessary to make

corrections for the effect of atmospheric attenuation, a factor which he determined empirically.

21 Published as volumes **50** and **54** of *Ann. Harvard Coll. Observ.* (Harvard, 1908).

22 Hearnshaw J B 1986 *The Analysis of Starlight: One Hundred and Fifty Years of Astronomical Spectroscopy* (Cambridge) section 5.1.

23 Parkhurst J A and Farnsworth A H 1925 'Methods used in stellar photographic photometry at the Yerkes Observatory between 1914 and 1924' *Astrophys. J.* **62** 179–90.

24 Zöllner J 1859 *Photometrische Untersuchungen, insbesondere über die Lichtenwickelung galvanisch glühender Plantindrähte* (PhD thesis). This was followed by a treatise on stellar photometry, Zöllner J 1865 *Photometrische Studien mit besonderer Rücksicht auf die physische Beschaffenheit der Himmelskörper* (Leipzig). For further biographical details, see *DSB* **14** 627–30.

25 Pickering, too, spent two years experimenting with variants of Zöllner's instrument before devising his meridian photometer.

26 For a discussion of the early Potsdam and Harvard observatories, see Krisciunas K 1988 *Astronomical Centers of the World* (Cambridge).

27 *DSB* **9** 563–4.

28 Typically 0.1–0.2 magnitude, or about 25% to 50%. See Lundmark *op. cit.* note 6 for detailed inter-comparisons of stellar catalogues listing magnitudes measured by visual photometry.

29 *DSB* **12** 95–6.

30 A spectrometer dispersed both the starlight and the light of a reference source, typically a flame, electric lamp or another nearby star of known characteristics. A region of the resulting spectra, located one above the other, was isolated using a slit, and the intensity of the reference band was adjusted to match the subject star.

31 The relative intensity as a function of wavelength was related to stellar temperature by blackbody formulae.

32 See Hearnshaw *op. cit.* note 22, pp 208 and 220–2.

33 He was subsequently one of the first to apply photoelectric methods to astronomical observations and developed recording photometers in the 1920s. The technology of astronomical photometry is discussed in chapter 6.

34 Astronomical photometry developed a larger academic component than did other versions, as evidenced by doctoral dissertations, e.g. that of Zöllner (note 24), Bennett A L 1928 *A Photometric Investigation of the Brightness of 59 Areas of the Moon* (PhD thesis, Princeton University) and Hall J S 1933 *Photoelectric Photometry in the Infra-Red with the Loomis Telescope* (PhD thesis, Yale University). See Hoffleit D 1992 *Astronomy at Yale* (New Haven) pp 131–40.

35 General histories of emission spectroscopy are given by McGucken W 1969 *Nineteenth Century Spectroscopy: Development of the Understanding of Spectra 1802–1897* (Baltimore), and Dingle H 1963 'A hundred years of spectroscopy' *BJHS* **1** 199–216.

36 See, for example, Newall H F 1910 *The Spectroscope and its Work* (London), which describes 'Principal' and 'Subordinate' spectral lines, the latter being 'fainter but sharper'. As in stellar photometry earlier in the century, spectroscopists used a rough estimate of intensity (usually into three or four ranks) to label lines.

37 Lockyer J N 1873 *The Spectroscope and its Applications* (London) p 51.

38 Lockyer cited the recent examples, too, of the discovery of the elements of caesium and rubidium in spring water by Bunsen (1860), of thallium by Crookes and of indium

by Reich and Richter in Germany. Despite this emphasis on mere *detection*, there was some interest in the potential for *quantifying* materials. A Mr Sorby, writing the same year, noted that he could measure the age of wine by the intensity of a particular spectral absorbance band. Using a 'microscope spectroscope' to examine vials of wine, he observed that 'the difference for each year is at first so considerable that wines of different vintages could easily be distinguished' [*Chem. News* December 17 1869 p 295].

39 Single-exposure photography was scarcely able to measure intensity ranges of 100:1, and this only when carefully calibrated.

40 For example, G G Stokes and others explored the ultraviolet spectrum in the early 1860s when quartz was found to make a suitably transparent prism. In 1865, Balmer discovered a simple numerical fit for part of the spectrum of hydrogen, supporting the contention that spectroscopy had a mathematical basis. New physical effects were discovered, such as the spectral perturbations caused by magnetic fields (Zeeman, 1896).

41 See, for example, Vogel H C 1892 'On the spectrographic method of determining the velocity of stars in the line of sight' *Astron. & Astrophys.* **11** 203–7. The precision of Vogel's spectrographic methods far exceeded that available by visual observations. For a further discussion of Vogel's work, see Hearnshaw *op. cit.* note 22, pp 77–89.

42 Parkhurst *op. cit.* note 14, p 8. The 'artificial star' was a lamp located behind a pinhole aperture, and collimated by a lens so as to appear to be located at infinity.

43 Liveing G and Dewar J 1892 'On the influence of pressure on the spectra of flames', *Astron. & Astrophys.* **11** 215–21.

44 Alglave E and Boulard J 1882 *La Lumière Électrique: son Histoire, sa Production et son Emploi* (Paris) pp 8–9, and Palaz A 1894 *Treatise on Industrial Photometry*, transl. G W and M R Patterson (New York) pp 111–18.

45 Williams T I 1983 *A History of the British Gas Industry* (Oxford). For an introductory history of gas lighting, see Schivelbusch W 1986 *Disenchanted Night: the Industrialization of Light in the 19th Century*, transl. A Davis (Oxford).

46 Clifton G C 1992 *Professionalism, Patronage and Public Service in Victorian London: the Staff of the Metropolitan Board of Works 1856–1889* (London) p 32.

47 *Ibid.*, pp 42–3. The MBW promoted bills in the 1860s and 1870s to allow it to supply gas or to purchase gas companies. These bills failed, but led to enforcement of stricter regulations of the gas companies by the MBW.

48 See Abady J 1902 *Gas Analyst's Manual* (London), in which the first chapters are devoted to photometric techniques.

49 Dibdin W J 1889 *Practical Photometry: a Guide to the Study of the Measurement of Light* (London) pp 181–2.

50 *Ibid.*, p 77.

51 *Ibid.* Dibdin became better known from the 1890s as a pioneer of biological sewage treatment. See Hamlin C 1990 *A Science of Impurity: Water Analysis in Nineteenth Century Britain* (Berkeley) pp 283–4.

52 Dibdin, *ibid.* pp v–vi. The book provides several examples of the legal disputes surrounding the intensity of gas lighting in Victorian London, and of the variety of hardware employed to resolve them.

53 The illuminating gas industry, on its part, consolidated expertise in photometry and other technical subjects by establishing the British Association of Gas Managers in 1863. It aimed at 'progress through the enlarged intelligence of its members to be brought about by the free interchange of opinion and experience' [Buchanan R A

1989 *The Engineers: a History of the Engineering Profession in Britain 1750–1914* (London) pp 95–6].

54 *Gas Works Clauses Amendments Act, 1871, schedule A, parts I and II.*

55 A P Trotter, quoted in Fleming J A 1907 *A Handbook for the Electrical Laboratory and Testing Room*, Vol II (London) p 240.

56 *Ibid.*, pp 238–55.

57 Thompson S P 1909 *Illum. Eng.* **2** 813.

58 National Physical Laboratory 1913–14 *Report* (Teddington) p 50.

59 Paterson C C 1907 'Investigations of light standards and the present condition of the high voltage glow lamp' *J. IEE* **38** 271–7.

60 *London Gas Referee's Notification for 1901*: 'The pentane is to be obtained from Light American petroleum by three distillations, at 55 °C, 50 °C and 45 °C in succession. The distillate at 45 °C is to be shaken from time to time, during two periods of not less than three hours each, with one-tenth its bulk of (1) strong sulphuric acid, (2) solution of caustic soda. After this treatment it is to be again distilled, and that portion is to be collected for use which comes over between the temperatures of 25 °C and 40 °C. It will consist of pentane, together with small quantities of lower and higher homologues, whose presence does not affect the light of the lamp.' The notification included mandatory testing of the product which comprised evaluation of density in both the liquid and gaseous state, and colour. In practice, pentane to be used in a Harcourt lamp for testing the illuminating power of town gas was prepared in bulk by the gas companies, and then tested by the Referees and supplied in sealed cans to the gas-testing stations, which were under the control of the chemical adviser of the London County Council. See Fleming *op. cit.* note 55.

61 Fleury P 1932 *Étalons Photométriques* (Paris).

62 Such national diversity in standards was the norm rather than the exception. The case of the resistance standard has been treated, for example, in Olesko K M 1993 'Precision and practice in German resistance measures: some comparative considerations', paper presented at workshop at Dibner Institute MIT 16–18 April 1993, and Hunt B J 1994 'The ohm is where the art is: British telegraph engineers and the development of electrical standards' *Osiris* **9** 48–63.

63 Trotter A P 1911 *Illumination: its Distribution and Measurement* (London) p 14.

64 For a particularly standardized measurement protocol, see Abady *op. cit.* note 48.

65 Alglave *op. cit.* note 44, pp 301–4; quotation p 303 (my translation).

66 Williams W M 1870 *The Fuel of the Sun* (London) ch 7.

67 By seeking to verify the 'countability' of intensity, the author was attempting to verify what Norman Campbell referred to as the third or most quantitative form of measurement. Lighting was generally accepted to be of the 'rankable, but not necessarily combinable' form (Campbell's class 2) at this time.

68 The decline of routine photometric testing of gas supplies was accelerated by a trend towards the simpler but not entirely equivalent technique of calorific testing, which 'quite a number of the leading companies' had adopted by 1910 [Gaster L and Dow J S 1920 *Modern Illuminants and Illuminating Engineering* (London) pp 72–3].

69 For general histories of the evolution of electric lighting, see, for example, Cox J A 1980 *A Century of Light* (New York) and Schivelbusch *op. cit.* note 45.

70 Palaz A *op. cit.* note 44, p 181. The widespread contemporary application of public electric lighting is illustrated by Alglave E and Boulard J *op. cit.* note 44; the Paris Expositions of 1878 and 1881 were important showplaces for the new technology.

71 Palaz *op. cit.* note 44 ch 2, describes over two dozen variants in considerable detail.

72 Bunsen R and Roscoe E H 1859 *Phil. Trans.* **149** 891.
73 In practice, this condition occurs only if the reflectance of the paper equals the transmittance of the grease spot. Practitioners overcame this difficulty by either equating the *contrast* of the spot on either side of the screen, or by causing it to disappear on each side and then averaging the resulting measurements.
74 Trotter *op. cit.* note 63, p 105.
75 National Physical Laboratory 1905 *NPL Report* (Teddington).
76 Ayrton M *J. IEE* **32** 206.
77 Walsh J 1926 *Photometry* (London) p 179.
78 *Himself*, because I have found no record of female photometric observers before circa 1905, when routine electric lamp measurements began to call for patient, careful and low-paid employees—commonly voiced attributes of women observers during this period. The requirements were similar to those at Airy's Greenwich Observatory, which had demanded 'indefatigable, hard-working, and, above all, obedient drudges' [S Schaffer, 'Astronomers mark time' *Sci. Context* **2** (1988) 120].
79 Abney W de W 1891 *Colour Measurement and Mixture* (London) p 79; author's italics.
80 Parkhurst *op. cit.* note 14, pp 2–3.
81 Stellar classifications had been increasingly refined over the previous decade by the examination of stellar spectra. Pickering, chairing the committee, was joined by Jöns Oskar Backlund (Director of the Pulkova Observatory); Karl Schwarzchild (Director of the Potsdam Observatory); Edwin B Frost (Director of the Yerkes Observatory) and Herbert Turner, an Oxford astronomer [Plotkin *op. cit.* note 12].
82 Pickering's *North Polar Sequence*, consisting initially of the photographic magnitudes of 47 stars, was used. The Sequence included 96 stars by 1912.
83 Helmholtz H 1924 *Physiological Optics—Vol I* transl. J P C Southall (New York) p viii.
84 Stiles P *Photometrical Measurements*, quoted in Walsh J W T 1926 *Photometry* (London) p vii.
85 Trannin H 1876 'Mesures photométriques dans les differentes régions du spectre', *J. de Phys.* **5** 297–304; quotation p 304 (my translation).
86 Bohle H 1912 *Electrical Photometry and Illumination* (London) p 82.
87 Walsh *op. cit.* note 77, p 316.
88 Trotter *op. cit.* note 63, p 67.
89 Abney's researches, widely cited, included: 'Colour photometry' (Bakerian Lecture, with E R Festing) *Proc. Roy. Soc.* **40** 238; Abney 1891 *Colour Measurement and Mixture* (London); Abney 1895 *Colour Vision* (London); and Abney 1913 *Researches in Colour Vision and the Trichromatic Theory* (New York).
90 Blondel A E 1894 *Electrician* **33** 633.
91 Blok A 1914 *The Elementary Principles of Illumination and Artificial Lighting* (London) p v.
92 Hospitalier M 1893 'Photometric fantasies' *L'Industrie Électrique*, reprinted in Hospitalier 1893 *Electrician* **32** 59–60.
93 The design of lighthouses had occupied such scientists as Michael Faraday and Augustin Fresnel earlier in the century. Fresnel (1788–1827) spent the last few years of his life devoted to work for the French lighthouse commission, which included designing stepped lenses to improve collimation and beam intensity. Some 65 years later, André Blondel followed him by being employed by the École des Ponts et Chaussées and by the Service Central des Phares et Balises. Blondel used

his experiences with lighthouse design and electrotechnics to devise the system of photometric units later adopted by international conferences. Because of the previous existence of national committees and an international association of lighthouse authorities, the otherwise influential Commission Internationale de l'Éclairage (discussed in chapter 7) steered away from this subject in light measurement and standardization.

94 Quoted in Cahan D 1989 *An Institute for an Empire: the Physikalisch-Technische Reichsanstalt 1871–1918* (Cambridge) pp 106–7.

95 See, for example, Langmuir I and Hall R N 1989 'Pathological science' *Phys. Today* **42** 36–48, an edited transcript of a talk given by Langmuir in 1953.

96 Blondlot R 1903 'Sur une nouvelle espèce de lumière' *Comptes Rendus* **137** 735–8. Blondlot was professor of physics at the Université de Nancy (hence the appellation 'n' rays), and a corresponding member of the Académie des Sciences. He was known for his previous investigations of x-rays.

97 Nye M J 1980 'N-rays: an episode in the history and psychology of science' *Hist. Stud. Phys. Sci.* **11** 125–56. For the more nuanced recent historiography see also Gelain C and Geoffrey H 1965 'A la pursuite des rayons "N"' *Ingenieur des Industries Chimiques* **41** 7–12; Lagemann R T 1977 'New light on old rays: N rays' *Am. J. Phys.* **45** 281–4; Ashmore M 1993 'The theatre of the blind: starring a Promethean prankster, a phoney phenomenon, a prism, a pocket, and a piece of wood' *Soc. Stud. Sci.* **23** 67–106.

98 Blondlot R 1904 'Sur une méthode nouvelle pour observer les rayons N et les agents analogue' *Comptes Rendus* **139** 114–15 (my translation).

99 Blondlot R 1905 *'N' Rays* transl. J Garcin (London) pp 82–3.

100 McKendrick J G 1905 'The "N" Rays' *Nature* **72** 195. See also Stradling G F 1907 'A resumé of the literature of the N rays, the N_I rays, the Physiological rays and the heavy emission, with a bibliography' *J. Franklin Inst.* **164** 57–74, 113–30, 177–99.

101 Blondlot 1905 *op. cit.* note 99, pp 61–8.

102 Anon. 1994 Editorial *Electrician* **54** 296.

103 Blondlot *op. cit.* note 99, pp 13, 17, 30.

104 Friedrich Paschen (1865–1947) found the wavelength of peak emission to be inversely proportional to temperature. Encouraged by the reliability of the data, theorists such as the Russian W A Michelson (1860–1927) and the German H F Weber (1843–1912) tried to fit formulas to them.

105 Histories of blackbody radiation research include Kangro H 1976 *The Early History of Planck's Radiation Law* (English translation, London) and Kuhn T S 1978 *Blackbody Theory and the Quantum Discontinuity* (Oxford). A good contemporary survey is Coblentz W W 1921 'The present status of the constants and verification of the laws of thermal radiation of a uniformly heated enclosure' *JOSA* **5** 131–55.

106 These characteristics were subsequently categorized by the American industrial physicist Irving Langmuir as 'pathological science' [Langmuir and Hall *op. cit.* note 95, p 44]. His symptoms of such a science are the following: (1) The maximum effect is produced by a causative agent of barely detectable intensity, and the magnitude of the effect substantially independent of the cause; (2) the effect is of a magnitude that remains close to the limit of detectability, or many measurements are necessary because of the low statistical significance of the results; (3) claims of great accuracy are made; (4) criticisms are met by *ad hoc* excuses; (5) the ratio of supporters rises initially and then falls continuously. Langmuir's points are questionable; symptoms 3 and 4, for example, are not particularly strong factors

in the ultimate rejection of observations. The definition of 'great accuracy' and '*ad hoc* excuses' could differ for supporters and opponents of the evidence. Even more tellingly, the number of supporters of a new phenomenon may vary for reasons other than internal scientific consistency or methodological rigour. Such sociological causes are ignored by Langmuir. However, his first and second points highlight the difference between a truly quantitative measurement and threshold detection. This single, crucial difference appears to have been central to the rejection of Blondlot's results and the acceptance of blackbody data. Intriguingly, Langmuir, who had used visual photometry during his incandescent lamp research, cited two cases of visual detection (n-rays and scintillation counting) as paradigmatic examples of 'anomalous science'.

107 One such case, published weeks after Blondlot's evidence, was Weiss G and Bull L 1904 'Sur l'enregistrement des rayons N par la photographie' *Comptes Rendus* **139** 1028–9. Repeating his experiment, they were unable to reproduce Blondlot's results: 'dans aucun cas nous n'avons pu obtenir de résultat positif'.

108 Franklin A 1986 *The Neglect of Experiment* (Cambridge) and Franklin A 1990 *Experiment, Right or Wrong* (Cambridge), discusses factors determining the acceptance of new experimental data in sub-atomic physics. He argues persuasively that the data and statistical evidence are a small part of the acceptance, and that other less tangible factors such as the reputation of the experimenter and the perceived complexity of the experiment are important factors.

109 Richtmeyer F K and Crittenden E C 1920 'The precision of photometric measurements' *JOSA & RSI* **4** 371–87.

110 Teichmuller J 1928 *Illum. Eng.* **21** 130.

111 Trotter *op. cit.* note 63, p 68.

112 Kennelly A E 1911 *Trans. Illum. Eng. Soc. (NY)* **6** 580.

CHAPTER 4

CAREERS IN THE SHADOWS

Measuring light was becoming an increasingly organized activity at the close of the 19th century. Promoting the new cultural values of quantification, standardization and control were new groups of career workers. Photometry was building its own technological networks, and becoming an important agent in what has been called the 'era of technological enthusiasm'[1]. What 'professional' alliances brought together its practitioners?

During this period, the measurement of light intensity was carried out in various milieus and by a variety of people. While the predominant users of photometry continued to be relatively unskilled inspectors, those responsible for the principal innovations in practice and technology changed during the period. These latter ranged from enthusiasts and amateurs during the 19th century to well-connected and influential career scientists active shortly before the Second World War. In Britain, at least, the subject of light measurement was profoundly shaped by individuals, both acting alone and giving purposeful direction to fledgling organizations. Britain was also the country exhibiting the greatest range of organizations involved with photometry in the first decades of the new century. This chapter therefore illustrates the organization of its practitioners by focusing on the careers of several Britons.

At least two social groupings of practitioners became established: engineers concerned with lighting technology, and a loose collection of scientists active in applied optics and instrumentation. By the end of the First World War, these communities increasingly were characterized by a growing self-awareness, identification of common aims, establishment of training programmes and interaction with other organizations. Technical societies united individuals active in the subject before other forms of organization became significant. This was to be augmented by direct employment in government and industry (chapter 5) and by the rise of delegated bodies (chapter 6).

4.1. AMATEURS AND INDEPENDENT RESEARCH

Peripheral to much of 19th century science, photometry was sustained by enthusiastic amateurs, a scientific type prevalent in Britain[2]. By championing an unpopular subject using private funds, they were able both to increase its exposure to particular communities and to nurture its development along individualistic lines.

William de Wiveleslie Abney (1843–1920) typifies the career pattern of a dedicated Victorian exponent of light measurement. Obtaining a commission to the Royal Engineers at the age of 18, he spent a decade in India. Invalided home in 1871, he was appointed as chemical assistant to the instructor of telegraphy at the Chatham School of Military Engineering, where he was able to pursue a boyhood interest in photography. Within three years Abney was responsible for a separate school of chemistry and photography there, and became Inspector of School Science at the Science and Art Department located at South Kensington. His career after this time devoted equally to education and science, Abney retired from the army in 1881[3]. In the same year, he introduced the first sensitive photographic emulsion based on gelatine. His interests, centring on scientific photography, extended to all matters photometric.

Abney published over 100 papers and a similar number of popular articles on photography, sensitometry, physiological optics and photometry—almost all connected with the measurement or perception of intensity[4]. Editor of *The Photographic Journal* (London) from 1876 until his death, he was a prolific contributor to numerous photographic, astronomical and scientific journals. He was active in scientific and technical societies, being elected president of The Royal Photographic Society four times between 1892 and 1905, president of The Astronomical Society from 1893 to 1895, and of The Physical Society between 1895 and 1897. For Abney, light measurement was an essential adjunct to scientific photography. He lamented that 'of 25 000 people who took photographs not more than one cared for, or knew anything about, the why and wherefore'[5]. With missionary zeal, Abney sought to convert the lack of scientific interest regarding photometric issues. During his presidency of the London Photographic Society in the 1890s, he transformed it into a scientific institution, prompting one commentator to remark that 'the meetings became still duller, and *The Photographic Journal* was devoted almost exclusively to scientific aspects of photography'[6].

Abney was central in laying the foundations for photographic photometry and unique in having a broad interest in light measurement as well as an unparalleled desire to understand the scientific basis of photography. The connection was not easy to popularize.

> The idea of measuring light is so unfamiliar to many quite intelligent people, that they confuse the word photometry with photography, and have neither the remotest idea that light can be measured nor how any operation of measurement can be carried out when no units of length, volume, weight. . . or time, or appreciable force or movement, enter into the question

complained one of his contemporaries[7]. Abney and his occasional collaborators studied the light sensitivity of photographic materials as a function of chemistry, wavelength of light and processing conditions[8]. He used photographic methods to explore subjects as diverse as the intensity of coronal light during a solar eclipse[9], the spectrum of electric lamps[10], the near-infrared spectrum[11], and numerous other topics of contemporary interest. Abney's contributions to photographic sensitometry, in particular, were much cited in contemporary texts. Drawing on his educational connections, he gave courses of public lectures on photography and colorimetry (both of which led to popular books). Abney's cross-fertilization of astronomy, physiology, photography and physics may well have introduced many of his scientific contemporaries to photometric approaches of investigation.

In a period when full-time scientific employment was still uncommon in Britain, William Abney was nevertheless more than the modern definition of an amateur. His investigations were careful and extensive, maintaining close connections with professional scientists. On the other hand, his research was usually divorced from the duties of his paid position, and he was active in several associations more closely linked with enthusiasts than to men of science. Apart from monetary remuneration, however, Abney was in most respects a career scientist.

Abney's research and occupational history were by no means unique. One of his near contemporaries, J Norman Lockyer (1836–1920), followed a similar career path in several respects[12]. Lockyer took up astronomy as a hobby while working as a clerk in the British War Office. His first observatory was set up in his garden at Wimbledon in 1862. Noting his interests, Lockyer's superiors assigned him to a succession of posts relating to scientific administration. These were followed by a grant for equipment to observe the 1868 eclipse, directorship of the Solar Physics Observatory which opened in South Kensington in 1879, and a professorship at the Royal College of Science in 1881[13]. He founded the journal *Nature* in 1869, editing it for 50 years, and was president of the British Association for the Advancement of Science in 1903. In the latter two roles he promoted the widespread application of science to social problems. By 1890, Lockyer was an influential figure, too, in British spectroscopy, for which he promoted photometric measurement.

Abney and Lockyer were typical of British investigators in photometry before 1900. Developing a strong amateur interest in a subject neglected by full-time scientists, they engaged in independent research, lobbied for support and popularized their studies by means of public lectures and books of general interest. The publicizing of scientific specialisms in this way was an effective method of gaining support in the late Victorian period, when lay-persons could and did read scientific journals and books. Neither Abney nor Lockyer had any success (nor expressed motive) in organizing scientists or engineers into special-interest groups. Rather, they attempted to rally other individual investigators to their cause by providing examples of its utility. Thus Abney preferred a cogent demonstration to a meticulous study, illustrating colour blindness, for example, by mapping the response of one subject's eyes to colour, rather than by examining a cross section of individuals. The result of this method of leading by example was that both Abney and Lockyer became respected members and officers of scientific and technical societies but never founded organizations of their own. Exemplars rather than leaders, their enthusiasms were not, on the whole, shared by their contemporaries, and these remained marginalized as minority interests in societies having broader goals.

The technique of mobilizing *popular* interest and secondarily entraining *scientific* attention was a tactic also employed by a separate group of individuals intimately concerned with light measurement: the 'illuminating engineers' ('illuminating', because, as several of the early engineers complained, the term 'illumination' was more closely associated with mediaeval manuscripts or fireworks than with lighting). In contrast to their seniors, Abney and Lockyer, however, the engineers proved remarkably effective in defining both a subject and a career structure for themselves.

4.2. THE ILLUMINATING ENGINEERS

In the first decade of the 20th century, illuminating engineering was a subject close to attaining a self-recognized career status, yet its practitioners were, for the most part, hesitant to call themselves professionals[14]. Their self-awareness sprouted in the span of scarcely a decade. Besides their impressive rate of growth, the utilitarian origins, too, of the illuminating engineers were quite separate from the more recreational scientific interests of Abney and his generation. Also in marked contrast to their predecessors the gas inspectors, the illuminating engineers promoted the scientific development of light measurement for utilitarian ends.

With the commercial availability of electric lighting in the 1880s, an atmosphere of rapid technological development and 'progress' had become widespread. Bright, steady light became not only a desired utility but a symbol of scientific advancement. The journal *La Lumière Électrique*, for example, founded in 1880, promoted every aspect of electrical technology and devoted a portion of its thrice yearly volumes to illumination and its measurement. Electricity would supply the light of the future, figuratively as well as literally.

Applying the new technology demanded more than just an engineering bent, however. The electrical enthusiasts who developed lighting systems found themselves faced with marketing, physiological and economic questions. How were they to convince purchasers of the *need* for more or better lighting? How could they compare meaningfully the competing light sources in terms of brightness, colour and efficiency? How much light *was* needed for various tasks, and how should lighting systems best be installed and employed? Increasingly, the measurement of the *illumination* of surfaces rather than the *luminance* of light sources was emphasized, raising concerns of fair pricing. 'If serious attention is to be given to the often recurring suggestion that the customers of lighting companies be charged according to the actual illumination secured and that street lighting be rated and paid for on a mean or a minimum illumination basis', noted one author, 'reliable methods of measurement are indispensable'[15].

The Illuminating Engineering Society was founded in New York in 1905 by a group of 25 who wanted a society dealing specifically with the art and science of illumination. As was to be mirrored in Britain, the society was preceded by a general-circulation magazine, *The Illuminating Engineer*[16]. Indeed, it appears that these publications preached the sermon of illuminating engineering before a 'common enterprise' was recognized, thereby hastening its advent. The idea had first been mooted by Louis B Marks, a consulting electrical engineer, and Van R Lansingh, an engineer at the Holophane Glass Co, who decided to contact interested persons, judging that 'six or eight men, if they are the right ones, would do for a starter'[17]. The society gained 93 members in its first year, and within two years the membership had swelled beyond 1000. Early prominent members included Thomas Edison and André Blondel, the principal French exemplar of intensity standards.

Despite its claimed interest in science, the new-born society's practical concerns were decidedly utilitarian. One proposed name was the 'Society for Economical Illumination'[18]. Indeed, the new members frequently stressed *economy* in their early rhetoric[19]. The motivations of this first Illuminating Engineering Society centred on the efficient usage of lighting. Its first president observed that lighting costs in the United States in 1905 were conservatively estimated at $200 000 000 per year, of which some $20 000 000 was wasted by the consumer 'by reason of his failure to properly utilise the energy supplied'. This 10% wastage rose to 25%, he continued, 'by improper disposition of light sources or unsuitable equipment of lamps, globes, shades, or reflectors'. The aim of the society was therefore 'to point out in what way the best illuminating result may be obtained from any source of light, be it electric, gas, oil, or candle'[20]. Relatively little mention of light *measurement* appears in its early publications. The 22 papers presented in the first year included two on photometry, both of them presented by British members[21].

Having branches in five north-eastern US cities, the society consciously sought members having a practical, rather than scientific, bent[22]. Their society did not attempt to attract scientists, instead including 'electrical engineers, gas engineers, architects and designers of lighting fixtures' among its members. Tellingly, 'the views not only of the engineer but of the practitician' were to be courted[23]. Significant support from industry is indicated by the income generated by advertisements in the *Transactions of the Illuminating Engineering Society of New York*[24].

The birth of a society dedicated to illumination was not welcomed by all. Some preferred that illumination and photometry be made the subject of sub-committees of existing electrical and gas societies. Moreover, it was argued, excessive technical development might make life more difficult for practitioners. One editorial noted that 'at present, commercial photometry is delightfully simple, and it is questionable whether anything tending to complicate it will be welcomed by practical men'[25]. Others felt that the subject was intrinsically unworthy of attention: 'Can illumination be measured with sufficient accuracy and with sufficiently simple apparatus to make it a practical basis for many matters?'[26] The writer concluded that it could not.

Figure 4.1. Provocatively defining a new movement: the front cover of the first volume of *The Illuminating Engineer*, 1908.

The situation in New York had several parallels with that in London. In both cities, competition in lighting systems was increasing, and growing numbers of self-trained specialists were acting as consultants on matters of illumination. Leon Gaster (1872–1928), a British engineer much impressed by this American example, promoted the foundation of a similar society in Britain[27]. He had become editor of a new magazine also called *The Illuminating Engineer* in 1908 (figure 4.1)[28]. The publication attracted 140 readers, drawn mainly from engineering and science, by the end of its first year. As with its American counterpart, the magazine also united many of them in a common interest. Writing for newspapers and other periodicals as well as his own, Gaster was a tireless proselytizer for the need of an organization concerned with illumination. His efforts paid off: at a meeting in a Piccadilly restaurant in early 1909, 26 interested individuals founded the Illuminating Engineering Society of London[29].

These two independent societies collected together a highly eclectic assortment of individuals interested in the practice and measurement of illumination. Unlike the economic and practical motives of the American society, however, the British version was to centre on scientific measurement and application[30]. Subtitling the magazine *The Journal of Scientific Illumination*, its editor strove to promote this orientation. At the founding meeting and in editorials, the London society made clear its objectives and laid emphasis on quantitative measurement. 'What is wanted, above all, is to make the measurement of illumination a practical and familiar practice', wrote Gaster, 'just as the measurement of electric current or gas is already felt to be'[31].

The 'Illuminating Engineering movement' (so-called by the founders on both sides of the Atlantic) was an uneasy collection of groups with narrower interests. Indeed, the titling of the periodical *The Illuminating Engineer* was a provocative attempt to define a hitherto non-existent community, because no such occupational identity was recognized even among practitioners. The society would encourage the cooperation 'of oculists, physicists, the optical industry, architectural profession and Society of Engineers in Charge'. There were, however, existing animosities to be overcome. One of the proposers noted that 'the bringing together of those representing gas, electricity &c. was a stupendous task'. The previous year, Gaster had written on this topic:

> At the time of his inception the illuminating engineer was hailed as a man likely to add to the gaiety of nations. It was freely prophesied, owing to the conflicting interests of electricity, oil, and gas, that a meeting of an illuminating society would have more the aspect of a beer garden than a sedate scientific assembly... but, as is often the case, the prophets have turned out to be windbags and the illuminating engineer, at least in America, is an established fact.[32]

The uncertain welcome of the illuminating engineer is suggested by figure 4.2: they were often viewed with mistrust by architects and lighting manufacturers in equal measure. Gaster was repeatedly to stress the *neutrality* of the journal and Society in questions of technological evaluation. Nor were the divisions restricted to engineers backing competing technologies. The disparate concerns of physiologists and engineers were remarked by an oculist: 'some attention has been paid to the subject [of the physiological effect of light] by the medical profession, but their views were not sufficiently impressed upon the engineers'[33]. In an activity so new, the scope of illuminating engineering itself was not yet circumscribed. Kenelm Edgcumbe, an instrument-maker, gave examples of the measurement of illumination later used for courtroom evidence, 'one illustration of the unexpected directions in which the need for light measurement was constantly being experienced'[34].

Despite Gaster's strenuous efforts to found the new society, he willingly accepted the position of Secretary and proposed a noted scientist as President. This served the dual purpose of linking the society to science and giving it a prominent figurehead. The founders sought 'one who is in sympathy with our movement and has taken a wide interest in light, illumination and illuminants generally'[35]. Rather than a scientific enthusiast like William Abney, they sought an established scientist having industrial connections, someone who had

Figure 4.2. The imperious illuminating engineer. 'The Troubles of the Electrical Contractor, No V. He receives instructions and technical advice as to the position of the fittings' *Illum. Eng.* **2** (1909) 763, reprinted from *Elec. Industries* September 21, 1909. The engineer holds a Holophane globe and *The Principles of Illuminating Engineering*.

made the subject his *business*. They found their man in Silvanus Phillips Thompson. Thompson (1851–1915) was a well known and respected educator and popularizer of science. His career until then had concentrated on electrical engineering and technical physics, having chaired the Research Committee of the Institute of Electrical Engineers, and been its President in 1899. During the 1890s he had researched x-rays and fluorescence and developed an interest in photometry, leading to the short work *Notes on Photometry* in 1893[36].

One of Thompson's acquaintances, the Engineer-in-Chief of the Post Office, William Preece, shared some of the qualities required of a candidate for leadership of the Illuminating Engineering Society. In 1893 he had organized a committee in England to act with a similar group in America to consider a standard of light and illumination. Preece had already been interested in photometry for over a decade, having been asked by the Commissioners of Sewers of the City of London in 1883 to prepare a specification for lighting part of the City by electricity, and granted a sum of £200 by them for experiments[37].

Some ten years before the formation of the Illuminating Engineering Society, then, Preece had asked Thompson, along with William Abney and John Ambrose Fleming, to serve on his committee[38]. Thompson, in turn, approached his acquaintance Hermann von Helmholtz, director of a new national laboratory, the Physikalisch-Technische Reichsanstalt, about German participation. As will be discussed in chapter 5, the Reichsanstalt was then completing research on a fundamental standard of light and felt little inclination to work with ill prepared collaborators. Nothing came of the committee other than Thompson's heightened profile both at home and abroad as an expert on photometry[39].

Barely eight years younger than William Abney, Thompson nevertheless followed a career path more effectively tuned to exploiting his subject in a rapidly changing society. Besides being a popularizer of science, Thompson was a promoter of better education and industrial links. In 1902 he began a campaign to organize an institute of 'opto-technics' (in analogy to the 'electrotechnical' training courses then becoming widely available). Elected President of the Optical Society in 1905, he organized the first Optical Convention at the sole British institution teaching technical optics, the Northampton Institute in London[40]. The Convention exhibited the work of the optical trades which, according to Thompson, employed some 20 000 workers in the London district alone[41].

With his background in electrotechnics and optics and his high public profile, Thompson proved an effective figurehead for the new Illuminating Engineering Society. He was vocal in his opinions about the current status of photometry and lighting: 'the ascertained facts are few—all too few; their significance is immense; their economics and social value great; but the ignorance respecting them generally is colossal!...To sum up, the work before us is *to diffuse the light*' (emphasis in original)[42]. During the four years of his presidency, Thompson promoted the Society and its governmental and international connections, continuing until shortly before his death in 1915[43].

The choice of President and Secretary was instrumental in crystallizing the goals and outlook of the Society and its members. The early publications mirrored the new society's self-perception. The founding members were not eager to claim professional status. Indeed, the very idea of illuminating engineering as a *profession* was actively derided. Leon Gaster noted that

> membership of such a society cannot, at the present time, be regarded as any claim to professional distinction. We naturally hope that in times to come, when the subject of illumination has been thrashed out in detail to a far greater extent than at present, 'expert illuminating engineers' will have a professional existence and will, even though few in number, be entitled to claim the distinction that the name implies...the number of experts in this country who are entitled to claim the title with any approach to justice are...few indeed.

The society was to be called not *The Society of Illuminating Engineers* but *The Illuminating Engineering Society*. 'This meant anyone interested in the subject of lighting could join the society but membership would not carry with it any professional status'[44]. The American society had agreed to a similar name for similar reasons; in both cases, the proposal for the name Illuminating Engineering Society prevailed, making it 'representative of an art' instead 'of a profession'[45]. In another editorial, Gaster again cautioned against defining arbitrarily the profession of illuminating engineer: 'any attempt to force his existence in name only, without the necessary qualifications, can only bring the title into disrepute'[46]. Both Leon Gaster and Silvanus Thompson voiced their desire to make the society a collection of non-professionals interacting like the participants at meetings of the British Association for the Advancement of Science. This tactic clearly had two benefits: it broadened the potential membership, allying the subject with more established fields and it promoted the synthesis of a new subject from components of the old. Gaster's co-founders agreed with his aims. One, seconding the motion to form the society, replied that he was 'much impressed of the responsibility in replying on behalf of a profession which [does] not yet exist'[47]. Yet as the first president of the society, Silvanus Thompson held a much looser and all-encompassing definition of their activities, stating that

> diverse and individual interests centre upon a common topic... *illumination engineering* [sic]. So far as this is their profession they are engineers—for is not the definition of engineering the art of directing the powers of Nature to the use and convenience of man?

The magazine and society were nevertheless directed at a specific audience, namely the Illuminating Engineering movement:

> In their movement, as in every movement, they must have a number of leaders before an appeal can be made to the masses. [Gaster] had, therefore, endeavoured in the journal to appeal to the scientists and to the better educated engineers, so that once there was agreement as to the necessity of spreading the knowledge of illumination, the public, who were the consumers, would gradually be educated by those pioneers who at the present formed the bulk of the readers of our magazine.[48]

The conscious rejection of professional status by illuminating engineers hinged on their recognized lack of qualifications or testing standards. While a few lectures were available, formal training was non-existent[49]. A physicist at Cornell University, F K Richtmyer, noted that photometry played a minor role in the education of physicists and engineers. 'Typically the photometrical measurements are only secondary', he remarked, 'the main point of the experiment being usually the study of some problem by the aid of photometry'. With so little formal training 'it would be presumptuous... to regard illuminating engineering as a separate entity in the great science of engineering'[50]. As a partial solution, he proposed a course of ten lectures for his students. The following year, the journal reported on a more elaborate course given at Johns Hopkins University in Baltimore. Thirty-six lectures were given, along with demonstrations and laboratory work, to 250 postgraduate teachers and other interested persons. A more permanent educational facility was set up at the Case School of Applied Science in 1916, which continued to give courses on illuminating engineering through the 1920s[51]. Unlike the academic courses provided for the older engineering specialties, such courses, presented in large part by the illuminating engineering staffs of large firms, presented a business-oriented view of the subject[52]. The Illuminating Engineering Society of New York, too, devoted attention to educational activities. An *Illumination Primer* was published in 1912, and other pamphlets and teaching materials were frequently produced for local chapters of the Society. Lectures were even published in book form[53]. In Britain, similarly, courses on illumination became more common after *The Illuminating Engineer* was launched. As early as 1908, lectures on illumination were held at two London technical institutes: the Northampton Polytechnic and the East London College, followed in 1909 by four Cantor lectures by Leon Gaster at the Royal Society of Arts during the month that the Illuminating Engineering Society was founded, and two years later at three London polytechnics[54]. The availability of the journal and lectures clearly promoted the formation of the society. The lighting industry played a major role in organizing courses, The Electric Lamp Manufacturers Association (ELMA), for example, holding annual series of lectures beginning in 1918[55]. In 1926 this educational drive was extended by a 'Home Lighting Course for Women', which included six lectures which were to 'take the audience by easy stages through the history of lighting, illustrating the demands of modern civilisation, and then explain, by the aid of numerous demonstrations, how the home should be wired and lighted'[56]. Despite such attempts by business and technical societies to instigate standards of training for practitioners and support increased awareness among the public, as late as 1936 one commentator was able to state that 'illuminating engineering still remains more of a trade than true profession'[57].

In spite of a reticence for claims to professionalism by both the British and American societies, by 1910 a well developed culture of illuminating engineering was established. The diffusion of state-of-the art knowledge is well illustrated by texts independently published by persons associated with the Illuminating Engineering Society of London around this time[58]. A spate of books appeared before the First World War in response to the growing organization of illuminating engineers. While discussing gas lighting, they generally sought to incorporate illumination and photometry into electrical engineering practice. Hermann Bohle, a South African practitioner, argued that photometry had previously been neglected,

> yet this subject is as important as, or even more important than, the design of dynamos and motors. It is useless to raise the efficiency of generators and motors by 1 or 2 per cent and afterwards to waste the power by improper illumination engineering.

This argument closely parallels an example given by the president of the New York society six years earlier:

> The electrical engineer goes to great lengths to gain a small percentage in the economy of his boilers, engines, generators and transmitting system; the illuminating engineer has a problem which is in many ways far easier, because he can take the bad conditions which

> prevail at the present time and can produce a much more considerable betterment in results than lies within the easy reach of the electrical engineer...it is very possible to gain very considerable economies quite as useful as the additional economies which are to be attained at the generating plant.[59]

The practitioners saw themselves as more than merely engineers of economy, however. The current president of the British society emphasized the multidisciplinary nature of his craft, writing: 'Illumination is not an exact science with well defined laws of what might be called illuminative engineering, but an art whereto an indefinable and incommunicable skill pertains almost as it does to the magic of a painter'[60].

The domain of the illuminating engineer indeed encompassed disparate skills. He was versed in lamp technology at a time when several systems were commercially viable[61]. Between 1880 and 1920, at least three technologies vied for dominance: (a) gas lighting, revitalized by efficient burners, incandescent mantles, and high-pressure operation; (b) filament electrical lighting and (c) arc lamps, for high-intensity lighting of public places. New, more reliable and economical systems were constantly being developed, such as the Nernst glower lamp. Between 1890 and 1910, the difficulties of incandescent lamp manufacture, and potential profits from more efficient technologies, motivated engineers to seek alternatives. During this 20-year period, both innovation and technical development blossomed. The great illuminating efficiency of the firefly was much discussed, and an electrochemical or luminescent analogue was actively sought. Yet Silvanus Thompson felt compelled to emphasize to its new members that the Illuminating Engineering Society would deal with *quantifiable* matters, and that 'our Society has as little to do with fireworks as with fire-flies'[62]. The illuminating engineer required a strong background in electrical engineering to appreciate the best operating conditions for lamps and their interconnection into electrical networks. Advertisements not infrequently called for an 'illuminating *electrical* engineer'[63].

Illumination expertise also included a strong component of human physiology. The illuminating engineer worked with detailed tables of appropriate lighting levels, itemized for type of work and buildings[64]. And less tangible qualities such as colour and mixture of natural and artificial lighting were also on the agenda[65].

Most pertinently, the illuminating engineer worked routinely with photometry, both in a practical and theoretical sense; it formed the sole experimental tool at his disposal and a theoretical model of his handiwork. This new community of practitioners rapidly became the principal vector of innovation, application and promulgation of photometry. As with gas inspection some decades earlier, technology and industry were closely linked. The characteristics of commercially available light sources increasingly were measured and tested in commercial production[66]. Numerous portable photometers were available by 1910, designed for either measuring the intensity of a light source or the illumination of a surface. Early portable illumination photometers measured the illumination in rooms or lighted streets by an extinction method, in which the operator sighted the illuminated scene and interposed graduated absorbers until it disappeared (figure 4.3). Unusually among his contemporaries, William Preece had in the 1880s urged the measurement of illuminated *surfaces* rather than of light sources themselves. In a paper presented to the Royal Society, he said:

> We do not want to know so much the intensity of the light emitted by a lamp, as the intensity of the illumination of the surface of the book we are reading, or of the paper on which we are writing, or of the walls upon which we hang our pictures, or of the surface of the streets and of the pavements upon which the busy traffic of cities circulates.... Hence, I propose to measure the illumination of surfaces quite independent of the sources of light by which they are illuminated.[67]

This shifted emphasis was to preoccupy the illuminating engineers and, somewhat later, investigators at government and industrial laboratories.

The growth of the 'illuminating engineering movement' in the first decade of the 20th century thus entrained technological and social change, and united a disparate collection of workers. Seeking to specialize in what appeared to be a readily exploitable subject, these practitioners began an active dialogue in their journals discussing all aspects of illumination and its measurement. Their expansion was attributable to a combination of practical need and scientific acceptance of an increasingly quantitative subject. One post-First World War practitioner commented that

> the rapid development of the lighting art, and its transference from the domain of pure empiricism to that of scientific method which has been a marked feature of the last decade of engineering progress, have tended to emphasize more and more the importance of this branch of photometric practice.[68]

The transition was accompanied by new sponsors and applications. The impetus that had been given to photometry over the previous half-century by gas lighting was now virtually spent. Electrotechnology promised to be the technology of the future for lighting and for light measurement. In turn, the emphasis on lighting applications caused mainstream photometry to develop increasingly in this direction.

When Leon Gaster died in January 1928, 20 years after his journal had started, the domain of illuminating engineering was more widely established as a stable endeavour. The field had been defined by a generation of practising engineers seeking to systematize the measurement of light. The career scientists and engineers now working in the field used the occasion to pay their tributes not only to Gaster and his Illuminating Engineering Society, but to bolster the subject itself. Alexander Trotter, a past President of the society, eulogized that in founding the journal and Society Gaster had 'had the courage to found in anticipation

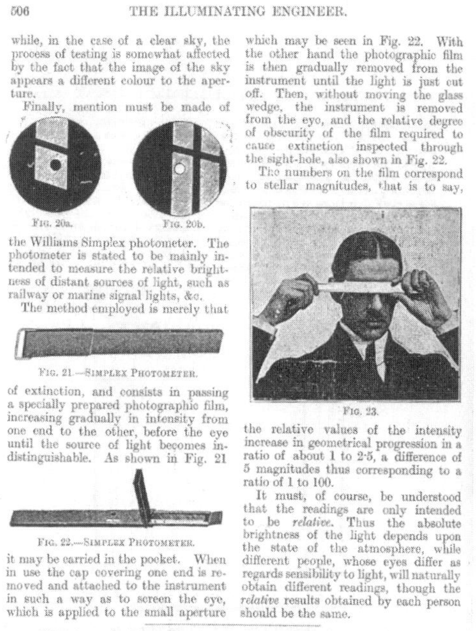

506 THE ILLUMINATING ENGINEER.

while, in the case of a clear sky, the process of testing is somewhat affected by the fact that the image of the sky appears a different colour to the aperture.

Finally, mention must be made of the Williams Simplex photometer. The photometer is stated to be mainly intended to measure the relative brightness of distant sources of light, such as railway or marine signal lights, &c.

FIG. 20a. FIG. 20b.

The method employed is merely that of extinction, and consists in passing a specially prepared photographic film, increasing gradually in intensity from one end to the other, before the eye until the source of light becomes indistinguishable. As shown in Fig. 21 it may be carried in the pocket. When in use the cap covering one end is removed and attached to the instrument in such a way as to screen the eye, which is applied to the small aperture which may be seen in Fig. 22. With the other hand the photographic film is then gradually removed from the instrument until the light is just cut off. Then, without moving the glass wedge, the instrument is removed from the eye, and the relative degree of obscurity of the film required to cause extinction inspected through the sight-hole, also shown in Fig. 22.

FIG. 21.—SIMPLEX PHOTOMETER.

FIG. 22.—SIMPLEX PHOTOMETER.

The numbers on the film correspond to stellar magnitudes, that is to say, the relative values of the intensity increase in geometrical progression in a ratio of about 1 to 2·5, a difference of 5 magnitudes thus corresponding to a ratio of 1 to 100.

FIG. 23.

It must, of course, be understood that the readings are only intended to be *relative*. Thus the absolute brightness of the light depends upon the state of the atmosphere, while different people, whose eyes differ as regards sensibility to light, will naturally obtain different readings, though the *relative* results obtained by each person should be the same.

Figure 4.3. The eye in the middle: portable visual photometers circa 1908. The lower two incorporate an electric lamp that can be adjusted in intensity to visually match the test source. *Illum. Eng.* **1** (1908).

of a demand, the enthusiasm to develop on scientific lines, the skill to balance between competing interests, and the satisfaction of producing so successful and attractive a form'[69]. Clifford Paterson, the then current President, noted that in the early days 'the need for the illuminating engineer was not appreciated and his *profession* only imperfectly understood'[70]. The members vaunted the future of illuminating engineering. John Walsh of the National Physical Laboratory echoed that he saw the subject as 'increasing...rapidly at present'. Elihu Thomson of General Electric in America even saw signs that illuminating engineering was expanding to encompass all forms of electromagnetic radiation:

> Just at present we find great interest in the production and application of rays which cannot be said to be illuminating, but which are of the same general nature. The usefulness of ultra-violet radiation has been thoroughly demonstrated, if we are permitted to use the term 'illumination' in reference to invisible rays... it is, indeed, difficult to assign limits to what can be done with this enormous range of wave frequencies, and, so far as illumination itself goes, many of the invisible rays are capable of exciting in special fluorescent materials visible light rays. I feel safe in predicting that the opportunities for usefulness for the *Illuminating Engineer* will not be diminished in the forthcoming twenty years.[71]

By 1935, illuminating engineering societies similar to the American and British examples and devoted almost exclusively to electric lighting were active in several countries. Representatives of the younger German and Dutch illuminating engineering societies applauded the international flavour of the journal, and traced its effect in influencing British legislation. Photometry was, in the early decades of the 20th century, a significant part of such organizations, which were principally tasked with the organization of standards, education and commercial promotion of lighting. Perhaps of most practical importance to a practising engineer, the subject also was receiving recognition from outside the fraternity. The 13th edition of the *Encyclopaedia Britannica* of 1927 had included an entry for illuminating engineering written by Gaster himself.

4.3. OPTICAL SOCIETIES

The linkage of illumination engineering with electrotechnology rather than with optics is attributable to the rapid expansion of electric lighting and the growth of a community of practitioners. By contrast, optics before 1914 involved a collection of disparate and unorganized practitioners much as illuminating engineering had done before the turn of the century. Despite the Optical Conventions of 1905 and 1912 in Britain which attempted to bring together all workers in optics, university scientists and optical craftsmen worked in different and almost mutually exclusive aspects of the field. There was little perception among them of optics being an activity of common interest, or of any potential benefit arising from organization, until the war changed their views. At that time government, industry and academia became acutely aware of the predominance of German commercial optics. This was particularly true in Britain and America, which had a dangerous reliance on German instruments and glass. The Department of Scientific and Industrial Research (DSIR) was founded in 1915 because

> many of our industries have since the outbreak of war suffered through our inability to produce at home certain articles and materials required in trade processes, the manufacture of which has become localised abroad, and particularly in Germany, because science has

> there been more thoroughly and effectively applied to the solution of scientific problems bearing on trade and industry and to the elaboration of economical and improved processes of manufacture.[72]

At the time, the UK was manufacturing less than a quarter of the types of optical glass being made by Germany, and a tenth of the requirements of the dyestuffs industry. There was an urgent practical need to design and manufacture optical devices and to develop national expertise in all aspects of optics for the war effort[73]. The DSIR and numerous national committees were set up to organize this. During and after the war, the new links that had been formed were maintained by the formation of optical societies. These professional groupings aimed to promote research and manufacture in an atmosphere of increased national awareness. Founded in 1916 principally by a group at Eastman Kodak, the Optical Society of America (OSA) brought together researchers and engineers concerned with all aspects of optics. This included photometry and colorimetry. Its *Journal of the Optical Society of America and Review of Scientific Instruments* (*JOSA*) became the principal English-language organ for scientific optics in the 1920s. Unlike continental journals, *JOSA* treated a much broader field than simply imaging optics. Along with lens design, it dealt with subjects such as colour measurement and the physical principles of light detectors. In England, the *Journal of Scientific Instruments* (founded in 1923) covered similar subjects, notably opto-electrical and opto-mechanical devices for measurement. Nineteenth-century optics was being broadened and redefined in terms of new technology.

The memberships, subjects treated and industrial linkages of the optical societies increased steadily through the 1920s. The economic depression of the following decade, however, caused a slump in the membership and publication rate of the Optical Society of America. Its flat membership rolls through the 1930s belied the number of new and extended activities of optical scientists in research, government and industry begun in that decade.

The turn of the century thus witnessed shifts in light measurement: a transition of photometric innovation from an activity of amateur scientists to career engineers; a transition from gas-lighting to electric-lighting firms; a transition from individual workers to groups organized in technical societies. Practice was appropriated by a new, self-aware community of illuminating engineers that increasingly became allied with the electric lighting industry. Coalescing first in America and Britain, the illuminating engineering movement championed the scientific development of photometry for utilitarian purposes. Optical societies encompassing the subject of light measurement joined in, particularly following the impetus of war-time shortages and organization, to enlist a broader range of career workers into the problems of light and colour measurement.

While providing a focus for common interests, the movement was ineffectual in carrying out research-oriented activities. Urging photometric standards and measurement practices, its members initially had neither the funds nor support needed from government and industry. Instead, the illuminating engineers relied upon a handful of interested scientists using make-shift equipment. The birth of the national and industrial research institutions greatly eased this impasse. Government- and industry-funded laboratories staffed by career scientists were now available, albeit having objectives distinct from those of the illuminating engineering movement. Organization of the subject by technical societies, industry and government brought new laboratories and a growing community of engineers and scientists concerned with light measurement.

NOTES

1 Hughes T P 1989 *American Genesis: a Century of Invention and Technological Enthusiasm* (New York).

2 D S L Cardwell has discussed reasons for the British condition of 'scientific amateurism' which persisted until the turn of the 20th century, ascribing it to the lack of a system of academic posts and of government commitment to funding scientific education and applied research [Cardwell D S L 1972 *The Organization of Science in England* (London) pp 179–84].

3 Abney's career, mixing service in the Royal Engineers with science teaching, was typical of the period. By the early 1870s, a lack of science teachers caused the War Office to allow officers of the Royal Engineers to supervise examinations of the Department of Science and Art. Abney told an 1881 Royal Commission 'the training and education of engineer officers renders them fit persons to be acting inspectors [of science classes]'; see Cardwell *op. cit.* note 2, pp 116, 136. He did not *share* the two roles, however: the War Office was informed in 1878 that his recall to his Corps would 'inconvenience the public service' [Departmental Minutes, quoted in Butterworth H 1968 *The Science and Art Department, 1853–1900* (unpublished PhD thesis, University of Sheffield) p 100].

4 In deciding reluctantly to promote him, his superior wrote in 1884 that he was 'never very sure of Abney, who had a strong liking for putting his name on original work'. Abney eventually succeeded him as Director of Science, and when the Department was reorganized in 1900 became 'Principal Assistant Secretary, Science and Art Dept.' and finally 'Head of the South Kensington branch of the Board'. He retired in 1903 but had continued contact with the Department almost until his death. See Butterworth *op. cit.* note 3, p 479.

5 Obituary notice: Anon 1921 *Proc. Roy. Soc.* A **99** i–v. Other biographical sources: *DNB* (1912–21) 1; *DSB* **1** 21–2 and Butterworth *op. cit.* note 3.

6 Gernsheim H and Gernsheim A 1955 *The History of Photography* (Oxford) p 256. Regarding the limited attention given to scientific investigation in the photographic industry, see Edgerton D E H 1988 'Industrial research in the British photographic industry, 1879–1939', in Liebenau J *The Challenge of New Technology* (Aldershot) pp 106–34.

7 Trotter A P 1911 *Illumination: Its Distribution and Measurement* (London) p 65.

8 Abney W 1874 'On the opacity of the developed photographic image' *Phil. Mag.* (4th series) **48** 161–5.

9 Abney W and Thorpe T E 1886 'On the determination of the photometric intensity of the coronal light during the solar eclipse of August 28–29, 1886' *Proc. Roy. Soc.* **44** 392.

10 Abney W and Festing E R 'The relation between electric energy and radiation in the spectrum of incandescence lamps' *Proc. Roy. Soc.* **37** 157. Festing knew Abney both during their time as Royal Engineers and later in his role as keeper of the Science Collection at South Kensington.

11 Abney W 1892 'On the photographic method of mapping the least refrangible rays of the solar spectrum' *Proc. Roy. Soc.* **30** 67, and 'On the limit of the visibility of the different rays of the spectrum', *Astron. & Astrophys.* **11** 296–305.

12 See, for example, Hearnshaw J B 1986 *The Analysis of Starlight: One Hundred and Fifty Years of Stellar Spectroscopy* (Cambridge) pp 89–94 and *DSB* **8** 440–3.

13 The publication of science books was also a significant source of his income. See Brock W H 1976 'The spectrum of science patronage', in Turner G E (ed) 1976 *The Patronage of Science in the Nineteenth Century* (Leyden) p 199.

14 Practitioners of light measurement generally eschewed the idea of a profession *per se*. Their goal was, rather, what has been called 'occupational upgrading' instead of 'professionalization' [Morrell J B 1990, 'Science in the universities: some reconsiderations' in Frängsmyr T (ed) 1990 *Solomon's House Revisited: the Organization and Institutionalization of Science* (Canton, MA) pp 51–64]. The term *profession* defies precise definition. Some of the characteristics commonly ascribed to professionals that the illuminating engineers *lacked*, however, were an educational process, recognition of status by the state and a self-perception of social duty. For a discussion of the 'impressive imprecision' surrounding the definition, see Buchanan R A 1989 *The Engineers: a History of the Engineering Profession in Britain 1750–1914* (London) pp 12–15. On scientific professionalization, see Morrell J B 1990 'Professionalization', in R C Olby *et al* (eds) *Companion to the History of Modern Science* (London) pp 980–9. For a discussion of the changing sociological definitions of professionalization and bureaucratization, see Torstendahl R 1982 'Engineers in industry 1850–1910: professional men and new bureaucrats. A comparative approach' in Bernhard C G, Crawford E and Sörbom P 1982 *Science, Technology and Society in the Time of Alfred Nobel* (Oxford) pp 253–70.

15 Wickenden W E 1910 *Illumination and Photometry* (London) pp 72–3.

16 E Leavenworth Elliott, the editor of *The Illuminating Engineer (NY)*, became the first secretary of the Society. The magazine retained its independent status, however, with *Trans. Illum. Eng. Soc. (NY)* becoming the Society organ.

17 Hibben S G 1956 'The Society's first year' *Illuminating Engineering (USA)* **52** 145–52. Marks had patented an enclosed carbon arc lamp as an undergraduate, and later worked for the Westinghouse Electric & Manufacturing company. The Holophane Glass Co, based in New York, specialized in the design and manufacture of novel prismatic lamp globes to control and redirect light, and employed a large proportion of the illuminating engineers of the area.

18 Hibben, *ibid.*, p 147.

19 See, for example, Wickenden *op. cit.* note 15, ch XIV: 'Engineering and economic principles in interior illumination'.

20 Marks L B 1906 'Inaugural address of the President', *Trans. Illum. Eng. Soc. (NY)* **1** 7–8.

21 Trotter A P 1906 'Errors in photometry' and Hyde-Cady M 'Lamp photometry', *Trans. Illum. Eng. Soc. (NY)* **1**.

22 The number of regional chapters increased to 14 during the 1920s, and to 21 by the Second World War.

23 Anon. 1906 'The organization of the Illuminating Engineering Society' *Trans. Illum.*

Eng. Soc. (NY) **1** 2, 8. Unlike their counterparts in London, the original officers and council of the Illuminating Engineering Society of New York were not closely connected with other developments in American photometry. This chapter therefore focuses on the British organization.

24 Anon. 1913 'Annual Report' *Trans. Illum. Eng. Soc. (NY)* **8** 683. Advertizing for the 1913 fiscal year provided $1097.14, some 13% of total income.

25 Anon. 1907 *Electrician* August 30, quoted in *Illum. Eng.* **1** (1908) 144.

26 Anon. 1907 *The Electrical Times* December 19 in *Illum. Eng.* **1** (1908) 144.

27 Gaster was born in Bucharest, and obtained a BSc in 1890. He worked for four years in electrotechnics under E H Weber at the Zurich Polytechnic, and moved to the UK in 1895. Gaster became a naturalized British subject in 1903, when he began to do consulting engineering. See Gaster L 1926 'Twenty-one years of illuminating engineering' *Illum. Eng.* **19** 12. The extent of his connections with the American society are unclear: Gaster had contributed a paper to its first year's *Transactions*, and was at least in contact with its officers. Although occasionally referred to as 'sister organizations', the two societies had no formal connection.

28 The financial backers of the Illuminating Engineering Publishing company and periodical are unclear, but did not include Gaster himself.

29 The German equivalent, the Beleuchtungstechnische Gesellschaft (Society for Illumination Technology) was founded in 1912 by the then director of the PTR, Emil Warburg. Its tardy formation may be attributable to the dominance of the Reichsanstalt in setting industrial standards and in centralizing action on questions of illumination and measurement. Illuminating engineering societies were organized later in several other countries: Japan in 1917, Austria in 1924 and Holland in 1926. Even in the USSR, which was less influenced by market forces, societies and research laboratories sprang up: in Leningrad in 1923, Moscow in 1927 and Kharkov in 1929.

30 The relative importance of British versus American scientists in 'authenticating' the new electrical technology at the turn of the century is discussed in Hughes T P 1983 *Networks of Power* (Baltimore, 1983) pp 53 and 234.

31 Gaster L 1909 'Editorial' *Illum. Eng.* **2** 796.

32 Gaster L 1908 'The illuminating engineer as specialist' *Illum. Eng.* **1** 175–7.

33 Parsons H 1909 *Illum. Eng.* **2** 156.

34 Kenelm Edgcumbe was co-director of Everett, Edgcumbe & Co, a firm specializing in the manufacture of optical instruments, particularly photometers. He was, in later years, a member and President of the British National Committee on Illumination, a delegate to the Commission Internationale de l'Éclairage, and chairman of the British Engineering Standards Association, in which capacity he set specifications for photometric instruments.

35 Thompson J S and Thompson H G 1920 *Silvanus Phillips Thompson: his Life and Letters* (London) p 274.

36 Thompson had considerable assistance in writing his *Notes on Photometry* from his friend Alexander Trotter, a London consulting engineer who supplied him with information on 'the very latest thing in photometers and photometry'. See Thompson *op. cit.* note 35, p 256. Trotter had also assisted William Preece in 1883–4 with his measurements on illumination.

37 Walsh J W T 1951 'The early years of illuminating engineering in Great Britain' *Trans. Illum. Eng. Soc.* **16** 49–60.

38 Thompson *op. cit.* note 35, p 273. J A Fleming (1849–1945) had been a consultant to the Edison Electrical Light Co from 1881 to 1885, and was Professor of

Electrotechnology at University College, London, for 41 years. His text on laboratory methods, published in 1907, included a chapter on photometry.

39 His Christmas Lecture of 1896 on 'Light visible and invisible' was translated into German by Otto Lummer of the Optics Section of the Physikalisch-Technische Reichanstalt.

40 For a discussion of its later-developing French counterpart, l'Institut d'Optique, see Paul H W 1985, *From Knowledge to Power: the Rise of the Science Empire in France, 1860–1939* (Cambridge) pp 310–13.

41 Thompson *op. cit.* note 35, p 264.

42 Thompson *op. cit.* note 35, p 275, quoting from Thompson's 1909 inaugural lecture as President of the IES (London).

43 In 1912, for example, he chaired a meeting of the London society and its American counterpart at the National Physical Laboratory to discuss photometric nomenclature.

44 Gaster L 1909 *Illum. Eng.* **2** 156.

45 Anon. 1906 'Organization of the Illuminating Engineering Society', *Trans. Illum. Eng. Soc. (NY)* **1** 1.

46 The desire among electrotechnicians and other engineers to replace unformalized knowledge by higher education in the 1880–1910 period is discussed in Torstendahl *op. cit.* note 14.

47 Dow J S 1909 *Illum. Eng.* **2** 158.

48 *Ibid.*, p 155.

49 This contrasts with the teaching standards of electrotechnics established by this time. See Gooday G 1991 'Teaching telegraphy and electrotechnics in the physics laboratory: William Ayrton and the creation of an academic space for electrical engineering in Britain 1873–1884' *Hist. Technol.* **13** 73–111.

50 Richtmyer F K 1909 *Illum. Eng.* **2** 851–2. Richtmyer (1881–1939) was active in early research into the photoelectric effect and its application to photometry. See, for example, Richtmyer F K 1913 'Photoelectric cells in photometry', *Trans. Illum. Eng. Soc. (NY)* **8** 459–69. He was also a promoter of purely photometric research in America, editing the 1937 text *Measurement of Radiant Energy*. See Ives H E 1943 'Floyd Karker Richtmyer' *Biog. Mem. Nat. Acad. Sci.* **22** 71–82.

51 The Case School courses were prepared principally by the staff of the Nela Research Laboratory (described in chapter 5). The two-term course for electrical engineering students covered 'all aspects of illuminating engineering as presently understood' in three lectures per week and laboratory work using Nela equipment. Lecturers included three Nela employees, five from the National Lamp Works of GE, an architect and representatives of two gas lamp manufacturers. See Anon. 1925 'Illuminating Engineering for Students and Engineers' *J. Sci. Instr.* **2** 365–7 and Cady F E 1920 'A cooperative college course in illuminating engineering' *JOSA* **4** 537–9.

52 The training situation in illuminating engineering had parallels with that in chemical engineering, a specialty that emerged in the inter-war period. See Divall C and Johnston S F 2000 *Scaling Up: The Institution of Chemical Engineers and the Rise of a New Profession* (Dordrecht), ch 4.

53 Illuminating Engineering Society 1911 *Lectures on Illuminating Engineering, Delivered at the Johns Hopkins University October and November 1910* (Baltimore), and IES 1917 *Illuminating Engineering Practice: Lectures on Illuminating Engineering Delivered at the University of Pennsylvania, Philadelphia, September 20 to 28, 1916* (New York). The former included Charles Steinmetz and Willis Whitney of General Electric as lecturers.

54 Walsh *op. cit.* note 37, p 53. In 1911, members of the Illuminating Engineering Society of London gave four courses, consisting of a total of 27 lectures.

55 In America, the National Electric Light Association was similarly occupied with 'propaganda lectures on illumination'. Equivalent organizations in France, Holland and Germany promoted public education regarding the benefits of good lighting.

56 Anon. 1926 *Illum. Eng.* **19** 144.

57 Moon P 1936 *The Scientific Basis of Illuminating Engineering* (New York) p 1.

58 These include: Fleming J A 1907 *A Handbook for the Electrical Laboratory and Testing Room, Vol II* (London) ch 3; Trotter A P 1911 *Illumination: its Distribution and Measurement* (London); Bohle H 1912 *Electrical Photometry and Illumination* (London); Bell L 1912 *The Art of Illumination* (London) and Blok A 1914 *The Elementary Principles of Illumination and Artificial Lighting* (London).

59 Bohle *ibid.* p v, and Marks *op. cit.* note 20, p 11.

60 Bell *op. cit.* note 58, p 336.

61 Invented by the chemist Hermann Walther Nernst (1864–1941), the lamp consisted of a solid bar of cerium oxide, and later zirconia and yttria, initially warmed by an external heater to reduce its resistance and then to incandescence by a controlled electric current. It was about twice as efficient as the contemporary carbon filament lamp (requiring about 2 watts to yield a candlepower of intensity), but proved only about half as efficient as the newer metal filament lamps which overtook it commercially. Another commercial disadvantage was the 10 to 60 seconds required for it to reach incandescence. See, for example, Anon. 1909 'A new high efficiency Nernst lamp' *Illum. Eng.* **2** 351, and Mendelssohn K 1973 *The World of Walther Nernst: the Rise and Fall of German Science* (London) pp 45–7.

62 Thompson S P 1909 *Illum. Eng.* **2** 815. The firefly example appears, for example, in Langley S P and Very F W 1890 'On the cheapest form of light' *Am. J. Sci.* **40** 97; in Thompson S P 1906 *The Manufacture of Light* (London); in Ives H E and Coblentz W W 1910 'The light of the fire-fly' *Illum. Eng.* **3** 496–8; in Pickering W H 1916 'Photometry of the West Indian firefly' *Nature* **97** 180 and in Ives H E 1922 'The firefly as an illuminant' *J. Franklin Inst.* **194** 212. Coblentz recommended mixing the greenish phosphor produced by the firefly with red and blue phosphors of other insects to yield an efficient white light source.

63 Anon. 1926 *Illum. Eng.* **19** 154; emphasis added.

64 Such tables had been empirically determined from the early 1890s using makeshift portable 'illumination' photometers. Later correlated with working speed and accuracy, the recommended office lighting levels increased fivefold over the period: 3–4 foot-candles (fc) in 1910 [Sunbeam Incandescent Lamp Co]; 4–8 fc [Bulletin 7C, GE Lamp]; 6–12 fc in 1925 [Bulletin 41B, GE Lamp] and 20 fc in 1935 [C E Wietz, ICS 2749A, GE Lamp] and rose by another factor of five by 1959 [*IES Lighting Handbook*, 3rd edn]. Higher values were set in the US than in the UK.

65 For example Clewell C E 1913 *Factory Lighting* (New York).

66 Dow J S 1906 'Glow lamp standards and photometry' *Electrician* **57** 855–7.

67 Preece W H 1883 'On a new standard of illumination and the measurement of light' *Proc. Roy. Soc.* **36** 270–5. The first 'illumination photometer' was constructed by Preece and Trotter at this time.

68 Walsh J W T 1926 *Photometry* (London) pp 6–7.

69 *Illum. Eng.* **21** 17. Trotter was arguably more influential in the British photometric community even than Gaster. Obtaining a BSc from Cambridge, he articled to an engineering firm where he designed lighting and photometric products. He met

William Preece in 1884, and began research in illuminating engineering with him. From that time until his later years, he maintained a 'private home laboratory devoted to photometry'. Trotter was briefly director of a dynamo factory, and then editor of *The Electrician* for five years. From 1899, Trotter served as electrical advisor to the Board of Trade, a capacity he filled for 18 years until his retirement. He also supported the formation of a photometry section at the National Physical Laboratory. See Anon. 1926 'Mr Alexander Pelham Trotter' *Illum. Eng.* **19** 77.

70 My italics. Paterson used the term *profession* loosely here, and never attempted to associate the more formal attributes of a profession with this community of engineers. See note 14.

71 *Illum. Eng.* **21** 19.

72 Anon. 1915 *Scheme for the Organisation and Development of Scientific and Industrial Research* (London), quoted in Melville H 1962 *The Department of Scientific and Industrial Research* (London) p 23.

73 For the war's effect on instrumentation companies, see Williams M E W 1994 *The Precision Makers: a History of the Instruments Industry in Britain and France 1870–1939* (London) pp 61–80.

CHAPTER 5

LABORATORIES AND LEGISLATION

The early 20th century shifted the domain of light measurement. Self-described illuminating engineers were calling for standards and scientific methods of measurement. The emphasis of photometry shifted from routine gas testing to the measurement of electric-lamp intensities and illumination. Visual methods became highly refined, and were joined increasingly by photographic and photoelectric photometry. Light measurement during this period was part of a broader trend towards quantitative methods, standardization and the growth of science-based industry[1].

The setting for these changes was a new environment of research and standardizing laboratories. National laboratories founded in Germany, Britain and America near the turn of the century, and the industrial laboratories that multiplied after the Great War, deemed light measurement a subject worthy of funding and attention. These new institutions nurtured the transition of photometry from the domain of isolated amateurs and consulting engineers to that of an increasingly influential body of career scientists and engineers—influential in that they affected government policy, international standards and the evolution of industries. The new social locus determined the problems engaged, the methods applied to their solution and the type of investigator studying them.

5.1. UTILITARIAN PRESSURES

Before exploring the changing methods and social environment of light measurement that institutions engendered, it is necessary to ask why photometry was transformed from a sideline of a handful of dispersed astronomers and engineers and a tool only of gas inspectors, into a technique of increasing importance that required the establishment of laboratories to exploit it fully. The answer lies in the increasing identification of practical reasons to measure light, coupled with a growing awareness of common aims.

By the end of the 19th century, engineers and scientists concerned with photometry agreed on its usefulness but bemoaned its lack of coherency. One text of 1894 described at least 13 current and proposed illumination standards, with the favourite standard varying from country to country, and industry to town[2]. Methods of photometric measurement were also varied. Some British gas engineers employed a simple variant of Bouguer's photometer, their counterparts in Germany favoured the Bunsen 'grease-spot' instrument and scientists increasingly used the considerably more precise Lummer–Brodhun device.

The rhetoric surrounding the Illuminating Engineering Movement suggests the frustration experienced by individual engineers when faced with the task of designing lighting installations using inadequate concepts and measurement methods. There were, moreover, the concerns raised by the financing of such installations. The electric lighting technology newly available at the turn of the century involved expensive and widespread replacement of gas in public spaces and in industry[3]. The power to control and to dramatically alter lighting was accompanied by expensive decisions, raising questions concerning the relative efficiency and cost of lighting systems. What *brightness* of illumination was required to write, weave or assemble products? Doubling the illumination levels in a factory or school could more than double the costs[4]. The *quality* of lighting was also of importance, even if difficult to quantify reliably. Lamp manufacturers such as General Electric in America, Siemens in Germany and Swan in Britain needed to verify the uniformity of the lamps produced. And, to make their products more competitive, they strove to produce as much light as possible from a given power input. Power generating companies, too, had an interest in lighting efficiency: illumination was the primary application of electrical power, and lamp designs could have a dramatic effect on the demands made of new power generating stations. Such questions of adequate illumination, product uniformity and efficiency thus concerned both government and industry. Institutional historian David Cahan has noted how 'scientists, industrialists and government officials had a common, pressing need to establish trustworthy measures for a score of electrical phenomena' including 'the amount of light radiated, the luminous intensity, the energy consumption and light-energy distribution of an illuminating source'[5]. Lighting systems were characterized by high costs of installation, some of which involved large outlays by governments at the local, regional or national level; the costs, in turn, were sensitively dependent on technological developments made by private industry. The granting of contracts for networks of street lighting and other large public works demanded input from impartial technical advisors.

Like the measurement of illumination, interest in the measurement of colour had strong utilitarian motivations. Dye production had expanded dramatically after the development of synthetic dyes in the second half of the 19th century. By the turn of the 20th century dye chemistry was a major industry, accompanied by the growth of research laboratories[6]. In the printing industry, colour printing processes had been much developed and were commonplace by the 1890s. Both of these applications demanded high-quality matching of colours and routine, rapid measurements. The demands from industry for colour standards for dyes and inks required research into the perception of colour, the effects of lighting, lamp characteristics and surface finish.

Such applications also provided great potential and risks for companies, increasingly competing on an international scale[7]. The situation led to a partial merging of government and industrial interests in a new form of institutionalized scientific research: the government standards laboratory.

Photometry was elaborated and systematized on an unprecedented scale at government institutions such as the Physikalisch-Technische Reichsanstalt in Germany, the National Physical Laboratory in England and the National Bureau of Standards in the USA. Each of these institutions was born around the turn of the century: the PTR in 1887, the NPL in 1899 and the NBS in 1901.

5.2. THE PHYSIKALISCH-TECHNISCHE REICHSANSTALT

Werner Siemens, head of the Berlin electrical firm Siemens & Halske, was a driving force in the foundation of the Physikalisch-Technische Reichsanstalt (the Imperial Institute of Physics and Technology, henceforth PTR or Reichsanstalt) in Berlin. Donating land to the Prussian government for a 'state institute in experimental physics' to promote the 'advancement of science and, thereby, also the technology closely bound to it', Siemens also encouraged the government to appoint Hermann von Helmholtz, the doyen of German physics, as director[8].

Unlike several others constructed by individual German states in the period, this was to differ in being an institution for all of Germany, in casting aside teaching duties for its employees and in promoting a mixture of science and precision technology[9]. The majority of members of the Reichsanstalt board were concerned with 'practical interests' and comprised chiefly experimental physicists, technologists and instrument-makers.

The PTR rapidly became the dominant German scientific institute by a combination of attracting first-rate scientists and gaining a voice in two journals. The editor of the *Annalen der Physik*, Germany's premier physics journal, agreed to publish all manuscripts from the PTR on the subject of pure physics. Similarly, the *Zeitschrift für Instrumentenkunde*, devoted to scientific technology and precision mechanics and optics, developed a close relationship with the Technical Section of the new Reichsanstalt[10].

The early Reichsanstalt was a closely organized and hierarchical institution. Helmholtz, its first and most charismatic leader, provided a strong sense of unity, making the rounds of the young workers 'like a doctor in a clinic... to see how his young interns were doing'[11]. While Helmholtz surrounded himself with capable young scientists, the style of work was quite unlike a university. Each scientist at the institution was directed to undertake particular projects, unlike their academic colleagues who were more free to choose the research topics they found interesting.

The study of heat radiation was one of the first successes of the PTR. Cahan has argued persuasively that

> the practical needs of the German illumination industry—better temperature measurements and better understanding of the economy

of heat and light radiation—provided the institutional justification and motivation for the Reichsanstalt's blackbody work.[12]

In 1888, for example, the Optics Laboratory of the PTR was requested by the Siemens company and the Deutscher Verein für Gas- und Wasser-fachmänner (German Association of Gas and Water Specialists) to develop photometric devices and reliable standards of luminous intensity. The German navy, too, was interested in improving the photometric design of its signalling devices[13]. From these initial utilitarian pressures, the researchers undertook a programme that led towards the understanding of the laws governing the radiation from a blackbody.

An early success was an improvement in visual photometers. Otto Lummer (1860–1925), head of the Optics Laboratories of the Scientific and Technical Sections, and Eugen Brodhun of the Technical Section, devised the photometer head described in chapter 3. The new photometer was an immediate success world-wide and, within a year of its commercial introduction, was being widely acclaimed as the best available[14]. Brodhun, a former assistant and doctoral student of Helmholtz, had moved with him to the new PTR, where he was to supervise all the running tests of the Optics Laboratory for the following 32 years. The routine investigations included certification of the Hefner standard lamp, testing the arc street lighting for Berlin, evaluating the relative performance of gas, kerosene, petroleum and electric lamps and making comparisons of coloured light sources[15]. In 1903 alone, they performed more than 600 photometric tests.

A reliable source of luminous intensity proved more difficult to develop. On the basis of prior theoretical and experimental work, a blackbody source seemed most likely to provide an absolute intensity standard[16]. By 1894 the Reichsanstalt scientists reported a luminous standard based on glowing tungsten, and measured by a sensitive bolometer detector. This entirely 'physical' method was nevertheless rejected by German industry and the international community: while it gave a reproducible measurement, the platinum-bolometer arrangement related poorly to human vision. It was an extremely hot source, appearing whiter than the commonly used gas lamps; the standard itself related so-called 'whole' and 'partial' radiations (i.e. comparing the entire radiant emission of the source, including invisible emissions, to an optically filtered portion) which was a meaningless criterion according to proponents of visual photometry and the standard was far from trivial to set up and maintain.

But despite the contentious practicality of the blackbody luminous standard, this linking of radiometric and photometric methods brought photometry a new prominence and respect. The tradition of quantitative measurement in radiometry now carried over to what the PTR scientists saw as its visible counterpart.

Alongside the environment of utilitarian research another PTR employee, Willy Wien, published 'unofficial' theoretical work on blackbody radiation. As his work fitted in with the practical investigations and promised to support a more direct definition of the unit of luminous intensity, the Optics Section, upon appeals from Wien, was instructed by the director to test the validity of Wien's theory. Lummer and Wien stated that the results would be 'as important for technology as for science'[17]. Work involved the experimental physicists of the Optics Section, theoreticians such as Wien and other scientists loosely associated with the PTR such as the infrared researcher Heinrich Rubens, employed at the nearby Technische Hochschule Charlottenburg, and Max Planck at the University of Berlin. This cooperative programme was substantially accomplished by the turn of the century, leading to Planck's formula for the blackbody distribution of radiation. Thus, motivated by utilitarian concerns, light measurement became associated with quantitative radiometry and played a central role in the emergence of quantum theory.

Cahan argues that the early successes in radiation research at the PTR were a consequence of its unique facilities and its willingness to undertake the necessary arduous precision measurements[18]. No less importantly,

> the Reichsanstalt and its physicists were motivated by a combination of pure scientific and utilitarian considerations...there existed utilitarian motives for pursuing this radiation research: such research would eventually advance the temperature-measuring needs of and contribute to the development of more energy-efficient lighting and heating sources for the German illuminating and heating industries.[19]

During its first 15 years, the Reichsanstalt embodied an admirably close-knit collection of German academics, technologists and industrialists concerned with light measurement. By their very concentration and unparalleled resources, they imposed working methods and standards that were to be retained in Germany for decades. Its workers also had a close connection with photometry. The original promoter of the PTR, Werner Siemens, had been manufacturing photometric devices from the 1870s. His senior engineer, von Hefner Alteneck, designed the intensity standard that was to be adopted by the German government. Helmholtz, the first director of the PTR, was renowned for his work in physiology and physics, having written an acclaimed three-volume treatise on physiological optics. Other German scientists such as Heinrich Rubens used the superior facilities of the PTR for their own related research, and freely shared their results with academic physicists such as Max Planck. Most of these scientists and technologists were to become board members of the Reichsanstalt, thus contributing directly to its management and planning. Owing to the institution's reputation for precision instrumentation, its close connections with German manufacturing and its direct publication organ the *Zeitschrift für Instrumentenkunde*, the photometric devices designed there received wide publicity and distribution. Indeed, the close links between industry and the institution made the selection of board members and subsequent directors awkward. The physicist Walther Nernst was rejected from the running for the directorship in 1905 owing to his investments in illumination manufacturing firms that sought Reichsanstalt certification for their products[20]. This highly integrated techno-scientific culture was central to the success and promulgation of the PTR's photometric research.

The unrivalled position of the Reichsanstalt during the last decade of the 19th century was to slip in following years. While serving as a model for other national endeavours it failed, in photometry at least, to make a sustained international impact. Despite the relative prominence and success of 'radiant heat' studies through the 19th century, the subject foundered at the PTR and the other national laboratories in the first decades of the 20th century. The workers at the Reichsanstalt ignored the implications of the new quantum physics, preferring to continue with experimental tests of radiation laws. As will be illustrated later, the German standards for intensity were not adopted by other countries and the relatively limited studies of colour were quickly overtaken by research elsewhere. Nevertheless, at the turn of the century, with its important successes in precision measurement, theoretical explanation of blackbody radiation and direct channels for self-publicity supporting it, the Physikalisch-Technische Reichsanstalt was a model for the achievements possible by concerted cooperation of government, industry and technology. Scientists and industrialists in Britain and America were soon urging for the formation of similar institutions in their own countries.

5.3. THE NATIONAL PHYSICAL LABORATORY

At the National Physical Laboratory in Britain, a rather different regime was to take effect[21]. Work and facilities comparable to those at the PTR were not established until more than a decade later. When government support was first urged in 1891 for a laboratory to do the research that industry could not do, a committee of the British Association for the Advancement of Science was formed 'to consider the establishment of a National Physical Laboratory for the more accurate determination of Physical Constants and for other *quantitative research*[22]. Oliver Lodge, an early promoter, noted that

> the further progress of physical science in the somewhat haphazard and amateur fashion in which it has been hitherto pursued in this country is becoming increasingly difficult, and that the quantitative portion especially should be undertaken in a permanent and publicly supported national physical laboratory on a large scale.[23]

Photometry was not among the handful of studies originally proposed for the NPL. By its second year of operation, however, requests were being received from industry for the testing of glow (incandescent electric filament) lamps, and for the establishment of standards of light and photometry. According to the authors of the annual report, these were 'impossible to carry out' owing to 'incomplete equipment of the laboratory'[24]. The Executive Committee observed that as 'the inception of new work involves additional expenditures, it will be difficult for the present staff to undertake the charge of a Photometric Laboratory'. Although they anticipated that testing fees would eventually cover the expenditure, this would take time. Nevertheless, the committee recognized 'the necessity for photometric work'.

Funding was a severe problem. For its first two years, the NPL had been allocated £3000 for equipment and fittings; this was supplemented by a further

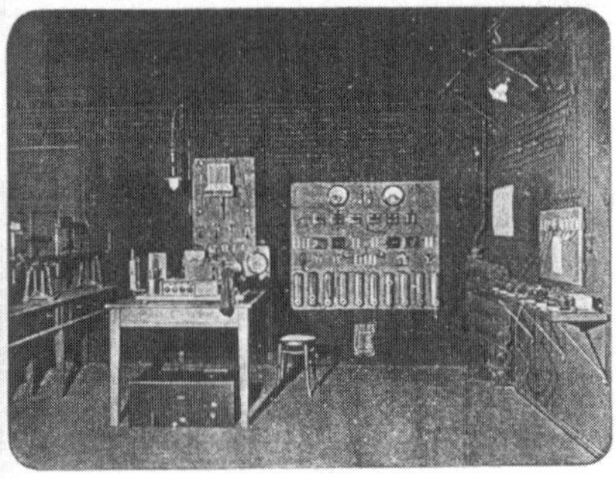

Figure 5.1. Shifting sponsors. An 'electrotechnical photometry' laboratory circa 1908, showing the photometric bench on the left, an oscillograph in the centre and an electrical distribution board on the right. From *Illum. Eng.* **1** (1908).

£4000 in 1903. By contrast, the annual allocation for 1902 was £40 000 at the PTR, £20 000 for the French *Bureau Internationale des Poids et Mésures* and £19 000 at the American National Bureau of Standards[25].

The solution came through donations. William Preece, whose earlier photometric work has been mentioned, donated a 'photometric outfit' consisting of a German-manufactured visual photometer bench of the 'Reichsanstalt pattern' and a Harcourt pentane lamp; the Electric Power Storage Company donated a 150-cell battery for powering electrical standard lamps and the consulting engineer Alexander Trotter donated another photometer. The following year, John Fleming provided 'three large bulb standard photometric lamps', with others donated by the Ediswan and Incandescent Lamp companies. The Gas Engineers Institute requested the NPL to make a comparison of the intensity standards of various countries, and donated Hefner and Carcel lamps. Alexander Wright & Co donated a flicker photometer, and £3 3s towards the NPL goal of a £2500 annual subscription[26].

With the help of such equipment donations and a meagre budget, the Electrotechnical and Optics Divisions were started in the summer of 1903 with Clifford Paterson engaged as Assistant and sole employee. Paterson undertook inter-comparisons of standard lamps with the PTR, the 'Electrical Testing Laboratories, NY' (which the director of the NPL visited) and the NBS[27].

Over the next five years, although the pentane burner was adopted as the NPL standard, incandescent electric lamps were receiving the most attention[28]. By then, photometry occupied a wing of the electrotechnical building, comprising 5000 square feet of floor space and including a battery room for photometry work (figure 5.1). Four staff were devoted solely to photometry, occasionally assisted by employees engaged in other work. At least two supernumerary staff were employed as photometric observers. The initial activities, dedicated almost wholly to lamp photometry, were later augmented by contract work for the Home Office Committee on Factory Lighting, of which Paterson was a representative.

Paterson left the NPL after the war to become research director at the General Electric Company (GEC). Facilities and projects kept expanding at the NPL with John Walsh as Senior Assistant responsible for photometry. Within a year of Paterson's departure and the war's end, other government departments were clamouring for various photometric researches to be carried out by the NPL. By 1923 over a dozen special projects had splintered the work of the Division, diverting it from its original task of standardization[29]. The meticulous cross-comparisons of the pentane standard with electric lamps and with the lamps of other countries which occupied nearly 15 years' work were completed and set aside; international agreement on the use of incandescent lamp sub-standards in 1921 meant that the pentane lamp was retained only for occasional national usage. Illumination and lighting studies now assumed great importance for the Division. A special 'illumination building' was erected in 1922[30]. Later, an additional 600 square feet of space was found in an old house on the laboratory grounds, and later still, 3000 square feet borrowed from the new high voltage research building. In 1936, the facilities in the four buildings were rehoused in a large new building which incorporated a 'physical photometry' room (for light bulb tests using photoelectric measurement), a spectrophotometry and illumination research room based on visual measurements, and a photometry room for the calibration of sub-standards[31]. To John Walsh, photometry was a branch of 'technical physics' to be pursued simultaneously on theoretical, experimental and practical grounds[32].

The growing organization at the NPL was not universal; an odd duality of purpose operated there through the 1920s. Unlike the PTR, where photometric measurements were the domain of the well-equipped Optics Section, photometric work at the NPL straddled two departments for its first few years. It was classified as Optics in 1904 and then as Electrotechnics the following year. The Optics Division, formed when Clifford Paterson joined in 1903 but taken over by another Assistant two years later, was evolving towards specialization in optical design and testing by the war. Paterson's own Electrotechnic Photometry Division concentrated on intensity standards. Unlike its German counterpart, the NPL Optics Division had little expertise and no mandate to engage in either radiometric or photometric research. By the early 1920s, however, both NPL Divisions were becoming involved with colour research. Special projects in the Photometry Division required the testing of railway signal lamps, as well as measuring dissimilarly coloured light sources[33]. On the other hand, the Optics Division had been donated a Koenig–Martins spectrophotometer, and an 'incomplete Hilger spectrophotometer developed during the war'. As early as 1911, in fact, the Optics Division had been designing visual spectrophotometers, although no object for or results from this work were mentioned[34]. With these instruments available but unused, the Optics Division stated its intention to begin colorimetry research in 1922[35]. The NPL annual *Record* documents completely independent but similar research by these two groups, with no cross-references or mentions of collaboration, throughout the decade. The overlap of work was considerable: in 1924, the Photometry Division began work on colour filters that had been undertaken by the Optics Division two years earlier; in the same year, the Optics Division did preliminary research on photometers for heterochromatic photometry already completed by their counterparts in Photometry[36]. In 1924, the redundancy of effort took a new turn when the Divisions undertook preliminary studies on the use of photoelectric cells in photometric research[37].

5.4. THE NATIONAL BUREAU OF STANDARDS

Photometric work at the National Bureau of Standards fell somewhere between the well organized early PTR and the under-funded, but ever expanding, NPL. In general, its organization closely mirrored that of its British counterpart. More than the other two institutions, however, and because of its direct administration as a government department, the NBS was efficient in proposing and imposing industrial standards.

The Bureau of Standards was founded by an Act of Congress in 1901[38]. The Photometry Division of the NBS was started in the autumn of the following year with a single laboratory assistant in a basement room of the Coast and Geodetic Survey in Washington, DC; the entire Bureau of Standards had only 14 personnel in its first year. By 1908, the Bureau could claim 110 employees and the Photometry Division five, three of whom were physicists. Their work was divided into the testing of lamps (for both commercial and Bureau use) and ‘investigation’[39]. The investigation was restricted to the evaluation of potential lamp standards for the first few years[40]. The first head of the photometry section was Frank A Wolff, Jr, formerly of the Office of Weights and Measures. Wolff, who had several acquaintances in Congress, had been instrumental in promoting the bill for the founding of the NBS. The Bureau itself was modelled on the Reichsanstalt, and its methods and standards initially drew heavily on its predecessor. In the initial pressure to establish laboratories of electrical and photometric references, Wolff was ‘obliged, as heretofore, to send to the national standardizing laboratories of Germany and England for verification the large class of alternating current measuring instruments, condensers, and photometric standards’[41]. His work was carried out in temporary headquarters in downtown Washington for three and a half years. By October 1904, the NBS was established in a purpose-built facility on the outskirts of Washington, DC. From the outset, photometric standards were part of the planned activities. Photometric laboratories occupied one floor of the mechanical engineering building and half an attic. The other, much larger, building housed the Physical Laboratory, which was to include a Photometric Standards Laboratory. This was, however, forced to give way to a lunch room, which had been omitted from the architectural design[42]. Upon completion of the new facilities, Wolff's work was turned over to Edward P Hyde from Johns Hopkins University in Maryland. The entire staff of the NBS comprised 58 persons at the opening of the new facility[43].

The American government soon made use of the NBS to ensure the quality of the products it purchased. The work of the photometry section was instrumental in persuading the government to move towards increasing industrial regulation. Incandescent lamps for Federal offices were, by 1904, being purchased at the rate of one million per year. When the purchasing agency sent a sample of light bulbs to the Bureau for tests for the first time that year, three-quarters were rejected because they failed the manufacturers' own specifications for luminosity. This success of the Bureau in weeding out unsatisfactory electric lamps was noted at government hearings on weights and measures, the incident leading to a wave of reform through the government service to set specifications and tests for items as varied as clinical thermometers, chemical glassware and mucilages[44]. In 1907, representatives of incandescent lamp manufacturers met with NBS engineers to adopt standard specifications. These detailed the power consumption required to produce a given illumination, and the minimum acceptable 'lifetime', defined as the time required to drop to 80% of their original light output. Ninety per cent of a test lot of bulbs was required to pass the specifications or the entire lot would be rejected[45]. The circular published by the Bureau called attention to the low illuminating efficiency of carbon filament lamps compared to the newer metal filament types. Avoiding outright mention of the brand name, another circular nevertheless made clear the marketing practices of the manufacturer: 'The tungsten lamp has been improved in quality and reduced in price to such an extent that no customer can afford to use carbon lamps, even if he were paid a bonus on each lamp for so doing. Many householders cling to the use of carbon lamps because they are usually supplied free'[46]. Such lamps required nearly three times more power than the Mazda tungsten lamp, a commonly available alternative[47].

The photometry of gas lamps similarly led the Bureau towards standards setting and regulation. In 1905, the Bureau of Corporations requested the NBS to investigate the illuminating power of commercial kerosene oils. When 40 such oils were tested the following year, the staff of the Photometry Section concluded that even the Hefner amyl acetate and Harcourt pentane standard lamps were inadequately stable. Citing the results of this preliminary work, the Bureau requested from Congress a special $10 000 appropriation for a two-year study of gas and oil illuminants in 1908. This was to be the first such specially funded investigation of the Bureau, a practice that was repeated almost yearly until 1936, when Congress began to lump special NBS research projects into general funds. The early special appropriations, being individually requested and granted by Congress, thus had a relatively high profile and gained both government and public attention.

As at the NPL, the early photometric work had an uncertain home. Photometry was decidedly *not* a branch of optics, however. A graduate chemist from the University of Wisconsin was hired and sent on courses in gas engineering, and then put in charge of the gas photometry investigation as a member of the Electrical Division. The work of his group over the next two years led to standards for illuminating and heating gas. In its circular on the subject, the NBS recommended that gas supplies be priced by their heating and illuminating power rather than by volume, as was the current practice in most cities[48]. This 'entirely advisory' information was disputed by the gas industry for a decade before agreement was reached to sell gas on this basis. The Electrical Division of the NBS continued to be responsible for gas photometry until the early 1920s, when the work was transferred to the Chemistry Division.

During the First World War, the photometry section switched priorities to searchlights and other forms of military illumination. The staff of the photometry section expanded to seven. After the war, the photometric work at the NBS was a notable part of a general crusade for standardization, which sought to simplify the variety and complexity of commercial products and thereby improve efficiency and competitiveness[49]. The standardization of electric lamps, gas purity and lighting systems were highly visible early successes.

Unlike photometry, radiometry at the NBS was a subject substantially uninfluenced by commercial pressures or government directives (it had, for this reason, played a minor role at the NPL). Perhaps as a result, the growth of light measurement responsibilities was rather *ad hoc* in the early years. For example, a promising young graduate who had done his PhD work in infrared spectroscopy was hired in 1903 to head the Radiometry Division. William Coblentz (1873–1962) kept this position, along with 'one or two minor assistants', for nearly 40 years[50]. In seeking practical justification for his post, Coblentz supplemented his radiometric research over the following years with work on visual response, ultraviolet filters and even the radiant heat losses of pig enclosures. During the depression, Coblentz worked on standards of ultraviolet radiation. Hospitals and several industries had sought means to calibrate the photoelectric dosage intensity meters used for measuring ultraviolet radiation. Around 1931, ultraviolet lamps became commercially available as 'household health aids'. The NBS produced a standard consisting of a quartz–mercury arc lamp calibrated in absolute units in 1936[51]. Unlike the PTR, which had sought to merge radiometry and photometry, the NBS enforced a distinction between radiometric and photometric work. Colorimetry and radiometry were subsections of the Optics Division, while photometry and illuminating engineering come under the Electricity Division[52]. Coblentz, responsible for radiometric studies principally in the infrared and later in the ultraviolet—bracketing the visible spectrum—was warned by his superiors to leave visible-light photometry to the Photometry Division[53].

As at the NPL, the work of the Electrotechnical Photometry and Optics Divisions began to overlap after the First World War. Both began investigations into colour measurement and standardization. The Photometry Division was motivated by extensions of 'white-light' photometry to lights of different tints. The Optics Division, on the other hand, felt that the design and evaluation of optical filters for signalling lamps fell naturally into its domain.

5.5. COLOUR AT THE NATIONAL LABORATORIES

The measurement of colour was a subject distinct from photometry in the early national laboratories, but one increasingly merged with it in terms of technique and measurement objectives.

By 1914 there was an increasing interest in, and demand from industry for, a general systematization of colour description. Industrial applications of colour matching were numerous, most having been developed in isolation to suit particular industries. The American, and then the British, national laboratories began to study colorimetry as part of the work of their Optics Sections. This work progressed independently of the radiometric and photometric activities of their electrotechnical laboratories, although there was occasional overlap of personnel and much commonality of technique. Interest in colorimetric research was considerably lower in Germany and France, where physical photometry retained most attention[54]. Although there was a large body of German work following the physiological optics research of Hermann von Helmholtz and Ewald Hering from the latter part of the 19th century, this made little impact in England and America[55]. The American investigators, with a growing body of recent studies behind them, were quick to denigrate foreign research. In a 1925 summary of advances in colorimetry, a reviewer from the American NBS mentioned Wilhelm Ostwald's *Farbenlehre* as typical of current German work, describing its author as 'very far from being abreast of current knowledge and practice'[56].

The NBS had begun its involvement with colour measurement in 1902[57]. From the beginning, it made use of existing empirical systems. The artist Albert H Munsell contacted the director of the Bureau soon after its formation in 1901, 'asking about color'. Munsell formed a company to market his colour charts, educational materials and books in 1917, the year before his death. Over the following decades, the Munsell Color Company under the direction of his son funded seven research associates at the NBS[58]. One of these, Irwin Priest, headed the Colorimetry Section from 1913 until his death in 1932, and was influential in the fledgling Optical Society of America, becoming its president in the late 1920s[59]. Priest provided considerable support in the planning and operation of the Munsell company. Another research associate at the NBS, Deane Judd (1900–72), was a central figure in defining colour standards that were eventually adopted by the Commission Internationale de l'Éclairage. Contact with the Munsell Company was close throughout the history of the NBS. Much of this centred on putting the original empirical system on a more regular footing. Attempting to *mathematize* or idealize human colour vision, the investigators used spectrophotometers, for example, to measure the reflectance of the various Munsell colours as a function of wavelength, and then adjusted the colour steps to follow a more regular mathematical sequence. A considerable amount of collaborative work took place at the Munsell Research Laboratory in Baltimore (founded in 1922), where seven individuals were assigned to mainly scientific work. Similar work in Britain was scattered through separate Research Associations, which published relatively little[60]. By contrast, the result of the more open American research was 40 collaborative papers before the Second World War[61].

Rexmond Cochrane has written that 'the field of research at the Bureau in which undoubtedly the greatest variety of industries and interests had a vital concern was the standardization of color'[62]. The NBS frequently served as the arbiter of disputes. In 1912, for example, representatives of the butter, oleomargarine and cottonseed oil industries requested help in colour-grading their products. Other queries dealt with the colour of paints, cement, porcelain, tobacco, foods and water purity. Irwin Priest, who had been hired in 1907 to conduct the Bureau's work in spectroscopy and applied optics, was moved to colorimetry. Investigating the use of spectrophotometric measurements for colour analysis, Priest was won over to this technique. By 1921, he was promoting colour standardization based on a carefully defined 'white light'. Based on a physical definition of colour, his ideas aimed at rendering the observer a minor and controlled part of colour measurement.

Work at the NPL in England was later in starting and more limited in scope than that in America. Unlike photometry, the study of colorimetry initially had no supporters from industry. Apart from the donation of an incomplete Hilger spectrophotometer during the First World War, British industry had little connection with the NPL for colour measurement. Before the war, in fact, there were only two recorded forays into colour measurement: one in 1908 concerning the measurement of the temperature of heated bodies by optical pyrometry, carried out in the Thermometry Division of the Physics Department[63], and the other from 1911 until the war, when a spectrophotometer was designed and built for testing the components used by the Optics Division[64]. Following the War, the Division decided that it would begin low-priority work on colour vision 'as occasion permits'[65]. The study initially involved a single observer, John Guild, who had previously been responsible for the testing of optical lenses. By 1921, however, interest grew because 'considerable attention has been devoted to it in America'[66]. The Division would do research on colour standardization by measuring 'a representative number of colours on various types of colorimeter, both scientific and commercial'[67]. Despite a slow start and limited resources, the research now had a clearly defined programme involving the development of a standard method of measuring colour and inter-relating different commercial instruments and practices. The NPL sought a consensus in British industry by aiming at 'a general coordination of the various colour systems...and their relationships to the fundamental facts of vision with a view to the evolution of a generally acceptable scientific basis for colour specification and standardization[68]. The first commercial system to be investigated was the 30-year-old scheme of Joseph Lovibond. Owing to the availability of only a single full-time investigator, progress was slow. The year 1923 was devoted to choosing a third colour between the standard green and red for railroad signal lamps, and 1924 to measurements of standard filters and instruments[69]. By 1925, however, Guild was developing a trichromatic measurement system based on standard colour filters, and collaborating with Hilger & Co in the manufacture of a trichromatic colorimeter. With the aid of other NPL staff and observers loaned from the British Woollen and Worsted Research Association in 1927, he was able to measure the vision characteristics of seven persons, from which he refined his colour measurement system and based a set of paint colours for the British Engineering Standards Association[70]. The Guild system of colorimetry found some application in British industry. The NPL assisted the Pharmacopoeia Commission in evolving colour specifications for cod liver oil, and to the Fuel

Research Station for standard colours for testing coal ash[71]. Guild's work amounted to a self-consistent body of research, but was not widely applied outside Britain[72].

Colorimetry in Britain thus began with desultory studies at the NPL around the time of the First World War, and picked up in response to American activity. Through the Research Associations sponsored by the Department of Scientific and Industrial Research, the NPL was the locus for research and development by the mid 1920s. This increasing national organization occurred in parallel with international developments to be discussed in chapter 7.

5.6. TRACING CAREERS

The employees of the national laboratories formed a community of practitioners distinct from their contemporaries, the illuminating engineers. Moreover, as previously discussed, the photometry departments of the national laboratories were allied more closely with the electrotechnical industries than with university scientists. During the first discussions of the role of the NPL, for example, the organizers had sought to extend their support by stressing 'engineering science and standards' rather than 'fundamental research'[73]. The members of the NPL departments were, nevertheless, recruited from universities. At the end of the 19th century, there were few permanent positions for physicists outside educational institutions[74]. The few individuals tackling industrial problems generally worked as consultants. 'When the NPL appeared at the turn of the century, it was an oasis in the vocational desert', writes Russell Moseley[75]. 'The profile of new recruits was remarkably uniform', generally men in their twenties often holding first class honours degrees and trained in physics. The NPL was organized into departments, each with a superintendent. In each department, a principal or senior assistant would be responsible for one field of activity. In accord with the NPL budget, salaries were low: in 1901, pay was about £100 per year for junior assistants, and £200–£300 for senior assistants. By the middle of the First World War, a proposal was tabled to increase salaries to £175–£235 for juniors, and £650–£750 for principal assistants. These 'by no means lavish' salaries were considerably lower than those available in industry[76]. In 1917, an advisory council recommended almost doubling them. Not surprisingly, the young graduates hired easily in the first decade of the century (when career prospects for physicists were particularly low) defected to industry when opportunities arose. Few made the move, however, from the NPL into academia. A good example of this industrial–national laboratory linkage, and academic exclusion, is the career of Clifford Paterson.

Clifford Copland Paterson (1879–1948), a close contemporary of the illuminating engineer Leon Gaster but a generation younger than A P Trotter, and nearly four decades younger than the scientific enthusiasts William Abney and J Norman Lockyer, joined the newly founded NPL as Assistant in 1903[77]. Unlike many others at the Laboratory, he had previously been employed in technical posts in industry. Having completed his sixth-form studies specializing in engineering and physics, he spent one year in a technical college training in electrical engineering. This was followed by apprenticeships with London and Glasgow companies, and then employment as a student assistant at an electrical manufacturer for two years. On installation projects in Switzerland and Italy, he became familiar with new technology as well as with industrial relations.

One of Paterson's first projects, the investigation of the effect of atmospheric conditions on the Harcourt pentane lamp, brought him into close contact with both British industry and the members of the newly founded Illuminating Engineering Society. Indeed, the equipment donations that made his Division possible had come from William Preece and Alexander Trotter, both of whom had known William Abney, Silvanus Thompson and Leon Gaster for over a decade. The personalities involved with British photometry, ranging from its amateur scientific aspects to illuminating engineering to government standards, thus all interacted around the turn of the century. Within a decade, though, Paterson, their junior, was a public figure and British authority on photometric standards and the NPL was the focus of national efforts on the subject. Paterson nurtured his connections with the members of the Illuminating Engineering Society in London and New York, and with representatives of the gas and electric lighting industries. Unlike his contemporaries, Paterson's post allowed him to develop a governmental and international perspective on the subject. As a representative of the NPL, he was an active member of the Commission Internationale de Photométrie from its second meeting in 1907, presenting papers on photometric standards in 1911. In 1913, he was appointed Secretary of the newly founded Commission Internationale de l'Éclairage, for which he had substantially drafted the statutes and constitution. He remained either its Honorary Secretary or Secretary until 1948, except for a period when he served as its president (1927–31). Paterson was an active participant on governmental committees, contributing to studies of factory lighting and sitting on boards responsible for ships' lighting and signalling lamps during the First World War[78].

Paterson was recruited after the war to become the first director of the GEC Research Laboratories, a position that he held from 1919 until his death in 1948. The period 1916–18 was a difficult one for the NPL, which had taken on a vast quantity of research and testing work during the war. The Treasury was unwilling to fund any more posts to ease the burden on the overworked employees or to significantly increase salaries. During the period, four senior staff members left for industrial posts[79]. When Paterson left in 1919, the funding crisis was in full swing. He took with him 'three valued members of the Laboratory Staff' to populate his new research facility. His transferred subordinates were B P Dudding, his second-in-command; Mark Eden, from Metrology; and Norman Campbell, the academic physicist and philosopher who had joined Paterson's department during the war[80]. Even Paterson's secretary and carpenter made the switch, swelling the payroll to 29 people by the end of 1919.

Paterson was thus involved centrally with British photometry in the first third of the century. He was the first investigator in the subject at NPL; he attained a wide reputation by serving on governmental committees during and after the war; he was a member of the Commission Internationale de Photométrie and of its successor the Commission Internationale de l'Éclairage; sometime president of the Illuminating Engineering Society and he was the first director of the GEC Research Laboratories, where he oversaw considerable work on photometry and commercial photoelectric light-measurement devices.

Paterson's career contrasts with that of John William Tudor Walsh (1891–1962), his successor at the NPL. Walsh had joined Paterson's group in 1913 at the age of 22 as Junior Assistant. He was promoted to Assistant in 1916 (with only women remaining Junior Assistants during the war) and Senior Assistant in 1921[81]. Unlike Paterson, and more typically of the now-established NPL, Walsh held an MA (Oxon) when he was recruited by the Laboratory, and subsequently earned a doctorate[82]. He spent his entire career at the NPL, gaining status comparable to that of Paterson in the photometric community. Walsh was less active than was Paterson in government committees, and had much less involvement with industry. He attained few of the honours that Paterson had gained. On the other hand, his professional reputation in photometry arguably reached a higher point, principally due to two books on the subject[83]. The dozen years between them witnessed a growing rigidity of career structure and integration within institutions.

A career regime much like that of the NPL operated at the NBS in Washington. There was a tendency to hire bright university graduates, often before the need for a Division had been demonstrated. One reason for the greater emphasis on recruitment of untrained university scientists rather than those with industrial experience was undoubtedly remuneration. Salaries at the new Bureau were considerably lower than in industry. In partial recompense, Stratton arranged agreements with several universities to accept research at the NBS as qualifications for advanced degrees. E P Hyde, the first investigator responsible for photometric research at the NBS, obtained his PhD in this way from Johns Hopkins University in 1906 for researches in photometry. With his improved academic credentials, however, Hyde was an attractive recruit for industry.

He left his position at the NBS to become director of the National Electric Lamp Association research laboratory[84]. While the NBS managed to retain a

Table 5.1. Heads of the NBS Photometry Section 1901–41.

Section Chief	Tenure	Period (years)	Next post
Frank A Wolff	1901–02	2	NBS Electrical Div.
Edward P Hyde	1903–08	5	Nela Research Laboratory
Eugene C Crittenden	1909–17	8	NBS Electrical Div.
A Hadley Taylor	1918–20	3	Nela Research Laboratory
J Franklin Meyer	1921–41	20	Retired

large fraction of its section heads for decades, others left to join industry (seldom academia). This tendency is illustrated by the Chiefs of the Photometry Section at NBS over its first 40 years (table 5.1). The short tenure of most of the Chiefs suggests that they saw the post as a stepping-stone to bigger and better things.

5.7. WEIGHING UP THE NATIONAL LABORATORIES

Photometric work in all the national laboratories grew rapidly in response to utilitarian responsibilities. The growth was spurred by, and contributed to, the increasing regulation of workplace illumination. Duncan R Wilson of the British Factory Department had surveyed industrial lighting, particularly in textile factories and printing works, between 1909 and 1911. As a result the Home Secretary in 1912 set up a Departmental Committee 'to inquire and report as to the conditions necessary for the adequate and suitable lighting (natural and artificial) of factories and workshops'. Richard Glazebrook, Director of the NPL, was chairman. A more extensive NPL survey was carried out in 1913, comprising 4000 measurements in 57 factories[85]. The Report of the Departmental (Home Office) Committee on Lighting in Factories and Workshops, issued in 1915, gave government guidelines. These guidelines had to be put into effect by engineers and verified by inspectors. Both groups required photometric standards, instruments and measurement procedures. In America, the Illuminating Engineering Society published a lighting code in 1910, which led to regulations for factory lighting in five states. During the First World War, the US National Defence Advisory Council Divisional Committee on Lighting issued a similar nation-wide code[86]. In Germany, the introduction of an illuminant tax law in 1909 burdened the PTR with routine photometric testing and certification of gas and electric lamps. The NPL and its counterparts in other countries made photometric standards a major part of their work.

While all three national laboratories responded to utilitarian pressures, the directions they took were different. At the PTR, requests for intensity standards were channelled into temperature research and radiometry. This choice of technical direction can be attributed both to the time and circumstances. In the early 1890s when the industrial requests were made, most practitioners of photometry believed the future lay in the Violle standard. This proposed unit of light, based on the radiation from one square centimetre of platinum heated to the melting point, was expected to promise the simplest and most fundamental of light sources[87]. Textbooks, engineers and scientists echoed this universal expectation[88]. Moreover, German investigators such as Heinrich Rubens were already engaged in research programmes to extend and measure light of increasingly long wavelength. The Reichsanstalt's embarking on the development of a primary standard and radiometry was thus the very activity that any well equipped and confident photometric laboratory would have undertaken at the time.

A decade later, when the NPL and NBS opened their doors, faith in a platinum standard had been shaken by the experimental difficulties encountered in stabilizing the temperature of molten platinum, maintaining a clean surface and measuring the intense white light. 'Like the mercury ohm, the Violle standard has been officially adopted again and again at International Congresses by people who have never tried to construct or even use one, and who were unaware that far greater accuracy may be obtained by less academical methods', wrote the peripatetic Alexander Trotter[89]. Despite several previous abortive attempts at realizing such a physical standard, it was nevertheless still the objective of the newly organized but inexperienced Photometry Division of the NPL[90]. In practice, the British and American laboratories found their funding inadequate for extensive scientific research, and relegated themselves to the pressing tasks of evaluating existing flame and electric lamp sources. With little time or experience in radiometric methods, they embraced visual photometry wholeheartedly and exclusively.

National differences affected the problems studied as well. By the 1920s, the NBS was directing its activities toward low-level applied science to benefit householders and small business[91]. Partly in response to criticisms of solving industrial problems at government expense, the NBS turned more towards academic science in the following decade. The NPL researches were motivated increasingly by projects for government departments, particularly those relating to lighting engineering[92]. The PTR turned away from both these trends, declining in international importance during this period owing to an increased emphasis on routine and test work[93].

All three laboratories nevertheless converged towards similar working practices in the inter-war years, largely owing to restricted resources and the rise of routine standards work. According to a historian of the NBS, 'because the national laboratories both here and abroad had fewer calls on them from industry, the depression years were remembered as a time of international conferences, of many inter laboratory comparisons and exchanges of data and equipment looking to new or improved international standards'[94]. All three photometric laboratories gradually lost control of their direction, yielding to an unplanned existence mediated by special requests from industry, growing routine work and increasing responsibilities for legal standards.

5.8. INDUSTRIAL LABORATORIES

Research into photometry and illumination was not restricted to government laboratories, even if it was concentrated there. The founding of industrial research laboratories, like government laboratories, was a distinctive feature of the early 20th century[95]. The GE Research Laboratory (NY) was founded in 1900; Kodak's was set up in 1912. One source puts the number of industrial research laboratories in America as 300 in 1920, and 1625 a decade later[96]. British firms also founded research laboratories in the inter-war period, and were conservatively estimated in the hundreds by the end of the 1930s[97].

As noted by Michael Sanderson for electrical innovation, the large industrial research laboratories 'came to replace the universities as the source of new

Figure 5.2. Enlightened industry: Nela Research Laboratory, National Lamp Works of General Electric, Cleveland, Ohio, where 'only pure research relating to the physics of illumination and its physiological and psychological effects on the human organism is conducted'. *Source*: Fleming A P M and Pearce J G 1922 *Research in Industry* (London) pp 127 and 160. Arthur P M Fleming was the Research/Teaching Director of Metropolitan-Vickers, British electrical manufacturers.

technology, and we cannot point to any set of achievements in the universities in this field in the inter-war years remotely comparable'[98]. The most relevant example is provided by the research laboratory created in the spring of 1908 for the National Electric Lamp Association[99]. The Nela was born in 1901, the same year as the NBS[100]. The member companies of the association emphasized its role in reducing competition. These semi-autonomous divisions were also aware of the need to develop products to compete with the more efficient metal-filament lamps being produced in Germany and Austria. In an environment of competition, marketing and government regulation the Nela Research Laboratory was conceived (figure 5.2)[101].

The first director of the Nela Research Laboratory, Edward Hyde, had begun his career as head of photometry at the NBS. He wanted to distinguish the laboratory as 'pure science' rather than as 'applied art'. Speaking at one of the first meetings of the Illuminating Engineering Society in New York, he observed that 'the future of this new science, and therefore the success of this new Society, will depend on the establishment of sound basic principles'. Putting behind him the ideas current in the national laboratories, Hyde believed that the future of photometry lay squarely on the shoulders of physical and *physiological* scientists: his laboratory would, he said, stress fundamental ideas before applications, with

> coordination of physics and physiology, the proper cooperation of the physicist, physiologist and perhaps the psychologist... Differentiation of science must be accompanied by a cooperation of the scientists if the great middle fields of science are to be adequately covered.[102]

The Nela Research Laboratory was not quite the cooperative industrial enterprise that it appeared. Although the National Electric Lamp Association consisted of nominally independent lamp manufacturers, in fact 60% of the stock at that time was owned by General Electric. Despite this, Hyde felt more freedom there than he had enjoyed at the NBS. 'Pure research is something of a hobby to me', he wrote to the director of the General Electric Research Laboratory, and for a dozen years he used his industrial laboratory as a place to exercise that hobby[103].

By its second year of operation, the Nela laboratory had seven people 'in a small one-storey and basement brick building recently occupied by the Buckeye Electric Co'[104]. The laboratory was re-housed on a green-field site in East Cleveland in 1911. Hyde wanted the facility moved away from smoke, gas fumes and disturbances—much as the NBS site had been selected some 15 years earlier[105]. Nela Park was, during and after the First World War, to carry out work much like that at the NBS and at the more commercially oriented General Electric Research Laboratory at Schenectady[106]. Following an anti-trust suit brought against General Electric, the National Electric Lamp Association was ended in 1911[107].

The name Nela, and the research laboratory itself, remained, although now clearly identified as the National Lamp Works of General Electric. Defections from the NBS continued, too. In 1921, A Hadley Taylor, at the time responsible for photometry and illuminating engineering at NBS, moved to the Nela Park Laboratory. In the same year, Ernest Nichols succeeded Hyde. Like his predecessor, Nichols saw the laboratory as favourable to basic research:

> The position offers complete freedom in the choice of research problems, and places at my unhampered disposal such human and material resources as no university I know of can at present afford.[108]

So unhampered were his options that Nichols renamed the facility the Pure Research Laboratory. Like Hyde, he directed its research over a range of studies from the physics of light sources to the physiology of vision. Upon Nichols' death in 1924, though, General Electric re-evaluated the function of Nela Park and reorganized it towards more direct industrial research. Its new director, Matthew Luckiesh (b. 1883), publicized the Laboratory's work in lighting research[109]. The Laboratory also undertook an educational role by organizing short courses on illuminating engineering, leading to its identification as 'the university of light'[110].

The large profits at risk encouraged other electrical manufacturers to launch research laboratories. The British version of General Electric set up a major laboratory to concentrate on lighting and thermionic valves[111]. The GEC Ltd Research Laboratory at Wembley was conceived in 1916, and first came into being early in 1919[112]. The formal opening of purpose-built facilities was in February 1923.

The company's aims were signalled by the research director it sought. Clifford Paterson's work in evaluating commercial incandescent lamps while at the NPL brought him into contact with the Osram Lamp Works, a company founded jointly by GEC and the German company DGA. Representatives of the company sought Paterson's suggestion of someone to organize a research department at Osram. Little came of the proposal for two years, but by the end of the war, Paterson's ideas about a research laboratory had developed and Osram had been bought outright by GEC from the Government Trustee of Enemy Property. Paterson himself took on the planning of a research laboratory for this enlarged company.

The first staff worked at a wooden building at the Osram Lamp and Valve Works at Hammersmith. Early work at the Laboratory centred on investigations of lamp design and manufacture. The first work on photometry appears to have been a proposal for a spherical integrating photometer, to be used to measure the total radiant output of lamps[113].

By the spring of 1920, at least nine GEC units were using or requesting the use of the Research Laboratories[114]. Among these were the Osram GEC Lamp Works and the Salford Instrument Works, a small company specializing in the manufacture of electrical measuring instruments. By the time of the opening of the new laboratory at Wembley in 1923, work was in progress in lamps, valves and photometry. Problems in lighting continued to receive attention. Paterson had been chairman of a British Standards Institution Committee on street lighting for many years. One of the GEC scientists, J M Waldram, took over the chairmanship later. Paterson also served on a Departmental Committee of the Ministry of Transport, on which Waldram was the member of an Experimental Committee[115].

Along with valves for radio broadcast, GEC researched photoelectric devices. Paterson took a direct interest in these activities, noting with satisfaction that his workers 'have probably devoted as much attention to photoelectric cells as any group of workers in the world'[116]. Although the photoelectric research received no mention in the official GEC history[117], it was a significant effort during the 1920s and 1930s. Norman Campbell and his co-workers publicized their work and products by publishing books on the practical usage of photoelectric tubes[118].

5.9. WARTIME PHOTOMETRY

A description of the institutionalization of light measurement would be incomplete without a discussion of the transformative organizational event of the early 20th century, the First World War. Unlike the Second World War, however, which profoundly altered the course of the subject, the influence of the Great War was of only indirect importance to photometry[119].

The PTR was the most affected of the national laboratories. Fully half of the personnel joined the German armed forces in the first months of the war. The reduced staff were occupied primarily in military-related work 'of a minor, testing nature'[120]. With 22 senior scientists absent, travel curtailed and research funds withheld, little research into light measurement was able to continue[121].

At the NPL, the hostilities were slow to affect the photometry and optics work. As late as the month before the war, representatives of the Reichsanstalt visited to compare standards. The war's first consequence was the increased workload caused by the quarter of NPL employees who had immediately volunteered for service. The loss of two observers and a laboratory boy burdened the remaining five photometry staff with additional work. By late 1915, the increase in investigations for government departments prevented more staff from volunteering. Disqualified men and female temporary staff more than doubled the size of the Physics Division, although the Photometry and Optics Sections were unaffected[122].

During the war, the activities of the Photometry Section remained evenly split between 'routine testing' and 'investigative, research and installation tasks'[123]. Among the 'several special confidential investigations' for government departments were studies of the intensity of luminous dials for watches and instruments and the development of a height finder for anti-aircraft guns[124]. The Optics Division reported a greatly increased workload owing to the routine testing of binoculars, theodolites and other war-related certification, and the urgent evaluation of optical glass manufacture.

The primary effect of the war at the NPL was organizational. In 1918, the newly created Department of Scientific and Industrial Research was given responsibility for the administration of the Laboratory. The DSIR funded research into building illumination after the war, an effort that demanded considerable resources. As already noted, dissatisfaction with salaries and workload caused several key employees, including Clifford Paterson, to leave in the last year of the war. His replacement, John Walsh, introduced the changes of administrative style that are inevitable in a small department. The increasing number of special projects did not slacken after the war, making the work of Walsh's Division considerably more fragmented than that of Paterson's.

The war had a comparable effect on light measurement at the NBS in Washington. Searchlight design and signalling lamps for ships demanded the resources of the Photometry Division, as they did at the NPL. Colour research, principally for camouflage design, also gained the attention of the Optics Division. In 1916, the director of the NBS requested government funding for special work on colour standards, noting that

> There never was a time in the history of the country when we should be looking at such matters as critically as at present. The items submitted—I think I can say all of them—are as fundamentally concerned with both industrial and military preparedness as any that will come before you.[125]

For the most part, however, the war was a temporary diversion for the photometry and colorimetry work at the NPL and the NBS. No crucial military applications of the subjects were identified as being worthy of post-war research[126].

Thus, at the PTR, the war hastened an already evident decline; post-war Germany would be unable to participate in international photometry[127]. For the victors, the chief effect of the war on these subjects was its demonstration of the benefits of organization for technological change. The consequent move towards increasingly planned research by technical delegations, and the effect of German exclusion from international photometry, are discussed in chapter 7.

5.10. CONSOLIDATION OF PRACTITIONERS

The first three decades of national laboratories thus witnessed a profound change in the social practice of photometry. The birth of national and industrial laboratories around the turn of the century marked a transition from a growing band of enthusiasts (the illuminating engineers and a handful of astronomers) to institutionalized photometric researchers. The light measurement work at the national laboratories was a direct outgrowth of industrial pressure for standardization and government-supported utilitarian research. These pressures provided the funding for a new class of scientist fitting imperfectly into either industry or academia, who wielded considerable influence on government purchasing, policy-making and international standards. These new career scientists and technologists, characteristic of the new century, were to direct the evolution of light measurement up to the Second World War.

The first quarter of the 20th century was a period of consolidation in the practice and research of light intensity measurement through institutions. It was also a time for constructing new alliances. By pursuing new methods and uses of light measurement, the new organizations had fostered a splintering into specialties[128]. The classification and subdivision of the subject, however, was specific to each laboratory: *radiometric* at the PTR, *optical* and *electrotechnical* at the NPL, *chemistry related* and *electrical* at the NBS, and *optical* and *physiological* at the Nela laboratory. By the 1920s, some practitioners were attempting to unite, or at least cross-fertilize, the various studies[129]. Illuminating engineers, in particular, were aware of the advantages of talking to optical experts. Leon Gaster, in large part responsible for the organization of illuminating engineering in Britain two decades earlier, said when addressing the 1926 Optical Convention in London:

> the use of light, whether natural or artificial, almost invariably involves consideration of problems from two distinct aspects; from the physical side, i.e. in regard to the most efficient utilisation of the luminous energy available, and from the physiological side, i.e. in relation to the effect of this energy on the human eye. It may truly be said, therefore, that optics and illuminating engineering are kindred sciences, and that there are many fields of work where experts in both can cooperate with fruitful results.[130]

It was, in a way, a compromise: an admission that photometry could not live up to its 19th century ideal of being an objective visual science. Instead, it necessarily straddled physics and physiology, and was not entirely part of either study. The new institutions researching light measurement could not successfully compartmentalize the field into radiometric, photometric and colorimetric components. Even with increasingly organized research, the standardization of light measurement proved difficult. The illuminating engineers, astronomers and institutionalized researchers remained separated by distinct technological approaches.

NOTES

1 For a broader perspective regarding these cultural changes, see Noble D F 1979 *America by Design: Science, Technology and the Rise of Corporate Capitalism* (New York).

2 See Palaz A 1894 *A Treatise on Industrial Photometry, With Special Application to Electric Lighting* ch 3. Adrien Palaz, born in Switzerland in 1863, studied electrotechnology under E H Weber at Zurich Polytechnic. He gained a position at the Bureau Internationale des Poids et Mésures at Sèvres in 1886, and edited the journal *La Lumière Électrique*.

3 Books on photometry began to emphasize the new illuminants, e.g. Stine W M 1900 *Photometrical Measurements and Manual for the General Practice of Photometry, With Special Reference to the Photometry of Arc and Incandescent Lamps* (New York).

4 In Britain, these questions led to influential committee reports by the Departmental Committee on Lighting in Factories and Workshops in 1915, 1921 and 1922.

5 Cahan D 1989 *An Institute for an Empire: the Physikalisch-Technische Reichsanstalt 1871–1918* (Cambridge) pp 17–18.

6 Homburg E 1992 'The emergence of research laboratories in the dyestuffs industry 1870–1900' *BJHS* **25** 91–111.

7 For an excellent study of the growth of electrical power systems, see Hughes T P 1983 *Networks of Power: Electrification in Western Society, 1880–1930* (Baltimore).

8 The chief source for this section is Cahan *op. cit.* note 5; quotation p 39. See also Pfetsch F 1970 'Scientific organization and science policy in imperial Germany, 1871–1914: the foundation of the Imperial Institute of Physics and Technology' *Minerva* **8** 557–80.

9 Cahan D 1985 'The institutional revolution in German physics, 1865–1914' *Hist. Stud. Phys. Biol. Sci.* **15** 20.

10 Cahan *op. cit.* note 5, pp 83–5.

11 *Ibid.*, p 71.

12 *Ibid.*, p 7 ch 4.

13 *Ibid.*, p 106.

14 For example Palaz *op. cit.* note 2.

15 Cahan *op. cit.* note 5, p 116.

16 A blackbody source is defined as one that absorbs all incident energy and, as a consequence, emits a characteristic spectrum dependent only upon its temperature. Silvanus Thompson facetiously complained in 1915 of the inadequacy of a language that required 'white' light to be defined in terms of a 'black' body. See Ryde J W 1949 'C. C. Paterson 1879–1948' *Obit. Not. Roy. Soc.* **6** 479–501.

17 Cahan *op. cit.* note 5, pp 147–9; quotation p 148.

18 Abney, when asked to carry his results to a higher degree of precision, not infrequently suggested 'leaving it to the Germans' [E H G-H (*ibid.*) 1921 'Sir W. de W. Abney, K.C.B.' *Proc. Roy. Soc.* A **99** v].

19 *Ibid.*, p 156.

20 *Ibid.*, p 179.

21 Pyatt E 1983 *The National Physical Laboratory: a History* (Bristol), provides an

overview of the institution, but almost entirely neglects the aspects treated here. The NPL annual *Reports* for the period provide details of staffing, finances, facilities and activities, both planned and accomplished.

22 Moseley R 1978 'The origins and early years of the National Physical Laboratory: a chapter in the pre-history of British science policy' *Minerva* **16** 222–50; quotation p 224 (my italics).

23 Moseley R 1976 *Science, Government and Industrial Research: the Origins and Development of the National Physical Laboratory, 1900–75* (PhD thesis, University of Sussex) p 41.

24 NPL 1902 *Report* (Teddington) p 5.

25 *Ibid.*, p 9. France did not form a national laboratory as did the other three countries. According to Harry Paul, the chief reasons were the reluctance of industry to make an investment in science and resistance by a significant number of purists to 'whoring for industry' [Paul H W 1985 *From Knowledge to Power: the Rise of the Science Empire in France, 1860–1939* (Cambridge) p 307]. See also Pestre D 1984 *Physique et Physiciens en France, 1918–1940* (Paris) pp 241–3.

26 NPL 1904 *Report* (Teddington) p 11. The flicker photometer had been invented in 1893 by Ogden Nicholas Rood (1831–1902), Professor of Physics at Columbia University, as a solution to colour photometry, following his observation that intensity changes, but not colour differences, were perceived when lights were rapidly interchanged. Practitioners quickly accepted it as the most precise instrument for heterochromatic photometry. The photometer was employed by first obtaining a visible flicker rate (too rapid a flicker was undetectable; too slow a rate appeared not as a flicker but as a colour interchange). The relative intensity of the two coloured sources was then adjusted to *minimize* the flicker visibility. See Whitman F P 1896 *On the Photometry of Differently Colored Lights and the 'Flicker' Photometer* and Tufts F L 1897 *The New Flicker Photometry*.

27 NPL 1904 *Report* p 17. The director of the NPL for its first two decades, Richard T Glazebrook (1854–1935) had worked at the Cavendish laboratory under Maxwell and Rayleigh, becoming its assistant director in 1891. As director of the NPL, he supported a combination of research useful to both science and industry. See *DSB* **5** 423–4.

28 Paterson C C and Raynor E H 1908 'Photometry at the National Physical Laboratory' *Illum. Eng.* **1** 845–54.

29 NPL 1923 *Report* (Teddington). The projects included work for the Commission Internationale de l'Éclairage, photometric studies of thermionic tube ageing for the Radio Research Board, ships' navigation lamps for the Board of Trade, motor car headlamps for the Ministry of Transport, miners' lamps for the Home Office and the lighting of the National Portrait Gallery and the House of Commons.

30 The illumination building was used for research conducted for the Illuminating Committee of the Department of Scientific and Industrial Research, of which both Paterson and Walsh were members. The DSIR, founded in 1915, formed the Illuminating Committee in 1923. For its early years, see Varcoe I 1970 'Scientists, government and organized research: the early history of the DSIR, 1914–16' *Minerva* **8** 192–217, and Varcoe I 1974 *Organizing for Science in Britain: a Case Study* (Oxford).

31 Walsh J W T 1936 'Photometry at the National Physical Laboratory' *Trans. Illum. Eng. Soc.* **1** 148–54.

32 Walsh J W T 1926 *Photometry* (London) p vii.

33 Railway lamps were internationally standardized in the mid 1930s; the first three-colour traffic lights were installed in London in 1932.
34 NPL 1911 *Report* (Teddington). The instruments were likely intended for measuring the transmissive properties of optical glass.
35 NPL 1921 *Report* (Teddington).
36 NPL 1924 *Report* (Teddington) p 77. Colour standardization work was carried out by the Optics Division for the Physics Coordinating Research Board; the work of the Photometry Division was motivated by employees' responsibilities as delegates to the Commission Internationale de l'Éclairage and as collaborators with the National Bureau of Standards in Washington.
37 The *NPL Report for the Year 1924* noted that photoelectric photometers had been in use in stellar photometry for a number of years, but that gas-filled tubes had been unreliable. The Photometry Section had, in fact, been characterizing selenium devices for industrial use since 1921, but these were generally employed as mere *sensors* rather than as quantitative detectors. See chapter 6 for further discussion.
38 The American National Bureau of Standards at Washington, DC, was officially entitled the Bureau of Standards for most of the period covered (1903–33) 'through an administrative whim' [Cochrane R C 1966 *Measures for Progress: a History of the National Bureau of Standards* (Washington, DC) p 332]. For consistency the abbreviation NBS is used here.
39 Evaluation of lamps as secondary standards continued for many years. The charges in 1916 were $3–$5 for 'seasoning' and standardizing lamps, $1 for candlepower tests and $2–$4 for tests of lifetime [Anon. 1916 *Circular of the Bureau of Standards 6: Fees for Electric, Magnetic and Photometric Testing* (Washington, DC)].
40 Hyde E P 1908 'Photometry at the United States Bureau of Standards', *Illum. Eng.* **1** 761–70.
41 Coast and Geodetic Survey, *Annual Report*, quoted in Cochrane *op. cit.* note 38, p 58.
42 Cochrane *op. cit.* note 38, pp 71–2.
43 The NPL, too, had a staff of 58 in 1904, two of whom were assigned to photometry.
44 Cochrane *op. cit.* note 38, pp 90–1.
45 NBS 1907 Circular 13, *Standard Specifications for Incandescent Electric Lamps* (Washington, DC).
46 NBS 1915 Circular 55 *Measurements for the Household* (Washington, DC).
47 General Electric, successor to the Edison company, owned the majority of manufacturing patents on incandescent lamps in America, which it licensed to at least 33 other companies.
48 NBS 1911 Circular 32 *State and Municipal Regulations for the Quality, Distribution and Testing of Illuminating Gas* (Washington, DC), and Anon. 1912 'Circular on regulations for illuminating gas' *J. Franklin Inst.* **173** 509–10.
49 On the American 'crusade for standardization' between the wars, see Cochrane *op. cit.* note 38, pp 253–63.
50 Meggers W 1967 'William Weber Coblentz' *Biog. Mem. Nat. Acad. Sci.* **39** 55–102.
51 Cochrane *op. cit.* note 38, pp 338.
52 Anon. 1925 'The National Bureau of Standards—its functions and activities', *NBS Circular No 1* (Washington, DC) p 2.
53 Meggers *op. cit.* note 50.
54 Political and social factors emphasized these technical divisions. Colorimetry drew increasing interest after the First World War, when German contributions to international science were restricted. French light measurement was dominated

by individuals who had already made an international mark on heterochromatic photometry and intensity standards, leads which were both actively pursued by university research. Coupled with a national self-absorption for French science, this success with physical photometry contributed to French scientists' neglect of colorimetry. For a discussion of the insularity of French physics in the inter-war period, see Pestre *op. cit.* note 25, especially chapter 5.

55 Part of the reason for this was the lack of English translations. Helmholtz's *Physiological Optics* was not translated until 1924, and Hering's *Spatial Sense and Movements of the Eye* not until 1942. For a good account of the internecine disputes between these two schools of German research, see Turner R S 1993 'Vision studies in Germany: Helmholtz versus Hering' *Osiris* **8** 80–103 and Turner R S 1987 'Paradigms and productivity: the case of physiological optics, 1840–94' *Soc. Stud. Sci.* **17** 35–68. For an earlier, positivistic history of colour science, see Bouma P J 1944 *Physical Aspects of Colour* (Eindhoven) pp 199–222.

56 Priest I G 1925 'Report of the Committee on Photometry and Radiometry for 1924–25' *JOSA & RSI* **11** 357–69; quotation p 366. Friedrich Wilhelm Ostwald (1853–1932), a Nobel-prize winning chemist, developed a colour system based on a triangle having black, white and pure colour corners. His system, first published in 1917, became widely known and was the basis of the Natural Colour System (NCS) later adopted in Sweden. He also wrote extensively on colour harmony through the 1920s, gaining considerable attention in the UK and America. See *DSB* **15** 455–69.

57 Kelly K L 1974 'Colorimetry and Spectrophotometry: a bibliography of NBS publications January 1906 through January 1973' *NBS Special Publication* 393 (Washington, DC).

58 'Research associates' were a response to inadequate funding at the NBS. In 1919, its director proposed to trade associations that 'where specific researches on important problems affect their industry, they send qualified men to the Bureau to do this research.' These research associates would be paid by industry, and their results published and made available to all by the NBS. See Cochrane *op. cit.* note 38, pp 224–5.

59 Ives H E 1932 'Irwin Gillespie Priest' *JOSA* **22** 503–8. Priest (1886–1932) joined the NBS in 1907 and was head of the Colorimetry Section from 1913.

60 Industrial Research Associations were promoted by the Department of Scientific and Industrial Research. Those concerned with photometry and colorimetry included the British Photographic Research Association (the first, set up in May 1918), the Scientific Instrument Research Association (1918), the Electrical and Allied Industries Research Association, the Research Association for the Woollen and Worsted Industries (1918), the Glass Research Association (1919) and the Research Association of British Paint, Colour and Varnish Manufacturers (1926). Some 31 such associations had been set up by 1931. The findings of the Research Associations were considered proprietary and for the exclusive use of the member companies; the DSIR could veto their communications to foreign individuals or companies. Such commercial secrecy inhibited dissemination of knowledge in colour measurement, and placed British workers at a disadvantage compared to their American counterparts. See Moseley *op. cit.* note 23, p 191; Varcoe I 1981 'Cooperative Research Associations in British industry, 1918–34' *Minerva* **19** 433–63; Varcoe *op. cit.* note 3, p 23 and Williams M E W 1994 *The Precision Makers: a History of the Instruments Industry in England and France, 1870–1939* (London) pp 123–39.

61 Nickerson D 1940 'History of the Munsell Color System and its scientific application' *JOSA & RSI* **30** 575–86.
62 Cochrane *op. cit.* note 38, p 270.
63 NPL 1908 *Report* (Teddington) p 20.
64 NPL 1911 *Report* (Teddington) p 64; NPL 1912 *Report* p 83; NPL 1913 *Report* p 76.
65 NPL 1920 *Report* (Teddington) p 54.
66 NPL 1921 *Report* (Teddington) p 73.
67 *Ibid.*, pp 71–2.
68 NPL 1922 *Report* (Teddington) p 75.
69 Similar work was being pursued independently at the NBS. See, for example, Gibson K S and Walker G K 1934 'Standardization and specification of railway signal colors' *JOSA* **24** 57.
70 NPL 1927 *Report* (Teddington) pp 78–80; NPL 1928 *Report* 93; NPL 1929 *Report* 96. See also the 1931 *British Standard Schedule for Colours for Ready-Mixed Paints* BSS 381.
71 NPL 1930 *Report* (Teddington).
72 Guild's researches are published in *Coll. Res. NPL* **20** (1928), and appeared originally in *Trans. Opt. Soc.*
73 Moseley *op. cit.* note 22, p 227.
74 In 1911, only 21 British firms employed graduate physicists, rising to 40 immediately before the war. Chemists were relatively better off, but still under-employed with respect to other countries. Some 1500 chemists, one-third with university training, were employed in British industry in 1902, contrasting with 4000 in Germany, of whom four-fifths had university training. See Varcoe *op. cit.* note 30 (*Minerva* **8**) 193.
75 Moseley *op. cit.* note 22, p 247.
76 Hutchinson E 1969 'Scientists and civil servants: the struggle over the National Physical Laboratory in 1918' *Minerva* **7** 373–98. The disparity between salaries of scientists and administrative staff continued when responsibility for the NPL passed to the Department of Scientific and Industrial Research (DSIR). See, for example, Hutchinson E 1970 'Scientists as an inferior class: the early years of the DSIR' *Minerva* **8** 396–411.
77 Biographical details are from Ryde J W 1949 'Clifford Copland Paterson', *Obit. Not. Roy. Soc.* **6** 479–501, and Clayton R and Algar J 1991 *A Scientist's War: the War Diary of Sir Clifford Paterson 1939–45*.
78 Paterson's obituary lists some two dozen offices he held. Among those related to light measurement were: chair of the Illuminating Committee of the Department of Scientific and Industrial Research; member of the Ministry of Transport Street Lighting Committee; Home Office Committee on the Lighting of Factories and Workshops. He was a founding member of the Institute of Physics in 1919, and helped establish its *Journal of Scientific Instruments* in 1922.
79 Moseley *op. cit.* note 23, p 166.
80 NPL 1919 *Report* (Teddington).
81 Walsh quickly assumed a prominent role in light measurement. He and Paterson had worked closely during the war, inventing an 'electric height finder' for which Paterson was awarded an OBE. Walsh dedicated his book [Walsh J W T 1923 *The Elementary Principles of Lighting & Photometry* (London)] to Paterson 'for an invaluable training in the study and practice of photometry'.
82 Walsh is listed in the NPL annual report as holding a PhD (London) from 1927.

Probably his sole obituary is Anon. 1962 *Trans. Illum. Eng. Soc.* **27** 214–15.

83 Walsh *op. cit.* note 81 and 1926 *Photometry* (London). The latter was updated as late as 1965, three years after Walsh's death. Walsh also wrote a textbook to be used for examinations of the Association of Public Lighting Engineers.

84 Hyde left his $2000 per year job at the NBS in 1908 to do similar research at the Edison lamp laboratories for $5000 per year. See Cochrane *op. cit.* note 38, p 98.

85 Walsh J W T 1951 'The early years of illuminating engineering in Great Britain' *Trans. Illum. Eng. Soc.* **16** 49–60.

86 Clewell C E 1919 'Industrial lighting' *J. Franklin Inst.* **188** 51–90.

87 For a technical history of the Violle standard, see Fleury P 1932 *Étalons Photométriques* (Paris) ch 4.

88 See, for example, Alglave E and Boulard J 1882 *La Lumière Électrique: son Histoire, sa Production et son Emploi* (Paris), and Palaz *op. cit.* note 2.

89 Trotter A P 1911 *Illumination, Its Distribution and Measurement* (London) p 8.

90 Plans for 1904, 1905 and 1906 mentioned in the NPL annual reports call for investigations of a 'primary standard of molten platinum'. See, for example, NPL 1903 *Report* (Teddington) p 7. When trials were finally undertaken in 1911 with the help of the thermometry division, they were shelved without publication of results.

91 Publications during the period included booklets on home maintenance, budgeting and efficient purchasing.

92 For views regarding the high proportion of government lighting projects carried out at the NPL compared to the NBS, see Walsh J W T 1929, 'Illumination research at the National Physical Laboratory' *Trans. Illum. Eng. Soc. (NY)* **24** 473–86.

93 See Moseley *op. cit.* note 22, p 256 for a discussion.

94 Cochrane *op. cit.* note 38, p 336. The effect of the depression on the NBS (with nearly half the staff furloughed in 1933) is described in Kevles D 1978 'Physicists and the revolt against science in the 1930s' *Phys. Today* **31** 23–30.

95 For the expansion of industrial laboratories, particularly in America, see, for example, Dennis M A 1987 'Accounting for research: new histories of corporate laboratories and the social history of American science', *Soc. Stud. Sci.* **17** 479–518 and Smith J K Jr 1990 'The scientific tradition in American industrial research', *Technol. Culture* **31** 121–31.

96 Dupree A H *Science in the Federal Government* p 337, quoted in Cochrane *op. cit.* note 38, p 218.

97 Sanderson M 1972 'Research and the firm in British industry, 1919–39' *Sci. Stud.* **2** 107–51.

98 *Ibid.*, p 135.

99 Another significant industrial laboratory that influenced illuminating engineering and photometry is the Westinghouse Electrical and Manufacturing Co in Pittsburgh. Photometry work at other light bulb manufacturers was more restrained. For the Dutch case, see Heerding A 1986 *The History of N. V. Philips' Gloeilampenfabrieken* (Cambridge). Another locus, influential in colorimetry research and in training career scientists, was the Eastman Laboratories of Kodak at Rochester, set up by C E Kenneth Mees in 1912.

100 The National Electric *Lamp* Association should not be confused with The National Electric *Light* Association formed in 1885. Initially an association of arc-lighting interests, by 1905 the *Light* Association represented 508 power generating companies and numerous individual and associate members from as far afield as Hawaii and the Yukon territory. Its stated goals were 'to advance the art and science of the

production, distribution and use of electrical energy'. The organization saw its role as primarily educational, however, and pledged not to become 'engaged in business'. It was reorganized as the Edison Electric Institute in 1933. See Wilkes J D 1973 *Power and Pedagogy: the National Electric Light Association and Public Education, 1919–1928* (unpublished PhD thesis, University of Tennessee) and Crickmer B 1993, 'Edison Electric Institute: the first 60 years' *Elec. Perspectives* May/June 46–66.

101 For an economic history, see Bright A A Jr 1949 *The Electric-Lamp Industry* (New York), especially ch VI.

102 Hyde E P 1909 'The physical laboratory of the National Electric Lamp Association' *Illum. Eng.* **2** 758–61.

103 Quoted in Wise G 1985 *Willis R Whitney, General Electric and the Origins of US Industrial Research* (New York) p 257.

104 One of the member companies. Quotation from Hyde *op. cit.* note 102.

105 Cox J A 1980 *A Century of Light* (New York) p 196.

106 During the war, for example, the laboratory designed signalling lamps and investigated optical glass, flares and camouflage, as the NBS was doing. This, along with 'many projects in testing and the creation of new light-measuring instruments, kept the staffs well occupied... at Nela Park'. See Keating P W 1954 *Lamps for a Brighter America: a History of the General Electric Lamp Business* (New York) pp 82, 122–3.

107 General Electric was the chief of 34 defendants in the suit, which disclosed the company's interests in the National Electric Lamp Association (by now owning 75%, with GE and Nela together producing 80% of American lamps). The court ordered that the National Electric Lamp Association be dissolved, that GE do business only in its own name and that it refrain from the price-fixing of incandescent lamps. See Hammond J W 1941 *Men and Volts: the Story of General Electric* (Philadelphia) pp 340–3, and Bright *op. cit.* note 101, pp 151–9.

108 Quoted in Wise *op. cit.* note 103, p 257.

109 *The Journal of the Franklin Institute* published research notes from both government and major commercial research laboratories, several of which were carrying out work in photometry. A number of individuals who were to become prominent in photometry and colorimetry in the following decade published early work in the journal, including Leonard Trolland at Nela, P G Nutting at Eastman Kodak, Irving Langmuir at General Electric and Harold Ives at the United Gas Improvement Company.

110 Noble *op. cit.* note 1, pp 122–3 and 171–3.

111 The General Electric Research Laboratory in America was much larger, but concentrated on incandescent lamp development and lighting arrangements rather than intensity measurement. The two companies had no financial connection except in the period 1928-34. See Reich L S 1985 *The Making of American Industrial Research: Science and Business at GE and Bell, 1876–1926* (Cambridge) p 104, and Wise *op. cit.* note 103.

112 Clayton R and Algar J 1989 *The GEC Research Laboratories 1919–1984* (London) ch 1. Much of the information in this section is based on information given in a talk and privately circulated article by Paterson, *A Confidential History of the Research Laboratories* (1945) and unpublished GEC reports quoted in the book.

113 A version of this device was commercialized a decade later: see Anon. 1929 'The 19th annual exhibition of the Physical Society and the Optical Society' *Illum. Eng.* **22** 42.

114 Clayton and Algar *op. cit.* note 112, p 45.
115 *Ibid.*, p 100.
116 Paterson C C 1932 'Photo cells: the valves which operate by light' *J. Sci. Instr.* **9** 33–40.
117 Clayton and Algar *op. cit.* note 112, p 47. Mention of phototubes is limited to in-house development of instruments to evaluate fluorescent lighting.
118 Campbell N R and Ritchie D 1929 *Photoelectric Cells: Their Properties, Use and Applications* (London), and Walker R C and Lance T M C 1933 *Photoelectric Cell Applications* (London). The work of the GEC laboratory is discussed in chapter 6.
119 The Second World War led to an identification of physical light measurement as a subject of military importance, particularly for aircraft and missile detection and for the analysis of materials by spectrophotometry, as discussed in chapter 9. The vision-based technology universal during the First World War largely precluded such military interest, although Alexander Trotter led a team studying flares and parachute lights (Walsh *op. cit.* note 85).
120 Cahan *op. cit.* note 9, pp 225–6.
121 In 1916, however, the PTR director awarded 2000 marks for constructing a blackbody radiator to be used as a unit of luminous intensity. See Cahan *op. cit.* note 9 226–7.
122 The 61 physics staff were joined by 89 temporary and volunteer workers, some 50 of whom were women.
123 NPL 1912, 1913–14, 1914–15 *Report* (Teddington).
124 NPL 1915–16 *Report* (Teddington) p 7.
125 Stratton J W 1916 Congressional Hearings February 2 991–2, quoted in Cochrane *op. cit.* note 38, p 171.
126 The wartime research was, however, popularized, for example in chapters on 'Lighting conditions in war time' and 'Searchlights and other appliances for the projection of light' in Gaster L and Dow J S 1920 *Modern Illuminants and Illuminating Engineering* (2nd edn).
127 In 1919, the International Research Council (IRC), sponsored by the Allies, advocated policies of ostracism for German scholars which excluded their participation in international meetings until the mid 1920s. See, for example, Kevles D J 1971 'Into two hostile camps: the reorganization of international science after World War I' *Isis* **62** 47–60.
128 This was also a general consequence of the increase in non-academic careers for physicists. After the First World War the existence of national and industrial laboratories promoted a schism between 'applied' and 'pure' physics. See Weart S R 1976 'The rise of 'prostituted' physics' *Nature* **262** 13–17.
129 For example, Fabry C 1925 'The connection between astronomical and practical photometry' *Trans. Illum. Eng. Soc. (NY)* **20** 12–16.
130 Gaster L 1926 'Illuminating engineering in relation to optics' *Proc. Opt. Convention* vol 2 (London) pp 297–304.

CHAPTER 6

TECHNOLOGY IN TRANSITION

With social organization came technological change. The inter-war period reshaped techniques and instruments. By the Great War, astronomers were increasingly adopting physical methods of light measurement, and laboratory spectroscopists soon joined them in converting to photographic methods. But engineering practice, wedded to visual methods, remained little changed from the 1860s until the 1920s for the vast majority of photometric work[1]. Photographic and visual photometrists had distinct outlooks after the war, leading to a divergence of practice between the communities. Only when all practitioners began to employ photoelectric measurement techniques in the early 1930s did practice again coalesce to a single technique.

This was not, though, a case of superior technology transforming practice. Instead, practice was socially shaped: the 'subjectivity' of visual photometry was first denounced; alternative physical techniques were then explored; and, with considerable difficulty, these new technologies were made to work. In this transferal of faith, the human component of the measurement process became minimized and the notion of the 'observer' was abstracted. Underlying the transition was a shift in cultural values.

This gradual process, repeated in each community, involved the recasting of photometry into less problematic terms. Nevertheless, the first decade of photoelectric instrumentation resurrected once again a concern of earlier periods: how reliable and reproducible were the measurements, and how did they relate to human perception? The new technologies proved, in their own ways, to be as troublesome as their predecessor. What were the contexts of the technological changes adopted by the scientific and engineering communities, and the specific problems surrounding those changes?

6.1. A FASHION FOR PHYSICAL PHOTOMETRY

The transition from visual to photographic, and subsequently photoelectric, methods could be portrayed as a natural evolution, replacing the eye by an alternative providing more sensitivity and convenience—indeed, this 'technological determinism' is the conventional view propounded by technical histories. However, there was a deeper motivation for the change relating to a growing scientific preference for physical methods. As other case studies have demonstrated, the adoption of new measurement technologies is seldom simple and frequently has a significant cultural component[2]. While espousing rational arguments for a physical detector of light, its proponents weighted their views with tacit considerations.

This transition began not with new inventions but with condemnations. By the First World War, nearly all practitioners—despite their disparate backgrounds and professional goals—sought a physical alternative to the eye. The ostensible reasons for seeking an alternative differed for each sub-culture of practitioners. But four principal motivations can be identified for the adoption of physical methods: perceptions of (1) objectivity, (2) precision, (3) speed and (4) automation.

6.1.1. Objectivity

The attraction of 'observer-independent' measurements was an important criterion for both scientists and engineers at the turn of the century. There were at least two aspects to this. First, human observations were increasingly labelled as unreliable; second, practitioners were placing greater emphasis on relating the perceptual property of *brightness* to the physical quantity of *energy*[3].

'Observer-independent' methods were expected to be free from the distortions and complications of human vision, influences that were suspected even if not entirely elucidated. By removing the human contribution from the chain of processes that converted a light intensity into a number, the quantification was rendered simpler and intrinsically more trustworthy[4]. In describing his first attempts to employ a physical photometer, for example, the astronomer Joel Stebbins at the University of Illinois noted that 'there is no evidence of a large difference in scale between my results and those derived from visual observation, but in any event it is my opinion that the selenium photometer *gives more nearly the absolute scale* than can be obtained visually'[5]. He was enunciating several views implicitly accepted by astronomers: first, that they should be concerned with measuring *physical power* rather than *perceived intensity*; second, that visual perception was a good approximation for what they sought; and third, that a *physical* detector was necessarily better at attaining astronomers' *physical* objectives of measurement. Yet Stebbins made no mention of the logical puzzles he posed: given only a visual and a selenium photometer, how could he judge one to give 'more nearly the absolute scale', and what, indeed, constituted an absolute scale? An implicit bias towards physical measurement and methods, without experimental justification, is thus revealed.

At the same time that physical methods divorced photometry from its association with human factors, they brought it into line with other specialisms in physical science where its proponents felt it more properly belonged. According to this view, the measurement of light intensity was merely a particular case of energy measurement. This appropriation and categorization of the subject is illustrated by the work of the Dutch physicist L S Ornstein (1880–1941), who spent much of his career defining methods of intensity measurement using photographic and reference-lamp methods, and working out a theory of spectral line intensities. Looking back from the perspective of 1933 to his professional beginnings around 1910, his colleagues noted the general enthusiasm of investigators for physical methods:

> They made use of instruments which had been planned and mounted in previous years in the very room now used for this investigation, viz. a thermopile and a galvanometer, the readings of which were recorded photographically. The complete objectivity of this method greatly impressed our neophyte; it satisfied his innate craving for accuracy and certainty, and the mere sight of these documents in black and white, fixing the results of the experiments as it were in a mathematical curve, must have delighted him too.[6]

The quotation may say as much about the newly entrenched ideas of experimentalists in the 1930s as it does of the transition period. The *complete objectivity*, *accuracy* and *certainty* were, however, recurring themes for the early promoters of physical photometry. By 1930, these characteristics had been associated with physical photometry in principle, if not entirely implemented or verified, by all practitioners. The term *neophyte* also suggests that a new generation of investigators was responsible for championing quantitative methods in light measurement.

The tendentious linkage between photometry and energy measurement was made increasingly explicit by physical scientists in the first years of the 20th century. The term 'mechanical equivalent of light' was commonly employed, in analogy with the term 'mechanical equivalent of heat'. This connection was problematic, however. To relate perceived intensity to physical energy, investigators were forced to *define* the average visual response, the light source and the viewing conditions[7]. Investigators glossed over this synthetic relationship in their enthusiasm to demonstrate a quantitative connection between light intensity and physical measurement.

The trend from visual to physical viewpoints overturned earlier scientific convictions. Not even the previously prevailing argument—that the intrinsically 'visual' characteristic of brightness demanded human observations—was reiterated in the growing mood of practitioners for physical measurements. The definition of photometry itself changed in the period from the turn of the century to the First World War: the centre of gravity had subtly shifted from the human eye to physical detectors. A new fashion, albeit one with convincing supporting arguments, had been adopted. The earlier physiological emphasis—the shared dogma of physical scientists such as Lummer and Brodhun as well as pragmatic engineers—was discarded in favour of a practical search for superior detectors. One of those converted was Leon Gaster, organizer of the Illuminating Engineering Society of London, who gave his support to physical methods:

> I agree... that physical photometers have great possibilities. Whilst realizing the difficulties that have yet to be overcome in connection

> with the use of photoelectric cells and similar devices, I hope that ultimately it may be possible to devise a direct-reading photometer based on their use. A reliable instrument of this type would be of immense value in illuminating engineering.[8]

At the very least, he suggested, the adoption of physical methods would distance these studies from the response of the human eye.

6.1.2. Precision

Researchers at the government standards laboratories stated the *precision* of physical methods as potentially their chief advantage. John Walsh, responsible for the NPL Photometry Division between the wars, secretary of the International Commission on Illumination, and author of the widely used text *Photometry*, became a proponent of the new photoelectric methods:

> The search for a physical photometer is as old as photometry itself.... In my opinion it is essential that photoelectric photometry should be developed. Visual photometry is adequate to meet most practical needs of the present day, but there is no doubt in my mind that a demand for much higher accuracy is inevitable sooner or later, and such accuracy is only attainable by physical methods. It has always to be borne in mind that increased accuracy in measurement means refinements in other directions, notably, as has been pointed out, in the design of electric lamps for use as standards. I feel sure that as soon as the need is indicated to lamp makers they will find a solution of the difficulties.[9]

While careful practitioners of visual photometry had been achieving measurement precision of 1% or better for decades, such results demanded the control of unpredictable human factors. These human factors were themselves unquantifiable. The degree of fatigue or the 'normalness' of an observer's response to light could not be related numerically to the precision achieved. Physical methods promised a way of grounding *all* aspects of the measuring process in details that could be quantified. According to this view, the effects of variables such as exposure time, developer concentration and temperature would be numerically and individually determined. Thus the uncertainties of the photometric reading could be decomposed into their component contributions. This, in turn, could allow experimental details to be separately improved to reduce their contribution to the net uncertainty. As a plan of action to improve photometric precision and to remove it from the conceptual mire of human visual response, this physical approach was attractive to scientists.

Yet this programme was based on faith rather than demonstrated potential. As discussed later, the NPL through the 1920s struggled to develop physical detectors that could equal the precision of visual photometry. Another justification was needed.

6.1.3. Speed

Where the astronomers made do with slow and technically difficult photographic methods, the engineers demanded speed and ease of use. Drawing an analogy with the popular Kodak cameras, for which the slogan since 1888 had been 'you press the button, we do the rest', one editor wrote in 1906:

> The apparatus which we describe this week also reduces photometry to the pressing of a button, while the selenium 'does the rest' and it can be used by unskilled observers.[10]

The urgency for rapid and convenient photometry rose as applications grew. At the Optical Laboratory of the Physikalisch-Technische Reichsanstalt in 1913, for example, scientists were encumbered with 700 photometric tests of lamps, requiring a significant fraction of their time[11]. A de-skilling of measurement would also promote mass production of standardized products such as light bulbs. Simplification was called for.

6.1.4. Automation

Closely allied to a desire for speed was a wish for the automation of measurements of light, part of a general trend towards automatic control in engineering and industry[12]. The meaningful employment of light intensity measurements frequently led to the need to acquire large bodies of data, whether of lamp characteristics as a function of angle, paint formulations versus wavelength or photographic emulsion transparency versus position. Even rapid measurements could require tedious work by patient instrument-minders. Following the First World War, such routine jobs were less attractive than formerly[13].

An early proponent of automated light measurement was the MIT physicist Arthur Hardy. He developed in 1922 the first recording photoelectric spectrophotometer to study the problems of colour printing, chiefly to acquire large numbers of data quickly:

> it seemed probable that a great mass of spectrophotometric data would be required.... The only escape from this situation seemed to lie in the direction of developing a more rapid method of spectrophotometry. There was little hope of decreasing the time required for a spectrophotometric analysis with instruments of the visual type. This type of instrument requires that the reflectance of the test sample be determined with high precision under illumination by homogeneous light of some thirty different wave-lengths within the visible region of the spectrum. Since at least five settings are usually necessary at each wave-length, the possibility that an instrument could be devised to determine these data and record them automatically seemed worthy of investigation.[14]

Hardy and colleagues devoted as much effort to automating their measurements as to improving their precision. Their labour provided an immediate pay-off: during its first year of operation, the spectrophotometer recorded over 1000 spectra, providing a wealth of information for colour scientists. Widely adapted, Hardy's device proved highly popular when commercialized some years later.

Automation symbolically removed the problematic observer from the measurement, making it an attractive and highly visible benefit of physical methods. By relegating the operator to interpreting graphs or numerical lists—an activity seemingly free of physiological and psychological factors—automated instruments appeared to redraw the boundaries to position photometry firmly within the realms of physical science. That such a demarcation entailed the adoption of new light detectors having their own complexities, and requiring a definition of how the visual sensation related to their replacements, was not initially an issue.

For different groups of practitioners, then, physical photometry promised distinct advantages: better objectivity, precision or speed than the eye could provide, and even the potential for removing the observer altogether. Along with these practical advantages, however, physical photometry required a change of philosophy. The new physical scientists who took it up saw photometry not as a common-sense procedure intimately tied to human vision, but as a branch of energy measurement. By interpreting light measurement in this way, they reclassified the eye to be one of the more unreliable detectors of radiant energy, rather than as the central element in a perception-oriented technique. This tailoring of photometry to the conceptions of physical scientists was to make it the dominant view for the first three decades of the century. How did this technological transition occur in the various technical communities?

6.2. THE REFINEMENT OF VISION

For engineers, the transition was a long time coming. Routine uses of photometry such as lamp standardization and testing had become commonplace after 1900. As a result, the techniques of visual photometry matured and were highly systematized in the first two decades of the century at the national and industrial laboratories[15]. This is not to say that these laboratories shunned physical techniques; rather, they saw their task as one of determining the brightness *as perceived by the human eye*. Bemoaning the difficulties, two engineers wrote in 1894:

> That we do have graduated slide scales in photometry means very little, for what we really want is a quantitative measure of the intensity of brain effect. And how can we do this with the brain itself? We are beset with physiological or, rather psychological, effects, and as yet there is no psychological unit which we can represent by anything concrete to give to the Board of Trade.[16]

The only option was to employ human observers. But the eye was not a detector of convenience; it was an intrinsic and central part of the apparatus. As Alexander Trotter observed, a photometer should merely furnish 'a development of our powers', and

> whatever results we obtain, however ingenious the apparatus used to arrive at them, and whatever the conditions we prescribe for carrying out the work, our measurements are of no value if they disagree with the common-sense estimate which anybody may make merely by using his eyes.[17]

This central role of the eye in photometry was accepted by contemporary physicists as much as by pragmatic engineers. The PTR physicists Lummer and Brodhun, inventors of the most popular visual photometer, noted:

> The purpose of practical photometry is to compare the total intensities of light sources as they are perceived by our eyes. In such a measurement of the purely *physiological* effect of flames only the eye can therefore be used; all other measuring instruments, such as the radiometer, selenium cell, bolometer and many more of the kind, are to be discarded in so far as these indicate *physical* effects of light sources.[18]

And Leon Gaster, representing illuminating engineers, echoed the physicists, observing that ‘all such “physical” apparatus, besides being inconvenient in practice, is open to the objection that it does not “see” the energy impinging upon it in the same way as the eye’[19].

Even though the intrinsic reliability of human observers was clearly poor, the laboratories sought to improve their results by carefully standardizing the conditions of observation and automating the observation process. In effect, the practitioners attempted to neutralise or compensate for the variable human aspects, making them as physically based as possible by restricting measurement to highly controlled circumstances. If the observer was to be a mandatory component of the apparatus, they reasoned, then the observer would be rendered as reliable as the rails, cranks and standard lamps that shared the room.

The strategy of standardizing viewing conditions yielded immediate gains. Investigators had found that results obtained using photometers employing differently sized illuminated areas gave incompatible results[20]. Another standardization was to restrict the range of illumination used, so that the Purkynje effect, an apparent colour change of weakly illuminated objects, was avoided[21]. By identifying ‘perturbing effects’ which caused deviations from the desired ‘linearity’ and by limiting the scope of measurements, quantification was thus made to appear increasingly plausible and, indeed, natural.

Besides controlling such instrumental and visual contributions to the measurement, serious practitioners reduced the variability of single observers by making multiple repetitions of measurements. Repeating a measurement hundreds or even thousands of times was not uncommon in precise work, and could yield repeatability of between 0.1% and 1%. If the starting conditions were suitably randomized (e.g. by beginning with the reference lamp at an arbitrary intensity with respect to the sample), multiple measurements could lower the uncertainty caused by observational factors such as fatigue or inexperience[22].

When differently coloured lights were to be compared, even this care was not enough. Because of the differences in the colour responses of different observers, no amount of repetition or control of viewing conditions could remove the inherent personal bias. For this reason, the comparison of the pentane standard with a carbon filament electric lamp (which had relatively yellow and white tints, respectively) at the NPL necessitated the drafting of all available technical staff as observers to obtain an unbiased mean[23]. Another approach to comparing light sources of different temperature (and hence colour) was the so-called 'cascade' method. To compare carbon-filament lamps with the newer (and whiter) metal-filament lamps when they became commercially available, a number of intermediate sub-standards were manufactured, designed to exhibit little or no colour difference compared to the sub-standards immediately adjacent[24]. The great advantage of the cascade method was that it required few observers, even if the colour sensitivity of their eyes was distinctly different from that of the average human eye.

Such systematization of observation could make an onerous task practicable. By 1908, Leon Gaster could wax optimistic:

> At one time, when such investigations had not yet been undertaken, the cumulative effect of unrecognised errors... was not infrequently ascribed to personal error; thus it came about that photometry came to be regarded as a hopelessly unreliable process, to the arbitration of which commercial matters could never be subjected. Now, however, the old sources of uncertainty are being one by one recognized and removed, and it must be recognised that photometry, well within the limits of accuracy imposed by commercial consideration, is possible.[25]

The other early 20th-century developments in visual photometry related to efficiency and simplification to suit the routine, high-volume measurements required by industry. The speed of observations could be remarkable. The process was made as routine as possible using human workers:

> In certain lamp factories, electric glow-lamps are tested by piece-work. This is generally carried out by girls working in teams of two, one seated in front of the photometer, adjusting it, making the observations, and reading the result either in candle-power at constant pressure [i.e. voltage], or in volts for a given candle-power; the other changes the lamps and marks them.

'With freely moving equipment a measurement can be made to an accuracy of 2 or 3 per cent in 5 or 6 seconds', continued Alexander Trotter[26]. Trotter gave much consideration to measurement errors, nearly all of which were related to human variations, citing ill-health, general fatigue and various forms of ocular fatigue as fatal to accurate measurement[27]. Indeed, 'ocular hygiene'—lighting to prevent general fatigue, eye strain and conjunctivitis and intended to promote speed and accuracy in fine work—was much mooted in industry at the time.

The standardization of visual photometry arguably reached its zenith in the establishment of legal specifications for visual instruments. An NPL staff member wrote in 1924

> the development of a cheap and accurate portable photometer is one of the problems of the moment. It is desirable that some standard of performance be specified for such instruments. A neutral glass is essential with most photometers of this description but many in use are far from being neutral.[28]

By the next year, the British Engineering Standards Association (BESA) had satisfied his wish, publishing a *British Standards Specification for Portable Photometers*[29]. This was followed four years later by another specification for integrating photometers, which defined attributes such as the surface reflectance, size of the reflecting sphere and diameter of viewing apertures[30].

The adoption of standardizing methodologies thus improved repeatability and went far towards legitimating the subject. But the regularization of the human factors in visual photometry illustrates the tantalizingly unattainable goal of the reliable measurement of a 'typical' human perception. An alternative approach, adopted increasingly by those scientists free of the pressures of utilitarian application, was to replace the complications of the human eye with what were claimed to be the more generally characterized vagaries of physical detectors of light. The best alternative at the turn of the century was the photographic plate.

6.3. SHIFTS OF CONFIDENCE

Despite the prevailing view that visual observation was essential for a meaningful definition of photometry, some physical scientists were willing to consider physical alternatives. William Abney, for example, interested in both vision and photography, predicted in 1893 that 'note-book records of photometric work would soon become obsolete, and that photographic records would become general'[31].

By the turn of the century, despite evolutionary improvements in visual photometers, *photographic* photometry began to make inroads among scientists. Part of the reason for this was analytical convenience. A photograph could record an intensity for later examination and matching by eye. This was particularly useful in astronomy, where a photographic record could be examined at convenience by one or more observers, rather than making a visual photometric reading by a single fatigued individual at the eyepiece of a telescope[32]. The ability to evaluate photographic records in an optimal setting was important to the acceptance of photographic photometry. So, too, was its ability to record the raw data. Visual photometry had no means of making a record of observations or to serve as an illustration for a publication. Photometric results had thus remained peculiarly individualized. The ability to record observations rendered the technique public[33].

To its first users, the conceptual difficulties of photographic photometry appeared minimal. Initially, at least, photographic methods of photometry simply replaced the eye by a photosensitive plate, the analysis of the resulting plates being carried out using the methods of visual observation[34]. The photographic record acted merely as an intermediary step translating the visual evaluation to a more convenient location and time. In a direct application of the visual methods of observation described in chapter 2, practitioners either noted the point of minimum exposure on a plate (extremum detection), noted the lack of exposure (thresholding) or equated the greyness of exposed plates (matching).

The cultural context was important in determining users' perceptions of photography. Photographic methods were taken up first by the community of astronomers and then by astrophysicists for determining stellar temperatures and for classification[35]; by the first decade of the 20th century, visual observations for stellar photometry had been completely superseded. For these astronomers, photographic photometry had unique advantages. For spectrophotometry in particular, visual methods proved simply too insensitive and time-consuming at the telescope. The photographic plate was clearly superior in this respect, being able to build up gradually an image over seconds or minutes to achieve a sensitivity far superior to that of the eye. In addition, fluctuations in brightness caused by atmospheric turbulence were averaged out by this integration process. Photographic recording also improved upon the measurement of the intensity of stars of different colour. The visual judgement of colour intensity in spectrophotometry was a process fraught with error. Photography, in contrast, yielded a monochromatic plate from which the density could be more straightforwardly judged by eye. The problem of colour sensitivity was transferred to the photographic emulsion, which could—with meticulous attention to emulsion chemistry and chemical processing—be rendered less variable than different human observers.

From the astrophysics community, photographic photometry spread to laboratory spectroscopists, who again found that the ability of the photographic plate to record a faint spectral image made it practicable where the human eye was not[36]. Again, the photographic plate averaged the irregular intensities produced by the flame or arc sources that were used for vaporising materials in spectral analysis. Photographic photometry had advantages over direct visual observation in two further circumstances, both related to spectrophotometry. First, when measuring the relative brightness of different portions of a spectrum when the light source is fluctuating, a method of simultaneously recording all wavelengths is required. Second, when observing the short ultraviolet wavelengths to which the eye is insensitive or blind, photography was unavoidable.

Applied to scientific measurement in the last decades of the 19th century, photography became the principal photometric method for scientists by 1920 and found its widest routine application in spectroscopic research. The complexities of the technology were well understood, and its methods rendered routine, by the mid 1920s[37]. This new technology remodelled photometry to emphasize features important to the astronomical community: instead of obtaining measurements linked to human perception, the practitioners stressed the ability to integrate weak images and to analyse records.

Despite astronomers' unproblematic exploitation of the seemingly straightforward analogy between visual and photographic methods of photometry, photographic photometry made no inroads whatsoever into industrial applications. Indeed, the use of photographic in preference to visual methods is a practical criterion for dividing engineering and scientific uses.

From the viewpoint of the illuminating engineers and standardizers of light intensity, there were good reasons to reject photographic photometry. First, it was impracticably slow and complicated. In the context of their work, the process of exposure, processing and subsequent examination of the plates by eye was pointlessly circuitous. As long as the eye served as the final arbiter of relative intensity, the only function of the photographic plate was to record the measurement. For an activity that generally did not have the leisure for subsequent analysis, photographic photometry offered no advantage. Moreover, the photographic method required standardized photosensitive materials and processing which introduced even more sources of error into the photometric evaluation. An understanding of the extraneous factors affecting photographic emulsions was only gradually becoming clear. By the First World War, then, engineers were becoming separated from scientists by technique as well as by motivations.

6.4. PHYSICAL PHOTOMETRY FOR ASTRONOMERS

A handful of astronomers formed the vanguard of an as-yet unelaborated physical approach, developing stellar photometry from a visual method to a technique based upon physical measurement. This conceptual development had three technological stages: first, photographic recording of the intensity, with subsequent visual analysis; next, photographic recording of the intensity with photoelectric analysis; and, finally, direct photoelectric measurement of stellar intensity. The photographic stage of the process has been discussed earlier; this section will deal with the technical difficulties associated with the photo-visual and photoelectric methods.

6.4.1. An awkward hybrid: photographic recording and visual analysis

Photographic recording of stellar intensities originated with William Bond at the Harvard College Observatory, who in the 1850s related stellar intensities to the diameters of the images they formed on photographic plates[38]. The technique, rendered reasonably precise by his successors, relied upon calibrating the relationship between the image diameter and apparent brightness. The image formed, although theoretically a minute point, in practice consisted of a dark centre surrounded by a halo of radially decreasing exposure, caused by the optical limitations of the telescope. The size of the image recorded also depended on the sensitivity of the photographic plate. Like Bond before them, David Gill and J C Kapteyn, who used photographic methods between 1895 and 1900 for their *Cape Photometric Durchmusterung* catalogue, simply measured the photographic diameters[39]. Because of the dependence of the image size on telescope optics,

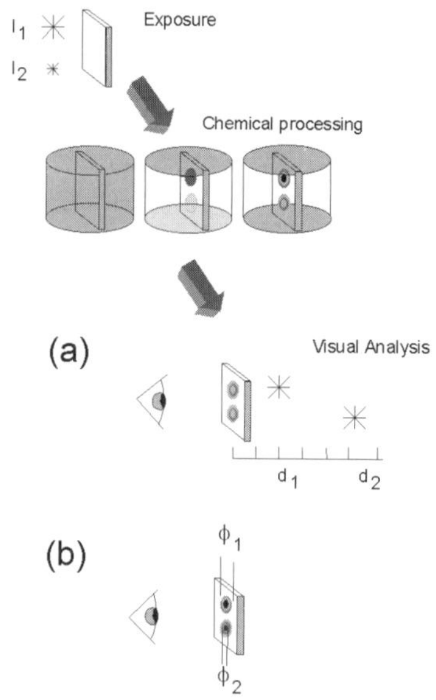

Figure 6.1. Steps in a photographic/visual measurement of intensity. The intensities I_1 and I_2 are ultimately related to either (a) the distances d_1 and d_2 of a reference lamp on a photometric bench that produce the same apparent brightness through the exposed plate (the *densitometric method*) or (b) the diameters ϕ_1 and ϕ_2 of the stellar images produced (the *size-of-image* method).

each instrument had to be individually calibrated—hardly strong evidence for the greater generality of the technique compared to problematic human eyes.

As the successors to Bond discovered, the brightness of a star affected not only the *diameter* of a photographic image, but also its optical *density*. To minimize the complexity of the effect, some investigators defocused the telescope to yield a blurred spot and measured its density. The relationship between the smudgey image diameter and intensity thus differed depending on the quality of the telescope optics, the type of photographic plate used, as well as exposure time, details of plate development and intensity range. The category of plate development alone included critical factors such as the chemicals used for development and fixation of the plate, development temperature, development time and agitation, with the precise method of agitation of the developing plate in the liquid significantly affecting the resulting density[40]. Measuring the diameter of the image had the advantage, however, that no estimate of intensity was needed. Photometry was again transmuted: the problems of photometric judgement were replaced by a mechanized process of exposure, chemical processing and metrology[41].

The alternative to this metric technique of photometry was a more conventional visual estimation of the greyness of the exposed plate. William

Abney, for example, compared the 'photographic values' of moonlight and starlight with a candle[42]. Unlike simple visual observation, the photographic technique involved several steps (figure 6.1). Abney first prepared a photographic plate having a series of stepped exposures to yield a gradation of density. He then used this plate as a neutral density filter through which his test lights shone to expose a fresh photographic plate. From the resulting exposures using moonlight and candlelight, he visually compared the grey tints of the stepped exposures to determine their difference[43]. The measurement of the greyness of point-like stellar images was difficult without microscopic examination. By either diffusing or defocusing the image, however, a larger, relatively uniform spot could be obtained which was more amenable to analysis. In some cases, observers used a combination of diameter measurement and grey-level matching for stellar photometry.

Photographic photometry benefited from the standardization of plates, chemical formulations and conditions of development. Using such methods for laboratory spectroscopy, the precision of a measurement by the inter-war period had attained typically 5–10%, or in optimal conditions about 1%[44]. Although this is somewhat poorer than the visual determination of standard lamps, the measurement of the unstable and weak spectroscopic sources was correspondingly more difficult. Claims of achievable precision could also be inflated. While 'under favourable circumstances results can sometimes be repeated to within one-fifth per cent', the American investigator C H Sharp gave 2% as the typical precision of commercial photometry, 'which is probably only approached in the best laboratories'[45].

6.4.2. A halfway house: photographic recording and photoelectric analysis

For astronomers, according to one historian, 'the development of recording microdensitometers, in some cases that could directly produce intensity records from the density, or blackening, in the nonlinear photographic emulsion, was the important instrumental development'[46]. Such densitometers, or 'microphotometers', some employing photoelectric detectors, were in common use before the First World War.

Before the turn of the 20th century, a photoelectric cell was almost invariably a compound of selenium. The electrical resistance of pure selenium falls when illuminated, leading to its description as a 'photoconductive' material. In combination with other substances, selenium can be made to yield a small voltage (thereby acting as a so-called 'photovoltaic' device) when illuminated. The causes of this photosensitivity were unknown, and indeed of little interest, to those seeking applications[47].

Another type of photosensitive effect was being actively investigated by the first decade of the century, however. The 'photoelectric effect' was the observation that certain materials, when used as a cathode in an evacuated glass tube, generated a weak electric current when illuminated with light[48].

The microphotometer was, in principle, simply a photometer incorporating optical elements to view a small portion of a photographic plate. The first such

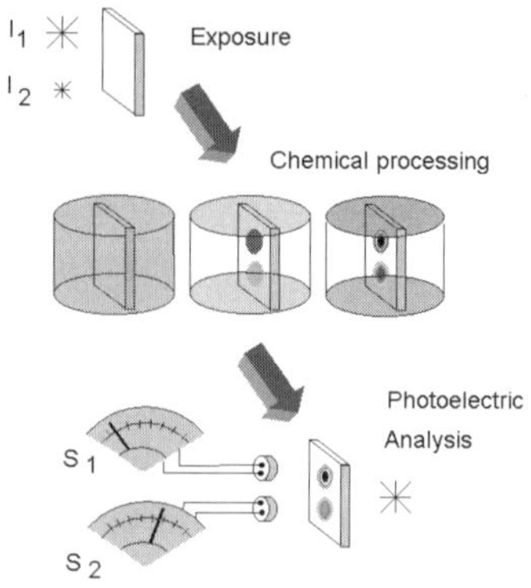

Figure 6.2. Steps in a photographic/photoelectric measurement of intensity. The intensities I_1 and I_2 are ultimately related to the signals S_1 and S_2 of the photoelectric detecting system. The diagram is schematic only; for example, the photoelectric cells were usually phototubes consisting of an alkali halide surface and anode connected with a large potential difference, surrounded by low-pressure gas and contained in a glass envelope. Intervening optical elements would be employed at both the exposure and analysis stages. The measuring instrument was typically an electrometer, or galvanometer operating on the null-balance principle.

instrument was designed by Hartmann in 1899 for stellar photometry[49]. This was a visual photometer employing a variable-density wedge as the reference against which the photographic plate was compared. Experimenters made attempts to replace the eye by a physical detector within a decade (figure 6.2). Koch, in 1912, used two sets of photocells, one illuminated directly by a small filament lamp, and the other receiving the light focused on and passing through the photographic plate. The ratio of the two signals, representing the fraction of light passing through the plate, was measured by a string electrometer. The replacement of the eye by photocells allowed Koch to automate the measurement process: the photographic plate was moved through the focused beam by a clockwork motor, which also moved a photographic film used to record the deflection of the electrometer. Development of this film revealed a tracing proportional to the optical transmission along the original plate[50]. Such a system made feasible for the first time the conversion of spectrograms, with their collections of dark and light bands, into a graphical display of intensity variations. The stability of such early photocell microphotometers was not adequate for routine work unless used with great care by their designers. Koch's electrometer was prone to interference from stray electrostatic potentials, and the sensitivity of his photocells varied with time and temperature. A more successful instrument that found wide application among astronomers was the Moll microphotometer. This device used a thermopile instead of a photocell, a detector that benefited from good stability and sensitivity, and a longer history of successful usage[51]. This instrument was perhaps the first physical photometer to justify claims of superiority over the eye. Such was its indifference to external disturbances that, while in use, it 'did not require any special supervision'[52]. The portion of the photographic plate viewed could be made as narrow as 0.02 mm by slits, allowing extremely fine detail to be measured. The microphotometer was used by Moll's countryman Marcel Minnaert to produce the Utrecht solar atlas in 1939. Such densitometer recordings of spectra revealed much more information than the photographic records themselves: Minnaert found it 'a continuous joy to "read" these records and to recognize many features, well known from verbal descriptions but now, for the first time, seen in graphical representation'[53]. He cited the ability to record variations of spectral intensity directly as an important advance in practicality and precision.

Spectroscopists and astronomers designed and used recording microphotometers increasingly from the early 1920s, with new designs being reported regularly in the journals[54].

6.4.3. A 'more troublesome' method: direct photoelectric photometry

The opportunities for propagating error in the multi-step process of photographic photometry were recognized by the astronomers who practised it. Some of them made attempts to measure stellar intensity electrically almost concurrently with photographic efforts[55]. Involving fewer components and processes, electrical methods promised better precision. Edward Pickering at Harvard College Observatory, who was to use visual techniques in his extensive astronomical surveys, performed some abortive trials using a selenium detector around 1877. In the early 1890s, George Minchin, an Irish professor of mathematics, experimented with photovoltaic selenium[56]. With William Monck, an amateur astronomer, he attempted in 1892 to measure starlight using a $7\frac{1}{2}$ inch refracting telescope without success, but they observed deflections of their electrometer due to the light from the moon, Jupiter and Venus[57]. Using more sensitive photocells three years later, Minchin reported observations on ten stars. Comparing the stars Regulus and Arcturus, he claimed favourable precision compared to the visual magnitude method. The size of the electrical signal was small, however: even for Regulus, a bright star, and employing the excellent light-gathering power of a 24 inch aperture telescope, Minchin measured a signal of only 20 millivolts at best, corresponding to a change of about 3% from the 'native' voltage of his photocell.

The experiments of Minchin and his collaborators went nearly unnoticed, and electrical detection of starlight was not attempted again until 1902, when Ernst Ruhmer in Germany observed eclipses of the sun and moon using a photoconductive selenium cell. Ruhmer's photoconductive cell was simpler than that of Minchin; it relied on the characteristics of selenium alone and so was not prone to oxidation of the liquid, which caused a consequent reduction in the magnitude and speed of electrical response. Five years later, Joel Stebbins (1878–1966) again tried selenium as a detector[58]. He reported that he had 'met some of the difficulties which confront everyone who tries to work with selenium. Other agencies than light affect the resistance, and apparently no experimenter has solved, to his own satisfaction, the mysteries of this particular element'[59]. Stebbins found that the sensitivity improved 20-fold when cooled, but the device was still relatively insensitive and the reading was prone to drift if exposed long to light or to air currents, which perturbed the temperature. The current used to measure the resistance of the cell also caused heating which decreased the resistance by some 10% after a half hour, 'of the order of 100 times the light-effect from a bright star'[60]. Stebbins was able nonetheless to measure the intensities of some bright stars to a precision of about 0.02 magnitude (about 5%) using a 12 inch aperture telescope, 'results which are considerably more accurate than have ever been obtained by visual or photographic methods'[61].

The experimental difficulties were nevertheless formidable. Despite Stebbin's claims, these early attempts with selenium were all unproductive compared to visual and photographic methods, and were largely ignored by the astronomical community. In 1910, however, Julius Elster and Hans Geitel, who had by then been experimenting with the photoelectric effect for over two decades, discovered a particularly photosensitive compound: potassium hydride. Two years later, Paul Guthnick at the Berlin Observatory used such a photocell to detect the light gathered by a 31 cm aperture telescope. With it, he was able to measure the intensity of bright stars reliably. And as Pickering had found with his earlier visual work, the *quantity* of data could serve as a tactic to sway doubters. By 1917, Guthnick and a collaborator had made 67 000 measurements on 50 stars and planets by this method, making a special study of variable stars. On the advice of his associate at Illinois, Jakob Kunz[62], Joel Stebbins, too, replaced his selenium photometer by a photoelectric version, noting a hundred-fold improvement:

> A comparison of the relative performances of the selenium and photoelectric instruments is somewhat difficult, but it is safe to say that with the new device, attached to the same 12-inch refractor, stars at least three magnitudes fainter can be observed than with the selenium photometer... the present measures of fifth-magnitude stars are better than the measures of any stars whatever with selenium.[63]

Such photoelectric observations were outside the domain of expertise of most astronomers. The German potassium hydride photocells were enclosed in glass tubes filled with low pressure argon, and supplied with a high voltage. Experimenters required expertise in chemistry, electricity and vacuum technology to make them. Operation was equally demanding. The output of the tube was measured by a delicate string electrometer suspended from gimbals, and mounted in a vertical orientation near the viewing eyepiece of the telescope where the photocell assembly was located[64]. Such mechanical detail, at least, was within the competence of the average astronomer. As to the measurement itself, the electrometer integrated the charge emitted by the photocell; the observer noted its deflection with a microscope and timed it with a stopwatch, and took the rate of deflection to be proportional to the brightness of the star[65]. The overwhelming practical difficulties associated with this technique are evidenced by the fact that most of the early publications concentrated on methods rather than science[66].

Guthnick used one of the first commercially available photocells; most other astronomers designed their own. In England, A F and F A Lindemann published the first account of the details of photoelectric apparatus and methods for astronomical photometry in 1919[67]. That the photocells responded differently to light than did the eye did not deter them; indeed, the Lindemanns marshalled it as a demonstration of the *success* for the new technology. They described the fabrication of photocells having potassium and caesium sensitive surfaces, noting that the two types could be used to measure a 'colour index' for stars. The potassium phototube responded most strongly to blue/violet light, while the response of the caesium type peaked in the yellow portion of the spectrum. The ratio of the two signals for a given star was an indication of the stellar temperature. Thus the astronomers recast the stumbling block of the illuminating engineers into a pedestal to extend their own observational grasp. They cautioned, however, that the new technology required some discontinuity with the past: because of the selective response to colour, they noted, 'it must be remembered that these magnitudes do not represent accurately either visual or photographic magnitudes, though they may be expected to approach the latter'[68]. The Lindemanns suggested a wide range of uses for photoelectric photometry, including measuring the variability of the sun, the albedo (surface reflectance) of the planets and brightness of the solar corona and sunspots.

Adequate sensitivity was a chronic problem. In 1920 Hans Rosenberg at Tübingen attempted to amplify the output voltage of his photocell using a triode valve, which allowed the electrometer to be replaced by a more robust galvanometer located away from the telescope. The poor stability of such early amplifiers, however, failed to convince other astronomers. Amplified photoelectric measurements did not become popular in the community until 1932, when a better design was developed by a member of Joel Stebbins' group[69]. This new amplifier was enclosed in an evacuated chamber to avoid sporadic fluctuations caused by cosmic rays, and amplified the photocell signal by over two million times. As one astronomer has written, 'the most successful early photoelectric photometrists were those who persevered with the intricacies of electronics at a time when electronic apparatus was generally absent from astronomical observatories'. He has noted also that the successful photometric astronomers before 1930 all collaborated with physicists who constructed or advised on the operation of their apparatus[70]. Stebbins, responsible for the first American group, complained in 1914 of the severe instrumental complexities to Harlow Shapley, who was considering taking up the technique:

> The whole problem is one of experimental physics, and our proportion of two physicists to one astronomer is about right. In fact I know of no man who has the requisite training to make a photoelectric cell, mount it on a photometer, and finally produce results on stars.[71]

Photometric astronomy was thus a distinct branch of astronomy demanding unusual skills.

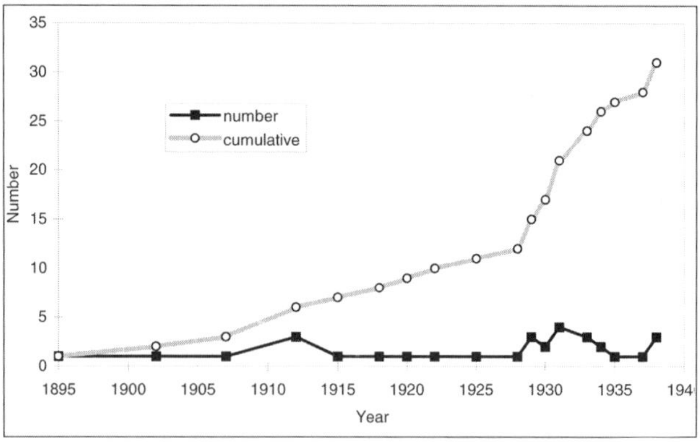

Figure 6.3. Experience with physical photometry. Number of astronomical observers using photoelectric methods before the Second World War. Source of data: Hearnshaw J B 1986 *The Analysis of Starlight: One Hundred and Fifty Years of Astronomical Spectroscopy* (Cambridge) p 19.

Despite the difficulties, interest in the photoelectric technique grew in the inter-war period, with over two dozen observatories in seven countries having attempted measurements by the end of the 1930s (figure 6.3)[72].

6.5. THE RISE OF PHOTOELECTRIC PHOTOMETRY

As with photographic photometry, the photoelectric techniques adopted by astronomers were generally ignored by other photometric practitioners[73]. One reason for this was that the astronomical and electrotechnical communities were dealing with different domains of light measurement. Astronomers measured angularly small and dim light sources. The measurements were consequently imprecise but could be used adequately to infer relative intensities, e.g. the fluctuations of variable stars. Electrotechnical engineers, by contrast, dealt with bright, large-area lamps. They demanded more precise measurements for comparing the technical performance of light sources. Also, as discussed earlier, the astronomers made an unproblematic transition from visual methods to physical photometry. For the purposes of illuminating engineering, however, the engineer was forced to consider the intensity as perceived by the eye; he was unable simply to dismiss the importance of the visual contribution. The difference in objectives between the two communities was reflected in their limited inter-communication. There were only occasional contacts between astronomers and engineering photometrists[74]. Most importantly, physical methods were rejected because they worked poorly in practice; only with the inclination provided by a strong bias against visual methods and faith in the unsubstantiated promise of photoelectric technology would a practitioner persevere.

Some engineers were, nevertheless, willing to consider measurement without the human eye. For those not deterred by the seemingly unavoidable human contribution to photometry, physical methods proved tempting, if elusive. One early illuminating engineer lamented the impracticality of quantifying light, observing that 'it will be evident in the first place that we cannot, at least at the present time, readily expect to measure [the illuminating power of a light] *directly* by the movement of a pointer or by any mechanical means, as in the case of electricity, for instance'[75]. Another wrote in 1894 that 'if there were any outside reliable effects in nature which were functions of the actual brightness of light, as we feel it, we would have a photometric principle'[76]. The same engineer nevertheless rejected the only photoelectric detector available, the selenium cell, observing that 'of all things to exhibit the total depravity of the inanimate this stands first. The variation of its resistance is truly a function of the brightness, but on a curve which changes totally from day to day'. Selenium cells had been proposed sporadically for general light measurement from the late 19th century, perhaps first in commercial form as a photoelectric photometer marketed by Werner Siemens in 1875[77]. The unexplained drift of the resistance of selenium was a serious problem for those eager to exploit it.

The drift problem was not immediately apparent to all investigators. Another early reporter on selenium cells was optimistic but not entirely accurate, reporting that 'light of all refrangibilities from red to violet is effective', and that 'a mere pin point of sensitive surface is as effective as a square centimetre'[78]. The convenience was also lauded:

> The use of the comparative or physiological photometer is irksome and demands some skill, while in the case of the selenium photometer the observation is reduced to the reading of a measuring instrument, and no special knowledge is required.[79]

Later investigators noted that such cells produced an inadequate voltage for deflecting an electrometer when illuminated with violet light. This made them unsuitable for colorimetric measurement, because researchers had established the importance of these extreme wavelengths to colour perception. Unable to respond to a colour to which the eye responded, selenium failed as a viable replacement for photometric applications. It still held some promise for physical measurements, though. A few die-hards remained enthusiastic, limiting their applications to the red end of the visible spectrum where selenium responded well:

> It has been established that selenium is capable of discovering differences of luminosity of the order of 1/100 per cent. This is an accuracy from 50 to 200 times that of the eye, and should add very greatly to the delicacy of all photometric processes. We have, therefore, tested the utility of selenium for discovering and estimating the difference in the amount of light transmitted by different glasses.[80]

Academic and national laboratory physicists familiar with radiometric methods began to extend their techniques to physical photometry. Like the illuminating engineers, there is little evidence that they had much contact with the astronomical community. Independently of astronomers, the physicists Nichols and Merritt devised a photoelectric photometer to analyse spectrographic plates. Speed was their motive: their instrument, incorporating a commercially obtained phototube from Germany, was used to make as many as 400 readings of plate transparency per hour[81]. Even more frequently than the newly available phototubes, thermocouples and thermopiles were used as detectors of visible light as well as heat.

Almost ignored by astronomers, the conceptual problem of adequately replacing the eye by an equivalent physical detector was broached by physicists. By the second decade of the century, the conjunction of a thermopile and a filter to screen out invisible radiation was being touted as an 'artificial eye'[82]. The central problem was to transform the spectral response of the radiometer (which responded almost equally to wavelengths over a very broad range) into a close approximation of the very uneven colour response of the human eye. Initial attempts employed liquid filters[83]. Practical problems, however, centred on the feeble response of such a system to visible light. 'The degree of sensibility required is very high', wrote one investigator, and hence the refinement of thermopile design and galvanometer sensitivity was severely limited[84]. He was to write 16 years later that 'the possibility of using some form of radiometer as a substitute for the eye in photometry has been a long-standing dream' and evidently one not yet realized satisfactorily[85].

The unreliable selenium cell was joined, in the second decade of the century, by the 'Thalofide' cell, a compound of thallium sulphide that changed resistance when illuminated, and the phototube, a thermionic valve having a photosensitive cathode[86]. The former found only limited use in photometry, however, because it responded to infrared radiation more than to visible light. Physicists were drawn to particular physical detectors for the same reasons that they rejected the human eye: because they could understand them more readily. Where the selenium and thalofide cells were unique flukes—unexpected discoveries—the phototube was based solidly on the photoelectric effect, which had been studied intensively from the first decade of the century. Contemporary theory was inadequate to explain the behaviour of selenium. Moreover, its characteristics were complex, depending on its purity, manner of preparation, type of electrical contacts and past exposure to light[87]. Norman Campbell, then designing phototubes, contrasted them with 19th century selenium cells:

> From its first discovery, the change in the conductivity of selenium when illuminated attracted the attention of the inventor rather than of the theorist, to whom it long remained an isolated fact of no special significance. The photoelectric effect, on the other hand, is one of the corner stones of physical theory; but until recently its practical potentialities were entirely unrecognized outside the laboratory, and insufficiently recognized within it. While the immense literature of selenium is directed mainly to its use, in the yet larger literature of the photoelectric effect its use receives scant attention.[88]

Photoelectric devices had to be elevated, suggested Campbell, from mere components for inventors to the subjects of scientific research. He and his contemporaries in the 1920s saw opportunities for merging theory with new applications.

Photoelectric cells were a part of the new physics, rather than outside it, but they were as yet subjects of study rather than components in scientific apparatus. The unexplored complexities resisted their being employed as unproblematic elements in instruments. Campbell himself used the new technology for colour matching, intensity measurement and spectrophotometry. At the National Physical Laboratory after the First World War, research into photoelectric photometry was considerably aided by collaboration with the GEC Research Laboratory, where former NPL staff were working. The director of the GEC laboratory, Clifford Paterson, had regular contact with his former subordinate John Walsh of the NPL through committee work. From 1924, when Norman Campbell at GEC headed a group developing photoelectric cells, the NPL Photometry Division was kept abreast of developments and received sample photocells to test. By 1925, this collaboration began to achieve results: the annual report mentioned

> use of photoelectric cells in place of the eye in a comparison of the light intensity of different sources; as a method of colour matching, the cell has been found, under suitable conditions, to give an accuracy ten times as great as the eye, but difficulty has so far been encountered in securing with the use of the cell the necessary sensitivity in the comparison of relative candle-powers of colour-matched lamps.[89]

Indeed, in the annual report the NPL staff expressed their indebtedness to the Director of Research at GEC, Clifford Paterson and his staff 'for much helpful cooperation in the early stages of the work' and for the production of 'suitable photoelectric cells'[90].

For straightforward photometry, the NPL investigators found the photocells to be 'no improvement' on the visual method, and definitely 'more troublesome'. Their initial researches used designs of test equipment and methods developed by Campbell and his group[91]. Despite being a 'corner stone of physical theory', photocells presented onerous practical problems. First, they suffered from 'photoelectric fatigue' caused by heating: the cells were one-tenth as sensitive at 50 °C as at 20 °C. Heating occurred when the cells were put into a reflective chamber (for measuring the integrated output of lamps) or even in a small unventilated room. Second, as astronomers had discovered two decades earlier, the photoelectric signal was small, requiring a sensitive (and delicate) electrometer to measure the emitted current. Various electrometers were tried, with the most successful being a design by Campbell. Attaining the necessary sensitivity and stability was difficult[92]. Third, the photocells did not produce a signal proportional to the intensity of light. This deviation from *linearity* of the devices depended on the wavelength of light, electrical supply conditions and other factors. The NPL workers avoided this problem by using photocells as they had the eye: the detectors were used merely to equate two light sources rather than to measure an intensity directly. Used in this way, only the *stability* of the response was important, and not the detailed proportionality[93]. The GEC group went further, developing a methodology to compensate for measurement drifts whether they were due to photoelectric phenomena or to the variabilities of human observation. Campbell emphasized 'establishing a scientifically accurate system of photoelectric photometry *in spite* of deficiencies of stability'[94]. The unreliabilities of the human eye were thus replaced by the different, but still considerable, variabilities of a physical detector. The problems of photometry were translated to a new, and as yet little explored, domain.

In the same year as the first success in the Electrotechnics Division, the Optics Division of the NPL was independently engaged in similar work. Its staff manufactured their own photocells to be used in a spectrophotometer. This was completed, and in regular use for colour standards work, by the following year. The stimulus for the research was the development of standards for the colours of railroad signal filters. In the post-war environment of restrained British innovation, this modest effort was appropriated as evidence for a burgeoning national optical industry: 'The work of the National Physical Laboratory is putting the whole subject of colorimetry and colour photometry on a firm foundation', boasted F Twyman[95].

Adoption of the new photoelectric technology appeared unlikely to the NPL staff in the mid-1920s. The Photometry Division used the cells produced by their Optics neighbours, and tried making their own as well as testing GEC products. The group was finding that, while photocells could detect minute differences between two nominally 'matched' colours, this very characteristic of colour sensitivity made them unsuitable for light standards work. Seemingly identical incandescent lamps could have slightly different colours owing to glass contamination or to slight temperature differences caused by insulation of the base. Campbell at GEC tried different cathode materials and optical filters in front of the photocells to make their spectral response more similar to the eye, with limited success. The NPL researchers tried filters of coloured liquids[96]. Campbell concluded that minor colour differences between nominally identical lamps would always unavoidably limit the precision of comparison to worse than 0.1%.

By 1927, the collaborators were experimenting with amplified signals, using thermionic valves. Even with cooled enclosures to reduce the 'photoelectric fatigue', drifts of the signal were troublesome, limiting precision to, at best, two to three times better than visual methods. In an attempt to improve this, they tried to switch rapidly between the reference lamp and sample lamp signals using two photocells, a commutator and amplifier[97]. The result was not a success, Walsh admitting that the best results still came from the 'original photometer' using a Campbell electrometer. Even so, 'in order to obtain results much better than those obtained with the visual photometer, every part of the apparatus needs considerable attention to ensure its perfect behaviour'[98]. The photometrist had been translated from meticulous observer to meticulous instrument minder.

By the next year, the group tersely reported that the 'flicker method of photoelectric photometry' was abandoned owing to 'commutator trouble', to be replaced by other more promising techniques. The NPL staff found that a 'thermionic balance' design, consisting of a photocell in a bridge circuit with a variable current source and detected by a micro-ammeter, could give precision of about 0.25%. The delicate electrometer still gave better results, however. Even so, they were able to report that 'much more confidence has been established in the reliability of illumination measurements made with photoelectric cells'[99]. Echoing Airy's attempt 70 years earlier, the NPL staff measured the change in illumination during a solar eclipse[100]. By the end of the decade, the staff were confidently designing more robust versions of their equipment for use in measuring the reflectance of surfaces and the diurnal variations of daylight[101]. The complications finally were being characterized and tamed.

By the end of the 1920s, the NPL group had enough experience with photoelectric photometry to cautiously support its gradual adoption[102]. Writing of the future of photometry in 1929, John Walsh predicted instruments and standards of greater precision and a simplification of apparatus. Photometric precision had been stalemated since the turn of the century by the reliance on visual observation. Improvements would be needed for progress in other fields:

> What is sufficient to-day may lag seriously behind even commercial requirements in ten or twenty years' time. Progress therefore is essential. Increased precision must be attained so that, in all that concerns the production and utilization of light, progress may not be hindered nor development retarded.

From a subject that had shown little real change during his career, Walsh must have been impressed by the transformations provoked by photoelectric technology. *Progress* was the keyword and it was linked firmly to physical photometry. 'Progress must necessarily lie in the use of physical methods'[103]. Walsh was not completely won over by the new light detectors, however. He saw the physical photometer as being analogous to a galvanometer, 'as a detector of minute differences, rather than as a measurer of integral illumination'[104]. Clifford Paterson, as head of the GEC research laboratory responsible for photoelectric photometry, was interested in promoting their commercial work even at the expense of denigrating his previous achievements at the NPL. Writing of the precision of visual methods he reminisced:

> If a greater accuracy than 2 or 3 per cent was wanted, even under favourable laboratory conditions, it meant several repeat readings with more than one observer. If an accuracy of one-half per cent were required one sat down for a good week's work.[105]

The handful of supporters of photoelectric measurement in the 1920s was to be swelled by many others a decade later, as commercial products began to appear. Straightforward replacement of the eye by a photoelectric cell in visual photometers was a common project through the 1920s[106]. The replacement was not without its difficulties, however; as at the NPL, complaints frequently surfaced that the new physical methods were not necessarily superior to the eye. One investigator warned that spectrophotometers 'must be pushed to the extreme possible limit in order to yield data truly significant in specifying color stimuli'[107].

6.6. RECALCITRANT PROBLEMS

As previously illustrated, early 20th-century photometry, like its 19th-century counterpart, was dogged by technical problems that limited its acceptance, impeded its application and restricted it to peripheral status. Where the experimental difficulties of the previous century had centred on the human observer, however, light measurement was now troubled by equally serious *physical* limitations. In contrast to the earlier hopes, light measurement could not be pegged straightforwardly to another physical quantity. For each community, the story of high expectations followed by the retrenchment of goals was repeated. To paraphrase sociologist Bruno Latour, the instruments resisted being 'black-boxed'[108].

6.6.1. Talbot's law

The use of a rotating sector disc to diminish the intensity of light found common use through the latter half of the 19th century. But as discussed in chapter 2, even Talbot saw no intrinsic justification for his law, although confirming that it worked in practice. By 1890, some experimenters claimed that the law failed for small apertures of the rotating disc—i.e. when the 'on' time was much shorter than the 'off' time. William Abney, who based much of his photometric and colour research on rotating discs, dismissed these concerns:

> it was admitted by this experimenter that with monochromatic light there was no error; it followed that what was true for each ray was true of the sum of them. [I will] not waste the time of the audience over such fallacies.[109]

Talbot's law came into question, too, in physical photometry. Unlike the eye, the photographic plate proved to be significantly affected by the rate of flashing, being relatively insensitive to slow flashes. The early selenium cells had been well-known to exhibit a similar exposure effect: typically a 10 second exposure to light would be followed by at least a minute of darkness, so that the cell recovered its full sensitivity[110]. Photoelectric devices proved even more troublesome than the eye in this regard, as the response time (and hence how the detector responded to rapidly changing illumination) depended on the type of device, its preparation, temperature and other details of the electrical circuits employed.

6.6.2. Linearity

An important concern regarding physical photometers was the relationship between incident intensity and the resulting signal. The *linearity* (or lack of it) of physical detectors was important for some types of measurement. When the intensity of light was to be inferred from the position of a galvanometer dial, for example, the measurement relied implicitly on the assumption that a dial position was proportional to the illumination. This assumption was frequently unjustified. The dial movement might rely, for example, on the precise winding of its electromagnetic coil, on the uniformity of the magnetic field of the surrounding magnet or indeed any other component in the chain linking optical energy to electrical signal to dial reading.

As with electrical phenomena, photographic recording had complications. The nonlinear nature of photography was explored in the last decade of the 19th century, principally by William Abney and the pair of investigators Ferdinand Hurter and Vero Driffield[111]. They showed that a photographic emulsion darkened as a result of chemical fogging and saturation of silver grains as well as by exposure to light. The result was a roughly S-shaped curve relating its opacity to the logarithm of light exposure. The mere recording of illumination could not, therefore, be used to infer intensity unless the photographic process had been calibrated carefully.

Some of the first post-war users of photoelectric cells believed that they had found a reliably linear method of recording intensity. 'The current produced is proportional to the amount of incident light... which renders photoelectric photometry so valuable for measuring in absolute units the light received from objects', wrote the Lindemanns in their account for astronomers[112]. Most astronomers, however, used their photoelectric photometers as comparators, interpolating an unknown stellar intensity between the intensities of two or more known stars. By the early 1920s, more extensive investigations of the characteristics of photoelectric tubes at GEC and elsewhere made it widely known that they could not be relied upon to yield a signal proportional to intensity except in very specific circumstances.

The usual method of dealing with problems of nonlinearity of response was to reduce the measurement to a process of comparison: the unknown quantity would be compared with a known reference. By simply observing the *balance* of two intensities—the equality of the instrument readings—factors such as amplification and the proportionality of the reading to intensity were avoided. As one industrial scientist put it:

> The traditional methods of making physical measurements... appear to imply that physicists as a body have a whole-hearted distrust of all types of instruments. Whenever possible, deflectional methods have been avoided and 'balance' or 'null' methods adopted so as to eliminate instrumental errors, and all essential instruments such as thermometers, or comparison standards such as boxes of weights or resistance boxes, have been calibrated with the utmost care before use.[113]

The criticism of nonlinearity was also levelled at early valve amplifiers. Since there was no guarantee that the output of an amplifier would be proportional to the input signal, distortion was the typical result. Amplifiers proved generally problematic for quantitative measurement. Again, compensation techniques were a partial solution. In describing a null recording colour analyser, a commentator noted that 'since equality of response to light from the two surfaces is indicated by no output from the amplifier, this method of recording is free from the usual objections which accompany the use of valve amplification for quantitative measurements'[114]. Another contemporary review reported a new instrument 'which combines the trustworthiness of the null method with the advantages of recording and rapidity of measurement'[115].

Yet, in photometry, new industrial applications made null methods too complex and tedious: a dial 'visible at a glance' was needed. Careful calibration of individual instruments also proved costly. The last available option was to create stable, linear instruments, in which a voltage or current was reliably proportional to light intensity. One approach was to carefully determine the characteristics of photoelectric tubes, noting the range of light intensities and supply voltages that yielded a reasonably linear output, and then designing an instrument to operate within these limits. Another strategy was to avoid any amplification of the signal at all. Photovoltaic cells, which produce a voltage when illuminated, or photoconductive cells, for which the resistance changes, could be used with sensitive electrometers. Finally, in situations where a non-proportional signal was obtained from an instrument, the dial reading could be calibrated by a nonlinear scale.

6.6.3. The spectre of heterochromatic photometry

The photometric problem *par excellence* of the 1920s was heterochromatic, or multiple-colour, photometry. Colour came pressingly to the attention of standards laboratories because of photometric standards. The availability of differently coloured light sources (gas flames, incandescent gas mantle lamps, carbon filament and other electric lamps) complicated the photometry programmes under way at the national laboratories. Owing to the unequal response of the human eye to different colours, it proved impossible to match the outputs or illumination provided by differently coloured lamps or to specify the colour of any object unless the light source, too, was specified. This problem provided an incentive to put colour measurement on a firmer footing.

The expansion of photoelectric photometry was limited, too, by complications related to colour response. Photoelectric cells did not respond to light and colour in the same way as the human eye did. While the eye's sensitivity peaked for yellow light, photocells could be produced to peak anywhere in the visible spectrum between red and blue. Secondly, while the eye had an approximately logarithmic response to light intensity, photocells could have a linear or markedly nonlinear response that varies with wavelength. This made the resulting signal not simply related to the either the subjective sensation or the energy content of light and colour.

An NPL physicist summarized the outstanding problems in photometry in 1924:

> The problems presented by the study of candle-power standards, flicker photometry, average visibility, and energy distribution must be solved before any further progress in photometry is possible, particularly as modern developments in high temperature radiations and spectral radiations seem likely to accentuate the existing difficulties to a very great extent. No reference has been made to physical photometry, as it seems that its basic problems are precisely the same as those of ordinary heterochromatic photometry, viz. average visibility, energy distribution, together with the technical problems of the sensitivity and reproducibility of whatever physical instruments take the place of the eye.[116]

Colour measurement and other problems thus plagued practitioners even while physical methods were being adopted. The physical method, he seemed to suggest, was a red herring and not a solution to photometry's problems. New technology was addressing new issues rather than facing the old ones.

The technologies of light measurement thus diverged and recombined between the turn of the century and the Second World War as practitioners hesitantly moved from a visual to a physical approach. Instigated by complementary convictions—that the eye was unreliable and that physical methods promised clear advantages—researchers sought a reliable method with limited success. By investigating photographic and then photoelectric techniques, they implicitly questioned the foundations of photometry and found them wanting. The defects of visual measurement were echoed in the complexities of photographic processing and of photoelectric amplification; the peculiar colour response of the human eye had its equal in the characteristics of photographic emulsions and photoelectric anodes. Despite the increasingly apparent analogy between visual and physical detectors, photoelectric methods rapidly came to dominate the subject. Nevertheless, the merging of technologies and the consequent programme to extend light measurement to new fields contained the seeds of problems. Colour could not easily be accommodated in a physical view of light. The definitions of light and colour would have to be renegotiated.

NOTES

1 The hiatus in technological interest is suggested by publication rates, which dipped after about 1912.

2 The case of the detection of ionizing radiation has been discussed by Hughes J 1993 'Making technology count: how the Geiger counter got its click', seminar, Oxford University; for radio astronomy, see Agar J 1998 *Science & Spectacle: The Work of Jodrell Bank in Post-War British Culture* (Amsterdam).

3 Or more accurately, *power density*, expressed as energy per unit time per area or per solid angle.

4 The importance of 'observation without an observing subject' as a precondition for non-subjective reasoning is discussed in Z G Swijtink 'The objectification of observation: measurement and statistical methods in the nineteenth century' in

Krüger L, Daston J and Heidelberger M (eds) 1987 *The Probabilistic Revolution* Vol I. (Cambridge, MA) pp 261–86.

5 Stebbins J 1910 'The measurement of the light of stars with a selenium photometer, with an application to the variations of *Algol' Astrophys. J.* **32** 185–214; quotation pp 205–6 [emphasis added].

6 Anon. 1933 *L. S. Ornstein: A Survey of his Work from 1908 to 1933* (Utrecht). See also Heijmans H G 1992 'The photometrical research of L. S. Ornstein 1920–1940' *Brit.-N. Amer. Joint Mtg. on the History of Laboratories and Laboratory Science* (Toronto) Paper 30.3.

7 The mechanical equivalent of light related the visual sensation to the energy, and was defined as the 'ratio of radiant flux to luminous flux for the frequency of maximum luminosity'. The value depended on the type of source employed, the definition of the colour response of an average human eye and the wavelength of greatest sensitivity. It was most commonly calculated for a blackbody source by multiplying the blackbody power by the relative sensitivity of the average human eye. See, for example, Drysdale C V 1907 'Luminous efficiency and the mechanical equivalent of light' *Proc. Roy. Soc.* A **80** 19–25; Ives H E 1924 'Note on the least mechanical equivalent of light' *JOSA* **9** 635–8; and Walsh J W T 1926 *Photometry* (London) p 296.

8 Gaster L 1926 'Illuminating engineering in relation to optics' *Proc. Opt. Convention* vol 2 (London) pp 297–304.

9 J W T Walsh, discussing Campbell N R and Freeth M K 1926 'Variations in tungsten filament vacuum lamps: a study in photoelectric photometry' *Proc. Opt. Convention* vol 2 (London) pp 253–74. As related in chapter 5, Walsh had been working with these GEC employees to develop accurate photoelectric methods of photometry since 1924. The term *accuracy* (agreement with reality) was less fitting than *precision* (variation from one measurement to the next) because physical methods had no obvious advantage for the former.

10 Anon. 1906 'Editorial' *Electrician* **56** 1037.

11 Cahan D 1989 *An Institute for an Empire: the Physikalisch-Technische Reichsanstalt 1871–1918* (Cambridge) p 214.

12 Stuart Bennett has written extensively on the history of automatic control. For an analysis of the attractions of automation in technical and popular culture, see Bennett S 1991 '"The industrial instrument—master of industry, servant of management": automatic control in the process industries 1900–1940' *Technol. Culture* **32** 69–81. For technical histories, see Bennett S 1979 *A History of Control Engineering 1800–1930* (London) and Bennett S 1993 *A History of Control Engineering 1930–1955* (London).

13 Stevenson J 1984 *British Society 1914–1945* (London) pp 182–202.

14 Hardy A C 1938 'History of the design of the recording spectrophotometer' *JOSA* **28** 360–4.

15 Until the early 1920s, when photoelectric techniques were investigated; see later. Commercially available photometer designs were essentially static between 1860 and 1900 in response to gas industry requirements. Compare, for example, illustrations in Dibdin W J 1889 *Practical Photometry* (London) and Abady J 1902 *Gas Analyst's Manual* (London).

16 Barr J M and Phillips C E 1894 'The brightness of light: its nature and measurement' *Electrician* **32** 524–7; quotation p 525.

17 Trotter A P 1911 *Illumination: Its Distribution and Measurement* (London) pp 66–7.

18 Lummer E and Brodhun E 1889 'Photometrische Untersuchungen' *Z. Instr.* **9** 41–50 and 461–5, quoted in Kangro H 1976 *Early History of Planck's Radiation Law* (London) p 152. The photosensitivity of selenium had been discovered by Willoughby Smith in 1872. The quantitative use of such electrical devices was made more practical by the development in 1882 of the D'Arsonval galvanometer.
19 Gaster L and Dow J S 1920 *Modern Illuminants and Illuminating Engineering* (London, 2nd edn).
20 By the turn of the century, photometer heads were frequently designed with a field of view of 2°, causing only the fovea near the centre of the eye to be employed.
21 'The Purkynje effect renders the photometric comparison of differently coloured lights at low intensities almost impossible' [Walsh *op. cit.* note 7].
22 See *ibid.* 175–80 for an account of the nature and control of personal errors in photometry.
23 NPL 1911 *Report* (Teddington) p 39.
24 At the NPL, a series of five such lamps was used. The observer used the standard techniques of visual photometry to compare each pair of lamps in the series. The difference between the two extreme lamps was the product of the ratios of the measurements on pairs. The measurement uncertainty was also increased in this technique, however, thus limiting the precision attainable.
25 Gaster L 1908 *Illum. Eng.* **1** 794.
26 Trotter *op. cit.* note 17, p 192.
27 *Ibid.* ch 9.
28 Buckley H 1924 'The field for international agreement and standardization in illumination', *Compte Rendu CIE* 412. From 1918, Buckley shared with John Walsh nearly all the photometric work of the Electrotechnic Division.
29 Edgcumbe K 1926 'The British Standards specification for portable photometers (No 230/25)' *Illum. Eng.* **19** 70–1.
30 Edgcumbe K 1929 'A standard specification for photometric integrators', *Illum. Eng.* **22** 106. The BESA specification was No 354, 1929. The integrating photometer measures the average intensity of a light source by receiving the light reflected from the interior of a diffuse white sphere or cube.
31 Anon. 1894 'Capt. Abney on photometry' *Electrician* **32** 625.
32 The application of photographic methods to astronomy was by no means straightforward, however. Some astronomers initially suspected that photographic recording of observations, while convenient for the 'automation' of observations, omitted detail evident to visual observers. Moreover, its use for quantitative measurements such as the transit of Venus was criticized for possible instability of the photographic emulsion, and for a dependence of the image size on exposure conditions. See, for example, Rothermel H 1993 'Images of the sun: Warren De la Rue, George Biddell Airy and celestial photography' *BJHS* **26** 137–69.
33 The ability to publicly witness experiments had been identified as a feature of good science since the 17th century. Photometry was thus marginalized by its requirement for closeted, individual observations.
34 Thus, for example, a photographic plate replaced the screen of the visual photometer and recorded two adjacent patches of light. The plate would be exposed to yield two blackened areas, the optical densities of which were assumed to be proportional to the original light intensities.
35 For example Wilson A E 1892 'A new photographic photometer for determining star magnitudes' *Astron. & Astrophys.* **11** 307–9.

36 The route for this technological exchange was undoubtedly through astrophysicists, who themselves employed laboratory spectroscopy to generate comparison spectra.

37 For surveys of the state of the art, see, for example, Conrady A E (ed) 1924 *Photography as a Scientific Implement* (London); Dobson G M, Griffith I O and Harrison D N 1926 *Photographic Photometry: a Study of Methods of Measuring Radiation by Photographic Means* (Oxford); Harrison G R 1929 'Instruments and methods used for measuring spectral light intensities by photography' *JOSA* **19** 267–307 and Harrison G R 1934 'Current advances in photographic photometry' *JOSA* **24** 59–71.

38 Norman D 1938 'The development of astronomical photography' *Osiris* **5** 560–94.

39 Waterfield R L 1938 *A Hundred Years of Astronomy* (London) pp 90–5; Ross F E 1924 *The Physics of the Developed Photographic Image* (New York) pp 88–107. Various calibration formulas were developed by, for example, Bond (1850), J Scheiner (1889), C L V Charlier (1889) and at Greenwich.

40 Dobson G M *et al op. cit.* note 37.

41 Some human judgement of intensity did remain, however: the stellar image generally appeared fuzzy, so that the measured diameter depended upon the grey level chosen as the true 'edge'. This uncertainty was sometimes reduced by employing 'hard' developers and plates which yielded higher contrast (and hence more sharply defined images), or by multiple copying of the plate to achieve this result.

42 Abney W 1896 'The photographic values of moonlight and starlight compared with the light of a standard candle' *Proc. Roy. Soc.* **59** 314–25.

43 By this technique Abney estimated that for Jupiter 'it would not be far wrong to assume that it is equivalent to a candle placed at 800 feet from the screen' and that 'moonlight is 44 times brighter than starlight when unabsorbed by more than 1 atmosphere' [*Ibid.* 324–5].

44 Dobson *et al op. cit.* note 37, p 14.

45 Gaster and Dow *op. cit.* note 19, p 221.

46 Hearnshaw J B 1986 *The Analysis of Starlight: One Hundred and Fifty Years of Stellar Astronomy* (Cambridge) p 419.

47 For an examination of early investigations of selenium, see Hempstead C A 1977 *Semiconductors 1833–1919: An Historical Study of Selenium and Some Related Materials* (unpublished PhD dissertation, University of Durham).

48 The research is described later in this chapter. Practical applications of the photoelectric effect, in fact, preceded its scientific explanation.

49 Hartmann J 1899 'Apparatus and method for the photographic measurement of the brightness of surfaces' *Astrophys. J.* **10** 321–32.

50 Koch P P 1912 'Über die Messung der Schwäzung photographischer Platten in sehr schmalen Breichen' *Ann. Physik* **38** 507–22.

51 The thermopile, a high-sensitivity variant of the thermocouple, had been in use since the middle of the previous century, and had figured in the precise blackbody measurements made at the PTR.

52 Moll W J H 1921 'A new registering microphotometer' *Proc. Phys. Soc.* **33** 207–16.

53 Minnaert M 1946 *Astrophys. J.* **104** 331.

54 For example: Toy F C and Rawling S O 1924 [British Photographic Research Association], 'A new selenium cell density meter' *J. Sci. Instr.* **1** 362–5; Gibson K S 1923 'Direct reading photoelectric measurement of spectral transmission' *JOSA & RSI* **7** 693–7; Baker E A 1924 'A convenient photo-electric photometer and densitometer' *J. Sci. Instr.* **1** 345–7; Dobson G M 1923 'A flicker type of photo-

electric photometer giving high precision' *Proc. Roy. Soc.* A **104** 248–51.

55 See Huffer C M 1955 'The development of photoelectric photometry' *Vistas in Astronomy* **1** 491–8.

56 Minchin G M 1896 'The electrical measurement of starlight. Observations at the observatory of Daramona House, Co. West Meath, in April, 1895. Preliminary report' *Proc. Roy. Soc.* **58** 142–54, and 'Observations... in January, 1896. Second report' *Proc. Roy. Soc.* **59** 231–3. His photocell consisted of a selenium coating on an aluminium plate immersed in (initially) acetone or (later) oenenthal in an air-tight glass tube.

57 Butler C J and Elliot I 1993 'Biographical and historical notes on the pioneers of photometry in Ireland' in C J Butler and I Elliot 1993 (eds) *Stellar Photometry—Current Techniques and Future Developments* (Cambridge) pp 1–12.

58 Stebbins *op. cit.* note 5, pp 185–216.

59 *Ibid.*, p 185.

60 *Ibid.*, p 187.

61 *Ibid.*, p 213.

62 Kunz (1874–1939) had obtained his PhD at Zürich, and was responsible for bringing Elster and Geitel's technology to American attention.

63 Stebbins J 1920 'The eclipsing variable star, λ Tauri' *Astrophys. J.* **51** 193–9; quotation p 194.

64 Minchin and his collaborators, unlike their successors, had used a quadrant electrometer located in a room below the telescope. The mirror mounted on the electrometer rotor reflected light to a scale seven feet away, and was said to give reasonably consistent results in the isolated observatory building. This was fortuitous considering that the very small signal from the photocell was transmitted by fine uncovered copper wires. For a detailed contemporary description of the design and operation of such devices, see Ayrton W E, Perry J and Sumpner W E 1891 'Quadrant electrometers' *Phil. Trans.* A **182** 519–34.

65 See, for example, Schulz W F 1913 'The use of the photoelectric cell in stellar photometry' *Astrophys. J.* **38** 187–91.

66 Hearnshaw J B 1993, 'Photoelectric photometry—the first fifty years' in Butler and Elliot *op. cit.* note 57, p 16.

67 Lindemann A F and Lindemann F A 1919 'Preliminary note on the application of photoelectric photometry to astronomy' *Mon. Not. Roy. Astron. Soc.* **79** 343–57.

68 *Ibid.*, p 351.

69 Whitford A E 1932 'The application of a thermionic amplifier to the photometry of stars' *Astrophys. J.* **76** 213–23.

70 Hearnshaw *op. cit.* note 66, p 18.

71 Letter of Stebbins to Shapley, June 11, 1914, quoted in De Vorkin D H 1985 'Electronics in astronomy: early applications of the photoelectric cell and photomultiplier for studies of point-source celestial phenomena' *Proc. IEEE* **73** 1205–20.

72 Hearnshaw *op. cit.* note 46, p 17.

73 One exception is the work of J Kunz at the Nela Research Laboratory: in an early paper [Kunz J 1916 'Photoelectric photometry', *J. Franklin Inst.* **182** 693–6], he noted that of the four lines of current research in photoelectricity (namely (i) the effect of frequency of light on electron velocity, (ii) the effect of light intensity on photocurrent, (iii) 'normal' versus 'selective' photoelectric effects and (iv) the influence of gases) the second had shown conflicting results by previous investigators.

Kunz investigated the photoelectric effect as a photometric indicator and concluded that, with caution, it was a reliable technique.

74 One tentative link with astronomers was made by Edward Hyde, director of the Nela laboratory, and W E Forsythe in papers describing photometric standards of high-temperature sources and how they related to stellar measurements. See, for example, Hyde E P and Forsythe W E 1920 'The gold-point and palladium-point brightness ratio' *Astrophys. J.* **51** 244–51, and papers in **36** (1912) 114; **43** (1916) 295; **58** (1923) 294.

75 Dow J S 1908 'The measurement of light and illumination' *Illum. Eng.* **1** 493–7; quotation p 494.

76 Barr and Phillips *op. cit.* note 16, p 525.

77 Siemens W 1875 *Nature* **13** 112. See also Siemens W 1875 'On the influence of light upon the conductivity of crystalline selenium' *Phil. Mag.* **50** 416. Siemens' photometer replaced the eye with a selenium cell and galvanometer. The cell, exposed briefly to the sample light source and the reference light source, was used to judge equality of brightness. Thus, despite the variation of its resistance with extraneous factors, it could be applied like the eye to the matching of intensities provided that the intensities were not too different and were available in close proximity.

78 Minchin G M 1892 'The photoelectric cells' *Astron. & Astrophys.* **11** 702–5.

79 Torda T 1906 'A portable selenium photometer for incandescent lamps' *Electrician* **56** 1042–5; quotation p 1044.

80 Fournier-D'Albe E E and Symonds E O 1926, 'Some new applications of selenium' *Proc. Opt. Convention* vol 2 (London) pp 884–93.

81 Nichols E L and Merritt E 1912, 'A method of using the photoelectric cell in photometry' *Phys. Rev.* **34** 475–6.

82 See Coblentz W W 1915 'The physical photometer in theory and practice' *J. Franklin Inst.* **180** 335–48 and Ives H E 1915 'A precision artificial eye' *Phys. Rev.* **6** 334–44.

83 One recipe for a 'luminosity curve solution' combined cupric chloride, potassium chromate, cobalt ammonium sulphate and nitric acid in water, contained in a 1 cm thick optical cell and kept at a constant temperature.

84 Ives *op. cit.* note 82, p 335.

85 Ives H E and Kingsbury E F 1931 'The application of photoelectric cells to colorimetry' *JOSA* **21** 541–63.

86 Case T W 1920 'Thalofide cell—a new photoelectric substance' *Phys. Rev.* **15** 289–91.

87 Hempstead *op. cit.* note 47, pp 100–5.

88 Campbell N R and Ritchie D 1929 *Photoelectric Cells—Their Properties, Use and Applications* (London) p v.

89 NPL 1925 *Report* (Teddington) p 6.

90 *Ibid.*, pp 6 and 107.

91 Harrison T H 1926 'Preliminary note on the use of photoelectric cells for precision photometry of electric lamps' *Proc. Opt. Convention* vol 1 (London) pp 245–52.

92 NPL 1925 *Report* (Teddington) p 123.

93 This obviates the need for Campbell's 'class 3' measurement by restricting observations to 'class 2' comparisons. The linearity problem is discussed at greater length below.

94 See Campbell N R 1925 'Photoelectric colour-matching' *J. Sci. Instr.* **2** 177–87.

95 Twyman F 1925 'The vitality of the British optical industry' *J. Sci. Instr.* **2** 369–80.

96 NPL 1926 *Report* (Teddington) p 132.

97 In this technique, a mechanically rotated switch (the commutator) alternately selected the reference and sample signals. The signal following the switch was thus a square wave with a 'peak' corresponding to the larger signal and a 'trough' corresponding to the weaker, and a frequency equal to the switching frequency. When the two components balanced, this fluctuating component disappeared. In principle, the amplifier could be 'tuned' to respond only to the commutator frequency and thus remove from the signal contributions caused by drifts and extraneous electrical noise.

98 NPL 1927 *Report* (Teddington) p 128.

99 NPL 1928 *Report* (Teddington) p 142.

100 See NPL 1927 *Report* (Teddington) p 137, and staff of the photometry department of the NPL 1928 'The variation of natural light during the total eclipse of the sun on June 29th, 1927' *Illum. Eng.* **21** 198–202. They found the minimum illumination during the total eclipse to be 0.18 foot-candles, compared to a full-noon value of 3000 foot-candles. The Illuminating Engineering Society of New York listed ten previous successful photometric observations of solar eclipses, dating from 1886. Half of these employed visual observation, one photography and the remainder photoelectric methods. Photoelectric observations of eclipses were subsequently extended, e.g. Sharp C H, Gray S M, Little W F and Eckweiler H J 1933 'The photometry of solar eclipse phenomena' *JOSA* **23** 234–45.

101 NPL 1929 *Report* (Teddington) p 143.

102 See, for example, Walsh J W T 1933 'Everyday photometry with photoelectric cells' *Illum. Eng.* **26** 64–72.

103 Walsh *op. cit.* note 7, p 8.

104 *Ibid.*, p 7.

105 Paterson C C 1932 'Some thoughts on the international illumination congress' *Illum. Eng.* **25** 89–99; quotation p 94.

106 For example Tardy L H 1928 'Remplacement de l'oeil par la cellule photoélectrique sur les spectrophotomètres visuels' *Rev. Opt.* **7** 189.

107 Priest I G 1929 'Note on the relative sensitiveness of direct color comparison and spectrophotometric measurements in detecting slight differences' *JOSA & RSI* **19** 15

108 Latour B 1987 *Science in Action* (Cambridge, MA) pp 2, 253.

109 Abney W 1913 *Researches in Colour Vision and the Trichromatic Theory* (New York).

110 Dobson G M B 1923 'A flicker type of photoelectric photometer giving high precision' *Proc. Roy. Soc.* A **104** 248–51.

111 Driffield V C 1903 'The Hurter and Driffield system: a brief account of their photochemical investigations and method of speed determination' *The Photo-Miniature* **5** 337–400.

112 Lindemann *op. cit.* note 67, p 344. There is evidence that the Lindemanns consistently underestimated the systematic errors in physical photometry. In the same paper, they optimistically wrote of a photoelectric photometer for measuring photographic plates, 'provided they are not overexposed in any part...there seems every hope that one could combine the two methods with advantage' [p 317]. In fact, as their photographic predecessors were aware, photographic recording of intensity is inherently nonlinear.

113 Moore H 1937 'The influence of industrial research on the development of scientific instruments' *J. Sci. Instr.* **14** 41–6.

114 Walker R C and Lance T M C 1933 *Photoelectric Cell Applications* (London).

115 C J H 1933 'A new microphotometer for the recording of the blackening of photographic plates' *RSI* **4** 553.

116 Buckley H 1924 'The field for international agreement and standardization in illumination' *Compte Rendu CIE* (London) p 408.

CHAPTER 7

DISPUTING LIGHT AND COLOUR

The locus of light measurement was changing. From small darkened rooms hissing with gas supplies, to busier rooms humming with transformers, to larger, well lit rooms buzzing with people: practice became tied to committees, commissions and delegations. This more public activity was different. Although drawing upon many of the same individuals as did the earlier associations and institutions, these new groupings fostered contention. The delegated bodies more often sought to operate by consensus than by hierarchical decision-making and were more goal oriented[1]. But as heterogeneous bodies bringing together different scientific and engineering cultures, they confronted differing worldviews.

Technical delegations came to dominate the subject in the inter-war period. Their goals were matched closely to the aims of the government, industry and technical associations that created them. They also proved appropriate for solving the type of problem then facing the subject. In the post First World War political climate, such technical panels embodied growing efforts to improve the cooperation of science and technology on a national and international scale[2]. The war had demonstrated the benefits of national organization in and between technologically intensive industries; after the war, these concerns shifted from military to commercial competition. The new committees sought the consensual solution of pressing industrial problems and the promotion of scientific activities by rationalizing standards. The situation for light measurement was a particular case of the increasing bureaucratization of international science.

The case of colour measurement highlights how this new bureaucratization operated. During the 1920s, the problem of quantifying colour came to the fore. The measurement of colour had previously gained little prominence within the communities concerned with light measurement, except where the photometric comparison of differently coloured lights was concerned. But coming to the attention of committees as a perceived hindrance to further progress in photometry, heterochromatic photometry opened the subject of colorimetry to different intellectual groups. Those most at odds proved to be communities of physicists and psychologists, which differed in their views on the nature, measurement and description of colour. A schism developed between proponents of physical measurement and supporters of a psychological view of perception. This was a recasting of the older, and seemingly completed, play of visual versus physical photometry for a new stage and new audience. The question of colour measurement was divisive for new associations of practitioners. Heterogeneous committees were forced to face these contentious issues soon after their formation.

The disagreements that developed around the subject, which could not be settled by the conventional methods of scientific closure, reveal the differing goals and methods of the protagonists. As sociologists Englehardt and Caplan have observed, 'one must establish by negotiation formal procedures to bring closure to a scientific dispute when more than one community of scientists exists...or when a conclusion has not yet been reached by sound argument and one intends to engage in common activities or undertakings'[3]. For colorimetry, those procedures involved appointing committees that included different scientific communities to examine the subject. The 'common activities or undertakings' which impelled the 'negotiations' were an abundance of commercial and utilitarian practices of colour matching and specification.

The initial attention of committees centred on the mundane questions of terminology. But the problems with colour were deeper than mere standardization of jargon. Their members found themselves grudgingly broadening the scope of discussions to consider a wider range of phenomena while simultaneously narrowing the definition of what 'colour' was to mean in quantitative terms. Underlying that definition was a particular conceptual foundation.

Committees proved to be central foci in the physical/psychological debate and in its eventual uneasy resolution. They brought together previously isolated communities to carry out a pragmatic agenda, namely the description and measurement of colour for industrial and scientific use. Colour measurement, then, was a problem substantially created and solved in the inter-war period by technical delegations. The solution, however, was a contentious one: colorimetry increasingly was appropriated and stabilized by physicists as a sub-category of photometry.

Commissions and committees are, more obviously than other forms of scientific interaction, a social response to social situations. They bring together decision-makers representing a range of expertise and opinion or the members of other social bodies. With the members of such groups drawn from one or more cultural milieus, their activities concern social questions in the broadest sense; the study of such organizations can probe the relationships between sub-cultures. Committees can also make explicit the connection between their subject and 'external' factors such as politics and personalities. The organization and membership of a committee depend on personal hierarchies and the status of various social groups. *Who* serves on committees, and *why*, can be as important as *what* they deal with, both for the results the committee achieves and for subsequent historical analysis. This is as true for scientific committees as for other types. Scientific commissions deal, in many cases, with the seemingly mundane topics of administration or regulation. But even such seemingly uncontentious agendas as measurement standards are influenced by social factors such as the domain of use of the measurement.

The product of a delegation is agreement on actions, reached by consensus or by the compromise of differing viewpoints. The decision-making bodies dealing with colour went beyond this conventional definition, however, in that they dealt also with *conceptual* questions. The commissions and committees defined not only nomenclature, but the very understanding and quantification of 'light' and 'colour'. Social and cognitive factors merged through the medium of decision-making bodies.

7.1. THE COMMISSION INTERNATIONALE DE PHOTOMÉTRIE

The first international body to concern itself with light measurement was the Commission Internationale de Photométrie (CIP). Its formation was triggered by an International Gas Congress held at the Paris Exhibition of 1900. Attended by some 400 gas engineers and industry representatives, the Congress included a paper entitled 'The photometry of incandescent gas mantles'. It excited unusual interest. The Chairman and President of the Société Technique de l'Industrie de Gaz de France, referring to the 'general and common interest of producers as well as consumers of gas to be exactly informed of the lighting power of mantles employed for incandescent lighting', proposed the formation of an international commission 'to fix the rules to be followed in photometric observations of incandescent gas mantles'[4]. Meeting later the same day, the officials of the gas congress decided upon a constitution for the new Commission. It was to consist of four members each from France, Germany and Britain—the principal representatives at the Congress—and one each from Austria–Hungary, Belgium, Italy, The Netherlands and America.

The meetings of the CIP were held in Zurich, and its proceedings published in French. At the first meeting in 1903, delegates agreed to investigate the luminous intensities of the various flame standards in use. The next meeting, in 1907, included representatives from the national laboratories of Britain (NPL), Germany (PTR) and France (La Laboratoire Centrale d'Électricité, Paris), specifically to organize the inter-comparison of flame standards. By 1909, the work on standards had led to the merging of the American, French and British candles into the *bougie internationale*[5].

This early success in international cooperation encouraged a further expansion of contributions to the CIP. At the third meeting in 1911, the Commission asked each National Electrotechnical Committee to nominate members, swelling attendance by about 50%. The extension of the membership indicates a broadening of scope from the restricted photometric questions of gas standards to other aspects of lighting. The new delegates also brought a new perspective: the dominance and interests of the gas industry in the CIP were weakened because of the pragmatic reliance that the national laboratories had placed on carbon-filament incandescent lamps as the most reliable light source for comparison with the flame standards.

The inclusion of electric lighting was followed by further calls to extend the Commission's mandate. During an International Electrical Congress held in Turin a few weeks after the CIP meeting, Leon Gaster, founder of the Illuminating Engineering Society of London, proposed the foundation of an international commission on *illumination*. The members of the CIP were polled, and they agreed to broaden the work of the Commission to include the new goals[6].

7.2. THE COMMISSION INTERNATIONALE DE L'ÉCLAIRAGE

The Commission Internationale de l'Éclairage (CIE) was formed in 1913. Instead of consisting of a few nominees of the national technical societies concerned with the photometry of gas engineering, the new Commission included representatives from any country willing to form a national committee that was truly representative of all organizations with a strong technical interest in lighting[7]. The change mirrored the commercial and technical shift in emphasis from gas to electrical illumination. Meeting every three years, the official languages of the commission were to be French, English and German. The object of the organization was 'to study all questions relating to the industry of illumination and to the sciences which are connected with it, and to establish, by all appropriate means, international agreements on questions of illumination'[8].

This early organization was stillborn. The outbreak of the First World War soon after the meeting caused the abandonment of the international work in progress and the suspension of CIE activities.

In 1920, E P Hyde, who had polled support for the formation of the CIE eight years earlier, made another European tour to gauge interest. Long prominently associated with American photometry, Hyde's career in many respects mirrored that of Clifford Paterson in Britain. Joining the NBS in 1903 to start its photometry department, he went on to head the newly established National Electric Lamp Association Research Laboratory in 1908. He was the principal organizer of the first regular university course on illuminating engineering, and was closely involved with the inter-comparison of flame standards. Hyde held the positions of representative of the CIP, President of the Illuminating Engineering Society of New York, and President of the American National Committee for the CIE.

The first meeting of the reborn and restricted CIE was held in Paris in 1921. The German-speaking countries were not invited to attend, and proceedings were printed only in French and English[9]. The lack of German participation was a consequence of the divided nature of international science after the war[10]. German attendance at international meetings and activities was boycotted. The membership broadened in the next meeting held in 1924, with Japan and Poland sending observers. The duties and attendance of the Commission sessions rapidly expanded (figure 7.1).

The Commission Internationale de Photométrie had limited the scope of its activities mainly to the measurement of gas lighting, and to about a dozen delegates from its member countries. The new Commission Internationale de

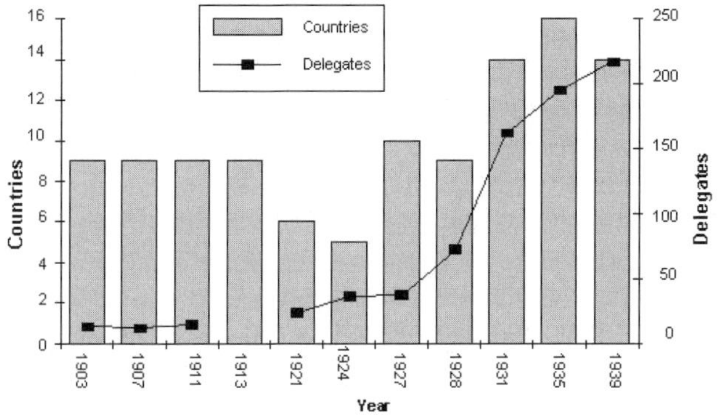

Figure 7.1. Attendance of countries and delegates at the CIP (1900–11) and CIE (1913–39) sessions. The 1913 session, dealing only with organizational matters, was never published. From 1928, the number of delegates per country was no longer limited to 10. Attendance at the 1939 session was reduced owing to the absence of Austria and Argentina. The Commission was dormant owing to wartime disruptions between 1939 and 1948. Sources of data: *Compte Rendu CIE* (1921, 1924, 1931, 1935 and 1939) and CIE 1990 *History of the CIE 1913–1988* (Geneva).

l'Éclairage took on a wider range of tasks, and opened its sessions to more national delegates and observers. The number of delegates quickly enlarged, particularly in the period 1928–31 when Germany was again represented. The number of topics covered also increased dramatically, although not according to a German agenda. Instead of organizing a few days of meetings chaired by the President as its predecessor had done, the CIE separated the discussions into various technical meetings chaired by delegates from the member countries. This structure was further refined in the 1927 meeting at Bellagio, Italy, when delegates agreed that the field of the Commission's activities be divided into several sections, listed in table 7.1.

The successor to the CIP thus maintained many of its original objectives. Photometric (items 1, 2, 8, 9 and 10) and colorimetric (items 1 and 6) subjects occupied six of its 13 topics of interest. Each of these sections was to be assigned to a National Committee of one of the member countries. The officers resolved that each National Committee should 'make a special study of its specific subject and be responsible for the reports which will be presented at the subsequent Commission meeting'[11]. The reasons for this division of subjects along national lines centred on practicality. According to N A Halbertsma, a Dutch illuminating engineer active in the CIE for several decades, this arrangement was formalized in 1927 because

> experience had shown that these committees of specialists from different countries had a low efficiency because the members could not meet regularly and had to rely upon corrrespondence. Therefore

Table 7.1. Subject areas for the CIE agreed in 1927.

1	Heterochromatic photometry
2	Definitions and symbols
3	Lighting in factories and schools
4	Automobile headlights
5	Street lighting
6	Coloured glasses for signals
7	Diffusing materials
8	Photometric test plates
9	Precision of photometric measurements
10	Light flux distribution
11	Daylight
12	Cinema lighting
13	Glare

> an important change for the work between the session was decided upon.... Each of the sections (or subjects) was assigned to the National Committee for that subject. It got the full responsibility for fostering on an international scale the study in that field and to maintain for that purpose contact with the other National Committees.[12]

The formation of national committees was modelled on the organization and practice of photometry in each member country. Membership on the Commission was open to those selected by their National Committees. Such committees generally chose a combination of individuals from those most active in the field, typically the presidents of national associations, academic scientists active in photometry or representatives from national laboratories. The British and American representatives were drawn primarily from the national laboratories and industry. In Britain, the Committee was generally a collection of representatives from the NPL, government departments, trade organizations, lamp manufacturers and instrument companies. Academic scientists were little represented[13]. These delegates represented the interests of commercial engineers, government scientists and standards organizations—a particularly productive mix that fairly sampled the active British light measurement community. But university scientists dominated the French committee[14]. Its 'Secretariat Committees', responsible for studying a particular problem assigned by the Commission, were generally based at universities. The later German delegates fell somewhere between the two extremes, with industry, academe and national laboratories represented[15].

The division of studies along national lines was to be crucial to the development of the subject of light measurement. Each Secretariat Committee was ostensibly responsible for fostering international study in its particular field and for maintaining contact with the other National Committees through experts that each appointed. These technical committees were intended to discuss contentious questions in the three or four years between CIE sessions, 'hors séance... les questions *en litige*'[16]. In practice, however, such cooperation was limited. The various technical committees were typically kept busy with their national responsibilities at government or university laboratories, and had relatively little time to travel or to manage international cooperative work. The communications were further hampered by the physical distance separating the various groups. At the 1924 CIE session, for example, the delegates agreed to hold the next session three years hence in America. Owing to other commitments and the long travel time, most of the delegates found the plan impracticable, and they met unofficially in Bellagio, Italy, instead. Even this unofficial meeting was productive, leading to *Comptes Rendus* running to 1250 pages. A meeting was held in Saranac, New York, the following year. Several of the delegates found the sea voyage and fortnight of American travel a useful and unaccustomed venue for further discussions[17]. Despite this exception, the relatively brief personal contact at the sessions usually made detailed collaboration between the committees difficult. Furthermore, the volume of work to be presented soon meant that there was no time for papers by individuals to be presented at the sessions. Instead, summaries were presented by National Committees. By the 1928 Saranac meeting, two or even three meetings of the technical committees met consecutively over the five days of the session. Contributions by individuals, when they were considered, were limited to semi-official venues. The host countries for some of the CIE sessions organized associated activities to demonstrate the state of the national industries, but which also promoted extended contacts between delegates and the sharing of information. At that meeting, 'in order to make the trip to the United States... attractive to the European delegates' there was an 'Illumination Congress' beginning three weeks before the official sessions with a series of technical visits to various American cities by chartered train, and culminating in the Annual Convention of the American Illuminating Engineering Society in Toronto, Canada. A similar Congress took place three years later for the Cambridge session of the CIE, with meetings and demonstrations held in Glasgow, Edinburgh, Sheffield and Birmingham. Coinciding with the centenary of Faraday's discovery of electromagnetic induction, it was a highly visible affair accompanied by the novelty of the flood-lighting of major buildings (flood-lighting had been employed at American war-time installations, and saw its first widespread commercial use in England in 1932). While the papers presented at these Congresses were published, they did not include the minutes of the discussion period as did the official proceedings. This arrangement of a series of meetings preceding the CIE sessions was an attempt to satisfy members interested in maintaining the CIE goal of providing 'an international forum for all matters relating to the science and art of illumination'. Nevertheless, the meetings for individual authors were dispensed with at the 1935 Berlin/Karlsruhe session: instead, five days were devoted to discussing the results of 25 technical committees. While the work of some technical committees may have been communicated informally before the session, preprints and formal papers were not circulated beforehand. This abbreviated format of the CIE sessions naturally limited the amount of discussion possible, and made the acceptance of the proposals of the secretariat committees all the more likely. By the 1930s then, if not earlier, the CIE sessions were restricted to merely setting the questions to be answered by the technical committees assigned to particular countries, and for ratifying their conclusions. So the *de facto* organization of the CIE had evolved towards shunting particular technical questions to individual countries. This national compartmentalization of problems was to be important to the foundation of colorimetric practice.

The officers of this *illuminating* commission were individuals closely associated with *photometry* in their own countries. The proposer of the CIE was Leon Gaster, founder of the Illuminating Engineering Society of London. The drafters of its constitution included Clifford Paterson, then responsible for the Photometry and Electrotechnical section of the NPL; Eugen Brodhun of the PTR, co-inventor of the universally used Lummer–Brodhun visual photometer; and Edward Hyde, formerly of the photometry section of the Bureau of Standards in America and then director of the Nela Research laboratory. Instrumental in gaining support for the Commission by visiting potential member countries, Hyde later gave up his seat on the founding committee to his former superior Edward Rosa (1861–1921), director of electrical research at the NBS, and a man with a strong hands-on interest in light measurement there. Photometry became an important part of the Electrical Division for the first 40 years of the NBS because of the attention gained by Rosa's early investigations of electric lamps for the US Government purchasing authority[18].

By its first technical meeting in 1921, Paterson, Secretary and now director of GEC Research Laboratories at Wembley, was joined by John Walsh, his successor at the NPL, in the role of Executive Secretary, and Kenelm Edgcumbe, director and chief instrument designer for Everett Edgcumbe and Co., as Vice President. The ascendancy of individuals on the national scene was mirrored in the positions they assumed on the CIE. Paterson became President between 1927 and 1931, and Walsh was eventually to succeed him for the period 1955–9. Although the CIE was based in Geneva, this British influence was significant and continuous. The British officials held more than one-third of the positions, and typically for the longest durations. And, unlike the CIP's French *Transactions*, the Commission's *Compte Rendu* was printed in England[19].

The officers of the CIE seldom were prominent in their national committees (figure 7.2). This was likely a choice by the individual for the higher-status and possibly less partisan international role provided by the CIE post. Paterson and Walsh of the NPL, for example, filled Commission posts, while members of British companies such as Edgcumbe were prominent in the British National Committee.

Cambridge 1931. Leaders of the Delegations and Officers.

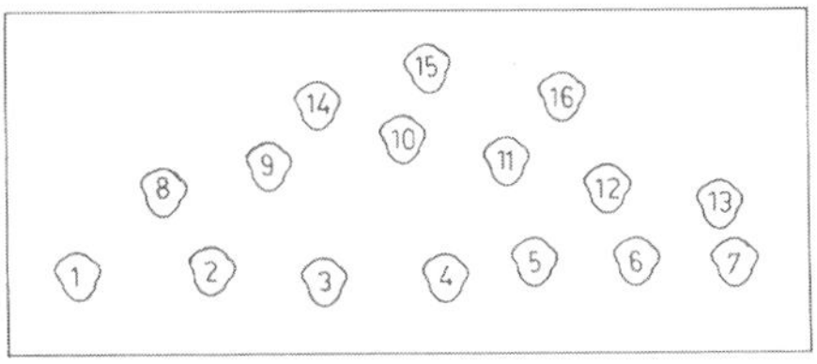

1. K Edgcumbe (GBR) - 2. W Dziobek (DEU) - 3. H Abraham (FRA) - 4. C C Paterson (GBR) - 5. A Filliol (ĊHE) - 6. E C Crittenden (USA) - 7. H de la Paulle (BEL) - 8. A M Baidaff (ARG) - 9. A Enstroem (SWE) - 10. N A Halbertsma (NLD) - 11. L V Rihanek (CSK) - 12. R Hiecke (AUT) - 13. T Czaplicki (POL) - 14. C Clerici (ITA) - 15. J W T Walsh (GBR) - 16. I Folcker (SWE).

Figure 7.2. Constructors of photometry and colorimetry. From Walsh and Marsden 1990 *History of the CIE, 1913–1988* (Geneva) p 12. Reproduced with permission of the International Commission on Illumination (CIE), Kegelg 27, A-1030 Vienna, Austria.

7.3. LEGISLATIVE CONNECTIONS

The work of the CIE was independent of, but loosely guided, legislation in its member countries. One of its first orders of business was to determine what laws or codes of illumination and light measurement were in effect. Although committees were active in several countries, only America reported specific legislation[20]. By 1921 lighting legislation existed in six American states. This consisted generally of a lighting code prescribing illumination levels for factories, schools and streets, but in at least one state included fines for non-compliance. France had set up a commission in 1912 to study factory lighting, and a similar committee in Britain grouped policy-setting representatives of the Post Office and the Ministries of Health and the Interior. The latter's mandate included providing the government with 'information on photometric and economic questions'[21].

The CIE organized committees to study technical questions that would allow international guidelines on illumination. These included committees on the lighting of factories, schools and mines; street lighting; aircraft and train signals. The need to specify intensities and colour demanded that even more urgent attention be given to photometric practice.

7.4. CONSTRUCTING COLORIMETRY

As table 7.1 indicates, the CIE placed the study and standardization of colour high on its list of priorities. The interest in colour by the CIE was a reflection of work already underway in its member countries, particularly America and Britain. Scientific investigation of colour measurement had been a recent development, however, dating barely from the First World War. The industrial need for colour metrics increased dramatically between the wars. In the British dyestuffs industry, for example, the production of dye colours rose fourfold between 1913 and 1927[22]. The scientific interest in the measurement of colour followed the establishment of professional societies, national laboratories and the organization of interested groups, especially in Britain and America. Between the wars, the subject was systematized and rationalized at these centres and formalized through the CIE.

Compared with radiometry and photometry, colorimetry proved far more problematic for quantification in the inter-war period. Owing to disagreement between the interested groups, the nature of colour was debated in an unusually public manner, and finally agreed by compromise and uneasy consensus near the end of the decade. In a very real sense, colorimetry was 'constructed' to suit the views of members of that debate. The events illustrate how technical delegations grew to influence not only colour but the more general field of light measurement during the inter-war period.

7.4.1. Colour at the CIE

Although there was considerable work in colour taking place at a variety of institutions, companies and societies in America and Britain, by the early 1920s an international nucleus was beginning to form through the CIE. Unlike its predecessor, the CIE tabled discussions of colour photometry from its first meeting in 1921, and faced the more fundamental problem of colour definition itself in its next meeting three years later. But unlike the national laboratories, the CIE was not initially concerned with questions of colour quantification. The commission was vitally concerned, however, with obtaining accurate *photometric* measurements, and practitioners now generally recognized these to be affected by questions of colour.

The first involvement began with a discussion of a sub-committee on the photometry of lamps, and the differing colours of various national intensity standards. The oldest extant standard, the German Hefner candle, had a distinctly red tint. The French, British and American light sources were intended as interim standards until they could be related to a more fundamental physical standard based on the light emitted by a platinum surface at the melting point (a standard itself adopted in principle at the 1884 International Conference on Electrical Units and Standards)[23]. This had proved difficult to achieve in practice, however, and so each of the national standards was based on electric lamps. The temperature of the filaments of these national sub-standards differed because the filament materials, construction and power consumptions had been differently specified by the individual laboratories. The result was a collection of national illumination standards of slightly differing colour. The investigators concluded that a comparison of differently coloured light sources was essentially meaningless unless the nature of the observer was also taken into account[24].

The problem of intensity standards thus devolved once more to the fundamental question of whether to specify light intensity and colour in terms of its physical power or in terms of its effect on a human observer. And, since human eyes varied in colour sensitivity, how could 'the human observer' be defined? The even greater difficulties of determining the intensities of different coloured lights had not been obvious to all investigators. Pierre Bouguer noted

> A comparison of two lights of different colours in the way that we prescribe is chiefly embarrassing in case it is necessary to do it with more care, that is to say, when the two intensities closely approach equality; but there is a point where one of two lights will certainly appear more feeble. We have then only to take the mean between these two limits.[25]

This technique of double-observation and averaging was unproblematically promoted by the first illuminating engineers. Alexander Trotter wrote

> It is true that with ill-devised apparatus and unsuitable methods some difficulties are experienced, but the judgement that two surfaces of different colours are of equal or of unequal brightness is an operation with which every artist in black and white or monochrome, and every engraver and etcher, is familiar.[26]

Yet the problem of differently coloured lights had been increasingly encountered with the advent of the incandescent and arc lamps in about 1880. Some practitioners made two photometric measurements, through red and green glass, respectively. But this simply displaced the problem: the standardization of these filters became necessary, with various schemes being suggested for preparing reliable coloured solutions or 'screens'. The early confidence in the ease of colour matching had been further eroded by the experiences at standards laboratories in the first two decades of the century.

The CIE committee initially minimized the scope of its enquiry by proposing the use of colour filters to restrict the wavelength range, and so avoid the problems of heterochromatic photometry[27]. The chairman deplored the lack of information, noting that 'the physicists are behind the photometrists' on the subject. Yet the delegates felt that the problems were not isolated to the study of colour. Discussion widened to the type of information needed. Would the description of colour be studied, or merely the physical question of the transmission of optical power by filters? The chairman admitted himself 'a little frightened at the size and difficulty of colorimetric questions'. A committee on heterochromatic photometry (based in Paris) already existed, having been formed at the previous CIE meeting in 1921; should this be expanded to include colorimetry, or should a new committee be formed? The president of that committee, Charles Fabry of the Université de Toulouse, wrote:

> The problem posed by colorimetry is, in some respects, the inverse of that of heterochromatic photometry, since, in [the latter] case, it is proposed to characterize intensity by a number with no allusion to colour, whereas in the [former], one seeks to define colour without concern for intensity.[28]

In his opinion, the Commission should concern itself with the physical side and ignore the psychology of colour. A Swiss delegate agreed, observing that colorimetry was too premature for international discussion. Instead, he suggested, the heterochromatic photometry group should first complete its study, then physicists in physical laboratories should 'precisely treat the questions which must constitute the bridge between colorimetrists and physicists'[29]. According to this view, *physicists* would define the concepts which other practitioners would then employ. The CIE delegates, consisting mainly of physical scientists and engineers, were not eager to complicate their work with questions of physiology and psychology. Were they not in the midst of putting the subject of photometry on a *physical* basis? Yet other delegates wanted to broaden the scope of the CIE work. John Walsh of Britain suggested forming a new colorimetry committee having the freedom to study all aspects of heterochromatic photometry, colour description and the establishment of a standard of white light. The American Edward Hyde concurred, calling it a 'question of high importance, and ripe for international investigation at present'. Rather than waiting to form a colorimetry committee '(which could find itself in contradiction to the heterochromatic photometry committee), it would be better to establish a collaboration between the two committees'[30]. Supporters of the two approaches separated into delegates involved with the existing heterochromatic photometry committee, based in Paris, and delegates from the Nela Research Laboratory and the NPL, who had little professional experience, but a strong interest, in colour measurement. Seeking compromise, the President noted that the two positions were 'well defined and not entirely incompatible'[31]. After deferring a decision until the final day of the session, the delegates unanimously voted to retain the narrow physical scope of the heterochromatic photometry committee but to form a new colorimetry committee having one representative each from Britain and America[32].

While narrowly escaping indecision, this episode was the first formal tabling of a conceptual question that would occupy the next 15 years, namely: Could a workable system of light measurement be constructed by treating colour as a purely physical phenomenon, or must the observer be an intrinsic part of the system?

The American contribution to the CIE colour committee was inevitable, an American committee already having investigated the subject. A Standards Committee on Colorimetry had been established by the Optical Society of America in 1919 to set forth terminology, summarize available data and to outline established methods of colour measurement[33]. Two years before the CIE meeting, the American committee had published a 69 page report attempting to formalize the measurement of colour. In it, they admitted to the provisional nature of what they hoped could become a science of colorimetry: 'the nomenclature and standards of color science are in an extremely unsatisfactory condition... manifest to practically all workers in this field'[34]. The work of the committee members had yielded a report which, 'being a more or less pioneer effort of its kind, must naturally be regarded as incomplete or tentative'. Indeed, the result was strongly disputed among the committee members themselves:

> The definition of the term *color* which is advocated in the present report is the result of very careful consideration and protracted debate between various members of the Committee.[35]

The *protracted debate* concerned not the experimental data but the concepts and language employed to discuss and understand it. The psychologists sought to express many aspects of colour perception that had hitherto been neglected. Different problems preoccupied the psychology and physics communities. The psychologists' efforts to determine inner mental relationships between stimuli and perceptions contrasted with the physicists' goal of employing the visual response to measure external phenomena. The psychological dimension approached that of the physicists most closely in the work of such 19th century investigators as Gustav Fechner (1801–87), Wilhelm Wundt (1832–1920) and Francis Galton (1822–1911)[36]. The physicists, on the other hand, wanted to concentrate on properties of colour that could reliably be rendered into numerical form, even if that meant simplifying or idealizing the complex characteristics of human vision. The American committee members were nevertheless more optimistic than the CIE committee to follow them:

> Practical colorimetry is... concerned with means for the unambiguous designation of those properties of objects and radiation which determine colour perception. Most of the means actually employed, however, utilize the visual apparatus as an essential element—in determining an equation of color—and hence the results are frequently not independent of the nature and special conditions of the apparatus. For this reason it is necessary, as in photometry, that the observers should be tested as average and normal.[37]

The very notion of an 'average observer', accepted without question by this time, was made possible by the 18th and 19th century realizations, particularly championed by Adolphe Quetelet, that human measures followed a normal distribution, and that 'l'homme moyenne' could be discerned from statistical analysis. Nevertheless, this trust had a narrow basis in scientific culture: the testing of groups or 'collective subjects' during the inter-war period was associated with applied, rather than academic, psychology[38].

In 1924, the CIE adopted data performed at the NBS on 52 individuals aged under 30, measured in 'good lighting conditions', as a definition of the 'normal visibility curve'. The Commission recognized that this adoption was rather arbitrary, since different data would have been obtained with other observers or the same observers measured under different conditions. By the late 1920s, several independent researchers had measured the 'visibility function' of human eyes, including Ives, Nutting, Coblentz and Hyde in America, Guild in Britain and Masamikiso in Japan. The CIE 'average' was a pieced-together combination of data from several of these sources[39]. Arbitrary or not, it was seen a useful construct that made possible further developments.

American interest in colorimetry had intensified after the 1922 OSA report. Helmholtz's *Treatise on Physiological Optics* was translated into English for the first time by the OSA; its second volume, devoted to colour perception, appeared in 1924. A reviewer noted that 'color vision at the present time is probably attracting a greater degree of attention both from the theoretical and practical points of view than ever before in its long history'. Describing its status, he also observed:

> it may be inferred that great difficulty has been experienced in completely harmonizing on any simple basis the extraordinary diversity of facts that must be explained consistently with each other.[40]

In Britain, John Guild at the NPL presented a one-man equivalent of the 1922 OSA committee report at the 1926 Optical Convention in London[41]. He echoed the American call for further research, and began to measure the colour response of human eyes. The Medical Research Council provided a grant to Imperial College for a research student, William Wright, to parallel and extend Guild's research. The good agreement between their results, which employed different apparatus and observers, convinced them and others of the feasibility of defining a 'standard observer'[42].

In 1931, the American and British work entered the international arena at the meeting of the CIE in Cambridge. Irwin G Priest of the NBS visited his co-member on the CIE colorimetry committee, Guild at the NPL. According to the NPL Annual Report, this 'enabled differences of view to be reconciled prior to the Cambridge meeting'[43]. The reconciliation was a hurried affair. Guild, having compared his and Wright's data late the previous year, had only recently finalized his ideas of a 'normal observer', i.e. an average human colour response. Seeking adoption of his methodology by the CIE, he lobbied members of the British and American committees by presenting a report to the Royal Society and sent copies to a few American researchers in the Spring of 1931[44]. Priest rallied by adapting the report and sending a written reply to Guild just two months before the CIE meeting. In it, he disputed that the British data were superior to earlier American results, but noted that he was willing to accept them. More importantly, the differences of view also related to the details of Guild's colour system, particularly his particular choice of three primary colours: 'not all countries... were prepared to adopt the NPL system of colour coordinates'[45]. The problem was that to produce certain colours, negative—i.e. unphysical—values of intensity were needed for one or more of the three component colours. Following a mathematical conversion to render all such sums positive, Priest accepted Guild's colour system. Because this agreement between the American and British committees occurred in the week before the CIE meeting, there was no time to print revised agenda papers and little opportunity for extensive discussion. Subsequently the CIE formally adopted the system, which included values for standard illuminants (coloured and 'white' light sources), numerical values for the visual response of a 'normal observer' and the mathematical relationships linking them. With these mathematical constructions, any colour could be expressed quantitatively.

The acceptance of the 1931 CIE standards thus can be seen as a result of conscious manoeuvring by the British and American delegates. Both Guild at the NPL and Priest at the NBS had restricted the subject of colorimetry to limit the importance of the human observer in the definition. Most aspects of colorimetry had *physical* bases: the definition of the 'white' and coloured illuminants; the method of calculating trichromatic coordinates based on the spectral transmission curves of the three primary filters; the method of converting between different trichromatic systems based on different colour filters. Only the highly artificial 'standard observer'—a table of numbers representing the response of a typical eye to the three reference colours—related this physical approach to visual perception. The acrimony in the subject through the remainder of the decade related to this restrictive *physical* definition of the subject.

The Commission's decisions on colorimetry were the highlight of the session, occupying 11 of the 24 pages of resolutions, and arguably have been the best known and most influential work of the CIE since. Industrial and national laboratories welcomed the standardization of a system of colour measurement, and began expressing colour information in the CIE terms. The activities of the Commission, however, waned for colour measurement. One highly likely reason for this is political. As noted earlier, the International Research Council's advocacy of policies of ostracism for German scholars between 1919 and 1926 had caused Germany to be unrepresented at CIE sessions until 1928, by which time the Colorimetry Committee had been assigned and work was well underway. France, too, was effectively excluded from participation in the colorimetry research by the decision of its delegates to support the opposing camp of heterochromatic photometry. As a result, while the British/American system of colour was accepted unanimously at the 1931 meeting, the German and French committees reversed their votes in the 'cooling off' period afterwards when National Committees examined decisions (enough other countries had nevertheless voted in favour for the system to become the international standard).

One participant later questioned 'why it was so much an Anglo-American concern', and decided that

> in the aftermath of the Great War... colorimetry cannot have had a very high priority in the European countries, and perhaps this helps to explain why France and Germany reversed their votes. They may well have felt they were being rushed into making decisions in a subject in which they were only just beginning to gain any practical experience of their own. They needed more time to think.[46]

So there was an impression that some countries were being railroaded into accepting an unsatisfactory compromise. Another reason for lack of effective action at the CIE after 1931 was its policy of rotating responsibility for Secretariat Committees. In sessions up to 1931, subject committees included representatives of several countries, even if most practical work was carried out in only one. In 1931 all committees were, for the first time, made the responsibility of individual countries. The subject of colorimetry was passed to Germany; colour specification and measurement were assigned to Japan. The American and British contributions were relegated to the lighting of factories and schools, and to the lighting of mines, respectively[47]. The lack of effective international cooperation limited the range of the work performed. Moreover, neither the German nor Japanese researchers benefited from the combination of industrial and national laboratory support for colour research that had sustained the American and British efforts. The next session in 1935 included no report from Japan, and a relatively brief contribution from Germany filling in omissions from the earlier American and British work[48]. The Colorimetry Committee was not reassigned at the session, and no programme of work was requested for the following four years. At the following session in June 1939, the proposals of the German representative were rejected by America and Britain because they would have required changes to the rapidly developing colorimetric practice[49]. The CIE then reassigned Germany the Colorimetry Committee but no work was begun before the outbreak of war. Thus active research in colorimetry returned by default to the ongoing national programmes in America and Britain.

By the early 1930s, then, a complex network had grown of institutions, committees and individuals involved in the standardization of colour measurement. In America, this network involved individuals working at large firms and at the NBS. The committees of the Optical Society of America served as the informal locus for this activity. In Britain, the NPL was the point of convergence for the DSIR-supported Research Associations (figure 7.3). Internationally, the CIE attempted to coordinate and disseminate these efforts to the less active programmes of other, principally European, countries.

The restrained international collaboration in colour research after the 1931 CIE meeting was not reflected in American work. On the contrary, bolstered by the international agreement, a second intensive phase of committee work started immediately afterward. A committee of its Illuminating Engineering Society was just then considering terminology and units for radiometry and photometry, and

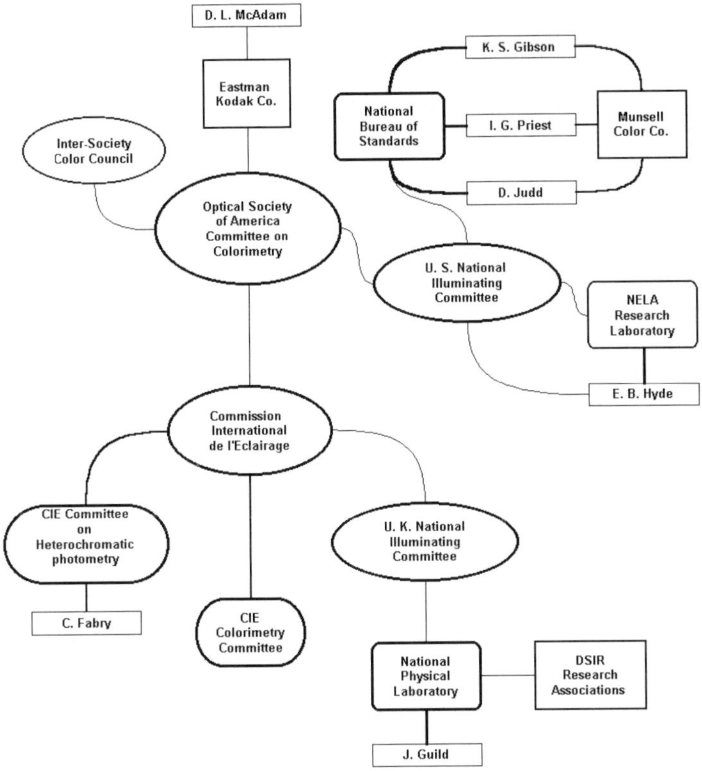

Figure 7.3. Networks of Anglo-American colour measurement between the wars. Thick lines indicate institutions employing individuals.

was extending this work to colour[50]. The American Committee on Colorimetry was also revitalized in 1932, when the Optical Society of America supported a more detailed examination of colour. The chairman, L A Jones, initially defined its purpose as being to 'introduce, advocate and facilitate use of the 1931 recommendations of the CIE'. Consisting 'almost entirely of industrial and government technologists', according to one participant, 'most members of the 1933–1953 committee had little experience with colorimetry'[51]. Another sign of continuing American activity was the birth of the Inter-Society Color Council (ISCC), set up in 1931 to define colour designations for drugs and chemicals[52]. Irwin Priest 'had most to do with the form which the council took', restricting its domain of interest to standardizing colour use in industry[53]. Not surprisingly, the ISCC defined its colours in terms of the Munsell colour notation, the product of the company that had sponsored NBS research associates. The *de facto* industrial standard for colour matching in America thus derived from the company that had so actively supported NBS activities[54].

Changes in personnel also played a part in revitalizing American colour research. In 1932, Kasson Gibson took charge of colorimetry at the NBS upon the death of Priest, who had dominated colour research at the NBS for nearly two decades. The success of committees belied the influence of individuals: while Priest spent 'many years of labor' on research into the specification of 'white' light, he 'left unpublished an exhaustive treatise giving the results of his studies and conclusions'[55]. His successor had a perspective less centred on the physical approach championed by Priest and adopted by the CIE, and was more amenable to studying the perceptual dimensions of colour vision. A shift of specialisms was occurring in the Optical Society of America, too. The original 1919–22 OSA committee was dominated by physical scientists[56]. Its original chairman, psychologist Leonard Troland, had been the only proponent of a psychological perspective. He died the same year as Priest, and was replaced by the physicist Loyd A Jones. Where Troland had fostered psychological research at the Nela laboratory and at Harvard, and applied his experience as the Research Director of the Technicolor Motion Picture Corporation, Jones specialized in the physics of photography[57]. The new 1933 OSA Colorimetry Committee included a larger fraction of psychologists than did its earlier incarnation. The increased visibility of the psychological perspective altered the very concepts of colour by the end of the decade.

7.4.2. Disciplinary divisions

The widespread acceptance of the CIE standards for colorimetry masked a deeper problem with colour measurement. The limited debates between proponents of 'colour as a sub-field of photometry' and 'colour as an independent subject' cloaked a deep, and worsening, conceptual rift. There were fundamental differences in the understanding of colour espoused by opposing social groups, drawn from physical science and psychology, respectively. The training, allegiances and experience of these 'core sets' determined the form of certified knowledge they produced[58].

The measurement standards and nomenclature adopted by the NBS and the NPL were, despite earlier disagreements with researchers in heterochromatic photometry, essentially *physical*. This was a reasonable consequence of their training in optics and applied science, and their answerability to industrial supporters. The CIE standards combined the responses of 17 British participants observing a 2–3° bright, plain visual field against a black background into a hypothetical 'average'[59]. This proved successful for simple colour measurements, such as the appearance of the light transmitted by colour filters. Psychologists argued, however, that the limited modelling of human perception made a wide class of colour measurement difficult. Surface texture, background interference, illumination level and a confusing assortment of other properties of coloured objects could influence the perceived colour.

The use of a committee structure at the Optical Society of America and the CIE to study colour was a consequence of their constitutions. But it also indicated an essentially confrontational standpoint and aura of compromise for the subject. Upon the formation of the American Committee on Colorimetry in 1919, discord between its members had soon surfaced. The difficulties centred upon the nature of colour itself. The assumption of a fixed relationship between spectral wavelength and perceived colour was implicit in the programme followed by these researchers and committee members. In the original 1922 report of the committee, for example, colour had been defined as

> all sensations arising from the activity of the retina of the eye and its attached nervous mechanisms, this activity being, in nearly every case in the normal individual, a specific response to radiant energy of certain wavelengths and intensities.[60]

Colour was thus defined as a subjective concept rooted in a physical phenomenon. Implicit in this was the assumption that, neglecting physical differences between the eyes of individuals, colour was an invariant sensation common to all observers[61].

The idea of *sensation*, however, was being criticized in the literature of psychology. As early as 1893, William James, professor of psychology at Harvard University, had argued that a sensation—a conscious response to a physical stimulus—could not be realized except in the earliest days of life, because memories and stores of associations clouded the response[62]. Instead, psychologists by the 1920s were expunging discussion of *sensation* and replacing it with *perception*, i.e. a stimulus interpreted by the brain in combination with other physical attributes[63]. This linguistic substitution represented more than mere terminology, but rather it was a conceptual shift away from attempts at measurement. Indeed, some psychologists sought to stem the tide by demonstrating that perceptions *could* be quantified:

> Psychology will never be an exact science unless psychic intensities can be measured. Some authorities [e.g. James] say that such measurement is impossible.[64]

Suggestions that colour be redefined in terms of perceptions caused complications. To the earlier definition in terms of the three attributes of *hue*, *saturation* and *brilliance* were added 'modes of appearance' such as *lustre*, *glow*, *gloss*, *transparency* and *body colour*[65]. The German psychologist David Katz concentrated on these perceptual aspects[66]. The Gestalt school of psychology included time-dependent effects such as *glitter*, *sparkle* and *flicker*. While such characteristics could be consciously experienced, they could not easily be reduced to physical terms.

7.4.3. Differentiating the issues

The disciplinary disputes can be summarized by observing that physicists tended to cordon off, or exclude, the importance of viewing conditions on colour perception, while psychologists focused and elaborated upon them.

The disputes between psychologists and physicists did not originate after the First World War, even if they escalated then. The issues being reopened had been raised earlier in a more localized and intra-disciplinary context. As discussed by R Steven Turner, the physicists' approach had been championed half a century earlier by Helmholtz, who, despite his close associations with physiology, found his ideas criticized as too 'physicalist' and simplistic by the proposer of an alternate system, Ewald Hering. Helmholtz's theory found stronger support among physicists, while Hering's was defended chiefly by physiologists and ophthalmologists. Turner notes resentment of non-physicists to the 'veneer of mathematics' in German colorimetry of the 1890s[67]. Indeed, the debates concerning the relation of colour to physical reality hearken to Goethe's criticism of the Newtonians in the first decade of the 19th century[68]. Such metaphysical overtones do not appear to have been a consideration in the American debate.

Psychologists were thus seeking to deconstruct physicists' colour to incorporate new and important phenomena. For them, 'decisions about the existence of phenomena [were] coextensive with the 'discovery' of their properties'[69]. The interpretation of colorimetry divided these cognitive communities; the move to restrict colour attributes was seen as progressive by physicists but *ad hoc* by psychologists. Physicists and industrialists believed the elucidation of 'modes of appearance' to be disruptive to standardization but psychologists took them to be cognitively essential. On another level, the technical divisions mirrored social organization; the desire to standardize units of commerce was favoured by physical scientists employed in intercommunicating national laboratories and industrial posts; psychologists, more frequently with academic affiliations, sought to bring new concepts and specialisms into both their study of colour and their broadening profession[70].

The interpretative flexibility in colorimetry existed at three levels. Most fundamentally, colour could be described either as a physical or mental entity. Second, the number of attributes required for a meaningful description of colour was open. Physicists generally opted for three, along with stringent viewing conditions. Psychologists either postulated more perceptual attributes or sought a deeper understanding for the dependence of colour perception on environmental context. Third, the precise definition of attributes—even when only three were invoked—was debatable. Thus colour systems could be based alternately on a partitioning of colour space into three additive (red, green and blue) or subtractive (cyan, magenta and yellow) components; or on less directly measurable quantities such as hue, saturation and brilliance; or on even more abstract entities such as chromaticity coordinates. The disputes between early colour systems, including the contentions surrounding the adoption of the 1931 CIE standard, operated at the last of these levels. The OSA committee discussions centred on restraining the interpretations at the first two levels.

Yet certain issues were closed for both physicists and psychologists. Observations themselves were generally accepted (although the scope of observing conditions differed for the two communities). Thus by agreeing at least on the results of experiments in artificially restricted conditions, the debate was constrained to a manageable number of issues and colour could be portrayed as a meaningful and replicable entity.

7.5. VOTING ON COLOUR

The difficulties of the OSA Colorimetry Committee from 1919 to 1939 centred upon the adoption of a physicalist, as opposed to a psychological, view of colour, and the consequences for the timing and content of a commercial standard for colour description. The even balance and differing philosophies of psychologists and physicists on the OSA Colorimetry Committee caused the meetings to be confrontational and stalemated. In a series of encounters through the 1930s, the committee members were split by their incompatible philosophies about the nature of colour.

The original OSA Committee Report in 1922 had opted for a definition of colour as a purely physical phenomenon—a definition that had carried through to the 1931 CIE standards. But when the question was re-evaluated in 1932, the majority on the new committee proposed considering the perception-based psychological concept to gain a more wide-ranging, and potentially applicable, system of colour description. When they heard the first discussion paper detailing this concept, however, the members were split down the middle. The majority of committee members rejected the addition of spatial or temporal colour characteristics, because the 'extra' attributes would be difficult to quantify or standardize. Instead, they attempted a return to the limited 'physical' definition of colour of the 1922 report, suggesting that it could be revised to make it acceptable to all members. Such a revision hinged on restricting the number of colour attributes to the original three—hue, saturation and brilliance—and in returning to the notion of colour as a 'sensation' or replicable and determinate physiological response to a physical phenomenon. This move simultaneously left the existing CIE system unmarred while disturbing the philosophical foundations of colorimetry itself, because 'sensations' were implicit and uncontentious in the physicalist version.

Such a definition was still unacceptable to psychologists, who increasingly subscribed to Gestalt precepts, maintaining that *perceptions* of colour were highly dependent on the viewing conditions. It was unacceptable for opposite reasons to instrument scientists, who saw colour as a physical phenomenon reducible to observer-independent data. The Committee as a whole agreed that neither perspective could be sustained; colour measurement, they decided, involved physical measurement and psychological factors which could, in the appropriate viewing conditions, be made adequately repeatable for standards to be practicable.

The stalemate between 'physicists' colour' and 'psychologists' colour' continued 'for more years than the chairman likes to remember' through 1937, when a proposal was published for nomenclature[71]. On this limited question, nearly unanimous agreement was obtained. Besides technical terms, though, the report attempted to relate the concept and measurement of *colour* to that of *light*. Colour was relegated to the psychological category, while light fell in the psychophysical category and radiometry in the physical category. Thus, for example, 'radiance' described a physical attribute (the amount of electromagnetic energy radiated per unit time into a unit solid angle), 'luminance' was the corresponding psychophysical unit and 'brightness' was the associated psychological unit. 'Slightly more than half' the committee accepted these definitions, with 'no one...particularly pleased with the outcome'[72]. This lukewarm agreement led the committee to explore a definition of colour as a psychophysical phenomenon.

7.5.1. *Configuring compromise*

The chairman of the original OSA committee, psychologist Leonard Troland, had earlier tried to marshal both the psychologists and physicists, writing:

> the term, light, is no longer used technically as an equivalent of radiant energy, whether or not the latter is 'visible'. Light consists in radiant energy evaluated in terms of its capacity for evoking brilliance, when it acts upon an 'average normal' psychophysiological organism. Consequently, if we are interested to formulate psychophysical laws which have exclusively physical terms on one side of the equation, we must avoid the photometric concepts and use those of radiant energy, pure and simple.[73]

And later:

> Light can neither be identified with brilliance nor with radiant energy. It has the properties of both, taken together.[74]

Troland, the sole psychologist among the physicists, had sought to establish a crucial link between perceived colour, physical measurement and mind.

According to Loyd Jones, the new committee chairman, the adoption of a psychophysical concept of colour was a matter of compromise. Initial reaction to a psychophysical concept of colour in 1934 had been 'quite unfavorable'. As described earlier, colour was associated with different phenomena and practical goals for physicists and psychologists. When a report on the consequences of a psychophysical definition was tabled in 1935 the reaction was 'not in the least enthusiastic', because, according to Jones, only 'a few had reached the point in their thinking where they felt that the psychophysical point of view should be considered...'. A second report was prepared to investigate these mixed physical–physiological–psychological definitions of colour more fully before they were finally rejected[75]. This had a more promising reception by the Committee, because the debate had moved slightly away from philosophical underpinnings (i.e. the nature of light) to workable schemes for merging physical phenomena (e.g. spectral distributions) with mental responses (e.g. awareness of brightness and hue). Again Loyd Jones appealed to various members to elaborate the psychophysical scheme. David MacAdam, a 28-year-old physicist at Eastman Kodak specializing in human colour vision, tabled a report based on a psychophysical scheme in 1938[76]. The content of MacAdam's report attempted to achieve a consensus by straddling both the CIE 1931 conclusions (based on the physicalist interpretation of colour) and concessions to the psychological perspective (in which the mental contributions to colour perception were acknowledged)[77]. This synthesis of two perspectives was not well received. 'A lengthy discussion indicated considerable dissatisfaction', but the committee members agreed to give it further consideration[78].

A key argument mounted by MacAdam and Jones was that there were only two options available: either (a) to reclassify light itself from a psychophysical to a psychological phenomenon; or (b) to reclassify colour from a psychological to a psychophysical phenomenon. Because of the prior work of photometrists (often associated with electrotechnical, rather than optical, specialisms), light had long since been interpreted as a psychophysical phenomenon, that is, a moderately repeatable mental response to a physical stimulus. The committee members generally agreed that light and colour were similar entities, and hence should either both be seen as psychological or both as psychophysical. But prevailing practice militated against redefining the concept of light; photometrists were content with their definition. As Trevor Pinch has persuasively argued for the detection of solar neutrinos, the attainment of consensus is tied up with the degree of 'externality' of debate, that is, by how widely the decision affects other 'facts' or cultural groups[79]. Applying Pinch's interpretation, the existing networks of photometry sustaining 'light as psychophysical' were too difficult to break, and so the concept of colour also defaulted to a psychophysical definition.

The large swings in committee opinion through the decade indicate the contention surrounding the subject and the difficulty in achieving consensus. In the end, the committee delegated Deane Judd, the principal spokesman for psychology, and Arthur Hardy, representing the perspective of physics, to give final approval to the report. MacAdam himself described the committee work as comprising 'long discussions, multilateral deadlock, and finally exhaustion'[80]. The result of this strained consensus was a definition of colour as a carefully delimited aspect of light, which in turn was interpreted as a physiological response to radiant energy:

> Color consists of the characteristics of light other than spatial and temporal inhomogeneities; light being that aspect of radiant energy of which a human observer is aware through the visual sensations which arise from the stimulation of the retina of the eye.[81]

7.5.2. An uncertain closure

The American committee took the hard-won psychophysical definition of colour and its colorimetric units back to the next CIE meeting in June 1939. At the international level, acceptance was considerably easier, with no significant dissension. A few reasons for this can be suggested. A psychophysical definition, originally inspired by German psychologists, was congenial to the German delegates. The British delegates had maintained a close working relationship with their American counterparts and generally supported their mixed units. Other nations were not immediately concerned with the conceptual points tied up in the new metrics and had fewer practical pressures to endorse any particular scheme. The psychophysical definition of colorimetric units was tabled as a discussion paper and quickly ratified. The psychophysical concept of colour thus suffused from an American committee into the international realm by way of the CIE.

The debates of the 1930s were never reopened by the formal committees. In America, though, there were open disagreements between the physical and psychological camps into the early 1940s. Physicists and psychologists continued to write about how they 'aimed at reconciliation of opposing points of view'[82]. The cracks were disappearing with continued effort. An OSA editorial soothed that the 'field of colorimetry will soon supply another example of cooperation among scientists'[83].

The subject stabilized further after the war[84]. When the OSA finally published its definitive book *The Science of Color*, the controversy was vanishing. Indeed, the book proved to have a role in capping the debate: the completed chapters, written principally by Jones and MacAdam, had appeared sporadically in the *Journal of the Optical Society of America* between 1943 and 1951. The first chapter, in which the debates of the 1930s were sketched, was followed by nine chapters in which colour was expressed solely and incontrovertibly in psychophysical terms[85]. The committee work of restricting colorimetry to a mathematical model and defining it as a shared property of mind and matter was complete. H D Murray summed up the situation in his book of the same period:

> Simplification of complex situations is a feature of all physical measurement and it has been nowhere more extensively applied than in subduing colour to the requirements of measurement.[86]

Subdued and yoked to its intended applications, colour measurement became less contentious. The philosophical basis of colorimetry no longer triggered controversy once the committees were disbanded and practical issues came to the fore. Key historical actors, ceasing to exist, no longer focused the issues. By emphasizing the utilitarian goals (standardization) over theoretical foundations (i.e. the physical, psychological or physico-psychological basis of colour), a mundane consensus was achieved for a broad technical community (delegates to the CIE). For Deane Judd, editing a collection of papers on the Munsell colour system, it proved difficult even to explain to a non-specialist readership the nature of the controversy. Psychological versus psychophysical concepts of colour had, he emphasized, either been seen as 'unproblematic' or as 'so utterly different in their concepts that there is no possibility of correspondence'. And, he cautioned, 'there are possible many psychophysical color systems'[87]. Things might have been otherwise. Similarly, the Inter-Society Color Council was careful to stress the limited nature of the agreement:

> These definitions of color, hue, saturation and brightness do not express a *unique* coordinate system, for they may be related to other sets of coordinates that may be more practically useful.... They represent a cultural development upon which there is reasonably general agreement.[88]

The social contingency of the standard may have been apparent to some of its key negotiators, but not to all their contemporaries.

Different constituencies of colour—disciplinary, practical and international—shaped the controversies in the subject and determined how they were eventually resolved. These factors embodied in the CIE system of colour are not all intrinsic in the science but arose from a range of historical situations, both in terms of the different conventions present in physics and psychology, and by inter-war politics. Colour measurement was a subject fashioned in a particular cultural and political context by heterogeneous committees. The decision-making bodies had a formal structure and rigidity, but this belied their transient compositions and contingent decisions. The distribution of the committee memberships shaped the dominant philosophical view and type of standard they adopted. Thus an evolved version of the three-colour theory of Maxwell and Helmholtz formed the basis of the international system because it was socially accepted as an operational concept by physicists and physiologists and, in restricted circumstances, by psychologists.

Committee-based colorimetry proved an ineffective method of reaching agreement. Disputes were both drawn out in the time between meetings and all too quickly debated in person. The dynamics of consensus were considerably more turgid than were debates between physicists alone, and not all constituencies were equally satisfied.

The history of colour measurement demonstrates the technical complexities and arbitrariness of definition faced in the inter-war period. Colour measurement evolved in a direction opposite to that of photometry and radiometry. While the networks of influence for light measurement (figure 7.4) are closely related to those for colour measurement—with both including several of the same individuals and institutions (the NPL, NBS, OSA, CIE and Nela research laboratory)—colorimetry entered the national laboratories with a fruitful history of empirical application and relatively little theoretical content, while photometry and radiometry struggled to adapt to the industrial problems faced between the wars.

The cases of photometric standards and colour measurement illustrate the central role played by technical delegations. The cultural schisms in colorimetry—technological versus scientific, Anglo-American versus German, physical versus psychological—made it peripheral for several communities and determined the method and shape of consensus. In such conditions, committees became the central, if fugitive, historical actors. For subjects whose scientific foundations were non-intuitive and contentious, committees defined limits and shaped content. Although goal oriented, the delegations did not maintain a fixed investigative course. Launched by particular interests (the CIP by the gas industry, and the CIE by government support for illumination standards), the Commissions nevertheless evolved in response to the experience of their delegates, the CIP shifting towards the photometry of electric lighting and the CIE undertaking colour investigations. And within these decision-making bodies, a handful of individuals proved to wield considerable power over the peripheral subjects they

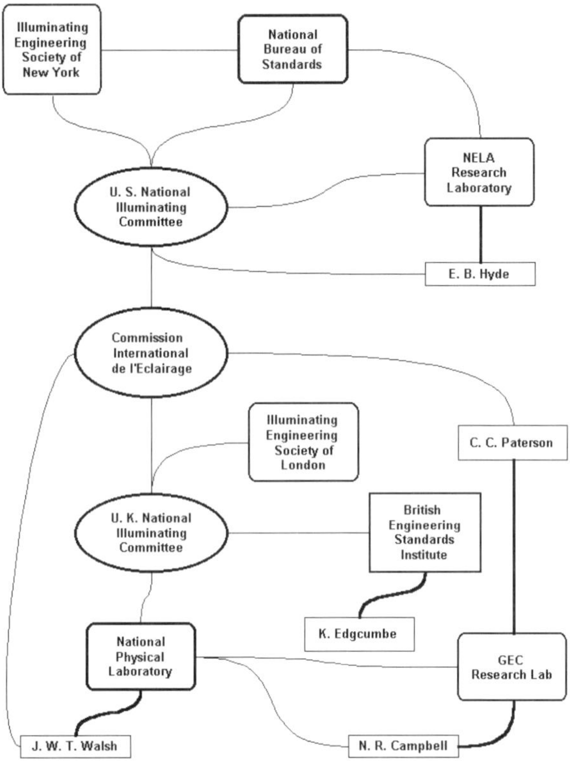

Figure 7.4. Networks of Anglo-American light measurement between the wars. Thick lines indicate institutions employing individuals.

constructed: Leon Gaster and Clifford Paterson in shaping the early CIE; John Walsh and Edward Hyde in proposing the international study of colour; and Irwin Priest and John Guild in devising the CIE measurement system. The goals and membership of the delegations moulded the subject as profoundly as did experiment and theory.

NOTES

1 *Committees* are, by definition, groups of people appointed to perform a specific task. *Commissions* are also groups charged with specific duties, but with the authority granted by a higher body, e.g. government.

2 For the rise in internationalism before the war, and 'international science without internationalism' after it, see Crawford E 1990 'The universe of international science, 1880–1939', in Frängsmyr T (ed) 1990 *Solomon's House Revisited: the Organization and Internationalization of Science* (Canton, MA) pp 251–69.

3 Engelhardt H T Jr and Caplan A L 1987 'Patterns of controversy and closure: the interplay of knowledge, values and political forces', in H Englehardt Jr and A L Caplan

(eds) *Scientific Controversies: Case Studies in the Resolution and Closure of Disputes in Science and Technology* (Cambridge) p 17. I use the term 'closure' in the senses they did, namely 'a bringing to a conclusion'; 'agreement'; or 'closing of a debate by competent authority' [p 2].

4 Quotation of T Vautier from Commission Internationale de l'Éclairage 1989 *History of the CIE 1913–1988* (Vienna) p 1 (my translation).

5 As noted in chapter 3, German industry and science had adopted the Hefner lamp as the standard of brightness, with the PTR attempting to promote it as the international standard. Its difference from the other standards (the Hefner being about 10% weaker) and its wide usage made the German-speaking countries loath to convert to the new international value.

6 John Walsh labelled the transition from measuring lamp *intensity* to *illumination of surfaces by lamps* as the beginning of the 'quantitative age' [Walsh J 1951 'The evolution of the lighting art' *Proc. IEE* **98** 309–15].

7 The requirements for membership of a National Committee were 'rather detailed', so the statutes were modified at the first meeting in 1921 to encourage the entry of new countries 'where it was difficult to comply fully'. For those countries still unable to ensure a representative committee, observer status was granted. See Walsh and Marsden *op. cit.* note 4, p 9.

8 *Ibid.* p 7 (my translation). The CIE numbered its meetings consecutively with those of its predecessor, the CIP. Neither published its minutes or findings until the fifth session in 1921. The fourth session of the CIP/CIE had been cancelled at the outbreak of the First World War.

9 The attendance during the 1920s was dominated by French and English speaking delegates. For example, the fraction of French, British and American delegates was 82% at the 1921 meeting in Paris and 63% at the 1924 Geneva meeting, but only 52% at the British meeting in 1931, when Germany and Austria together fielded 16% of the delegates, and other European countries were more strongly represented.

10 Following the First World War, Germany and Austria did not send delegates to the CIE until 1928. The exclusion enforced by the International Research Council was in effect during the formative years of the CIE, but was short lived. German attendance at commissions such as the CIE, almost nil early in the 1920s, increased to about 85% of international meetings by 1926, when the IRC lifted its bar against the Central Powers. This correlates with the appearance of German delegates at the CIE meetings of 1928 and afterwards. See Crawford E 1992 *Nationalism and Internationalism in Science, 1880–1939: Four Studies of the Nobel Population* (Cambridge) p 50. The political climate of international science between the wars is also discussed in, for example, Kevles D J 1971 'Into two hostile camps: the reorganization of international science after World War I' *Isis* **62** 47–60, and Forman P 1980 'Scientific internationalism and the Weimar physicists: the ideology and its manipulation in Germany after World War I' *Isis* **64** 151–80.

11 Walsh and Marsden *op. cit.* note 4, p 10 (my translation).

12 Halbertsma N A 1963 'CIE's golden jubilee' *Compte Rendu CIE* **15** 25.

13 'The National Illumination Committee of Great Britain is constituted by the Illuminating Engineering Society of Great Britain, The Institution of Electrical Engineers, The Institution of Gas Engineers, and the NPL, in cooperation with industrial, technical and professional associations and government departments interested in the subject of illumination' [Anon. 1928 *Illum. Eng.* **21** 106]. In 1927, 18 organizations and government departments were represented.

14 Despite the formation of the Institut d'Optique and its journal *Revue d'Optique Théorique et Instrumentale* in 1920, the industrial–scientific–governmental linkages in French optics were weaker than in Germany, although training was better organized than in Britain. The inter-war period saw a succession of government agencies tasked with the promotion of science and technology. See Paul H W 1985 *From Knowledge to Power: the Rise of the Science Empire in France, 1860–1939* (Cambridge) pp 311–12 and 340–53, and Williams M E W 1994 *The Precision Makers: a History of the Instruments Industry in England and France, 1870–1939* (London) pp 139–44.

15 The figures for the two years for which delegate affiliations were listed are as follows: for the 1924 session, France sent six delegates, all but one academic; the British sent nine—seven from industry and two from the NPL; the US sent seven, of whom five were from industry and two from the NBS. In 1931, Germany sent 16, 14 representing industry and one each from the PTR and university; France sent 29, eight of whom were academics, four from government and 17 from industry; Britain sent 32, five representing government departments and two the NPL. For a discussion of the 'rapports inéxistants' between the physics community and industry in France in the inter-war period, see Pestre D 1984 *Physique et Physiciens en France, 1918–1940* (Paris) pp 238–41.

16 CIE 1921 *Compte Rendu CIE* 5th Session (London) p 10, emphasis added.

17 Clifford Paterson, the President of the Commission, wrote, 'You will... appreciate how valuable is such an experience when illuminating engineers from all countries are thrown together for several weeks in informal relationship for study, instruction and recreation' [Paterson C C 1928 'Some notes on the meeting of the International Commission on Illumination in the United States', *Illum. Eng.* **21** 337–8]. Another delegate wrote: 'The sea trip from Southampton to New York gave time for recreation and for the final organization of the British delegation. Mr Good [the President of the British National Committee]... probably curtailed many delegates' social programmes by dividing the party into groups responsible for various subjects, whose members met, often several times a day, to decide on their course of action at Saranac' [Anon. 1929 'A review of the proceedings of the 7th session of the International Commission on Illumination and the International Illumination Congress in the United States in 1929' *Illum. Eng.* **22** 167].

18 See Cochrane R C 1966 *Measures for Progress: a History of the National Bureau of Standards* (Washington) pp 110–11 and Coblentz W W 1936 'Edward Bennett Rosa', *Biog. Mem. Nat. Acad. Sci.* **16** 355–68.

19 The 1913 plan for the CIE had called for the central office to be based at the NPL in Teddington, for which secretary and office space were being arranged at the outbreak of war.

20 Marks L B 1921 'Législation de l'éclairage aux Etats-Unis' *CIE Compte Rendu* (London) pp 22, 204–21.

21 CIE 1921 *Compte Rendu CIE* 6th Session (London) pp 23–4.

22 Brightman R 1934 'The dyestuffs industry in 1933' *Indus. Chem.* January 18–21. The tonnage of all colours was 4069 in 1913, 17 604 in 1927 and 22 045 in 1932.

23 The original suggestion had come from Jules Louis Gabriel Violle in 1881 and was taken up by Waidner and Burgess at the NBS. See, for example, Wensel H T, Roeser W F, Barbrow L E and Caldwell F R 1931, 'The Waidner–Burgess standard of light', *Bur. Stan. J. Res.* **6** 1103–18.

24 For example, an eye or detector sensitive mainly to red light would judge the relative intensity of a pair of light sources, one bluish and the other reddish, differently

compared to an eye sensitive mainly to blue light.

25 *Traité d'Optique sur la Gradation de la Lumière*, transl. W E K Middleton 1961 (Toronto) p 49.

26 Trotter A P 1911 *Illumination: its Distribution and Measurement* (London) p 68.

27 CIE 1924 *Compte Rendu CIE* 6th session 28–38.

28 Fabry C *Compte Rendu CIE* 190 (my translation).

29 Joye M *Compte Rendu CIE* 31 (my translation).

30 Hyde E P *Compte Rendu CIE* 32.

31 *Ibid.*, p 32 (my translation). Although Fabry, chairman of the Heterochromatic Photometry Committee, retained this position for an unusually long period in the CIE, the American contributions (from Crittenden of the NBS, and Hyde and Taylor of Nela) outweighed his reports by three to one. The differing views for a new committee cannot be seen, however, as a simple desire of the existing committee to retain control. Rather than wanting to explore all aspects of colour in an expanded version of the Committee, the members wished to omit all question of colour measurement until they, and other physicists, had cautiously investigated practical techniques for removing its effect from photometric measurement. The two positions amounted to either including or excluding colorimetry from the study of photometry.

32 Three members had been sought, but only two were proposed. The appointed members were Irwin Priest of the NBS and T Smith of the NPL. Smith, the head of the Optics Division, was not present at the CIE Session. The proposers were unaware of the work already begun by John Guild of the Division, who performed all colorimetry work at the NPL until Smith collaborated in the early 1930s.

33 Colorimetry Committee of the OSA 1920 '1919 report of the Standards Committee on Colorimetry', *JOSA* **4** 186–7. Copies of the unpublished 50 page report were provided to parties who had expressed an interest in colour measurement, namely researchers at the NBS, Nela Research Laboratory, Cheney Bros, Johns Hopkins University, Du Pont de Nemours & Co, Columbia University, Carnegie Geophysical Laboratory and the Corning Glass Works.

34 Troland L T 1922 'Report of Committee on Colorimetry for 1920–21' *JOSA & RSI* **6** 527–96; quotation p 528.

35 *Ibid.*, p 531.

36 See, for example, Ladd-Franklin C 1893 'On theories of light sensation' *Mind N.S.* **2** 473–89. For a social constructivist history of psychology discussing the drive for quantification and the resulting 'methodolatry', see Danziger K 1994 *Constructing the Subject: Historical Origins of Psychological Research* (New York), especially chapter 9. Regarding the simplistic metrology of human characteristics from an anthropological viewpoint, see Gould S J 1981 *The Mismeasure of Man* (New York).

37 Troland *op. cit.* note 34, p 574.

38 See Obserschall A 1987 'The two empirical roots of social theory and the probability revolution' in Krüger L, Daston L J and Heidelberger M (eds) 1987 *The Probabilistic Revolution* (Cambridge, MA), Vol 2 pp 109–11; Lazarfeld P F 1961 'Notes on the history of quantification in sociology—trends, sources and problems', in Woolf H (ed) 1962 *Quantification* (Indianapolis) pp 147–203 and Hacking I 1990 *The Taming of Chance* (Cambridge); Danziger *op. cit.* note 36 ch 8.

39 See, for example, Kaiser P K 1981 'Photopic and mesopic photometry: yesterday, today and tomorrow', in *Golden Jubilee of Colour in the CIE* (Bradford) pp 29 and 31–2.

40 Anon. 1925 'Helmholtz's treatise on Physiological Optics Vol. 2', *JOSA* **11** 369–74.

41 Guild J 1926 'A critical survey of modern developments in the theory and technique of colorimetry and allied sciences', *Proc. Opt. Convention* vol I (London) pp 61–146.
42 Wright W D 1981 'The historical and experimental background to the 1931 CIE system of colorimetry', in CIE *Golden Jubilee of Colour in the CIE* (Bradford) pp 2–18.
43 NPL 1931 *Report* (Teddington) p 15.
44 Wright *op. cit.* note 42, pp 13–17.
45 *Ibid.*, p 105.
46 *Ibid.*, pp 2–18.
47 CIE 1931 *Compte Rendu CIE* 8th Session (London).
48 CIE 1935 *Compte Rendu CIE* 9th Session (London). The Japanese delegation of seven persons did not table a paper or participate in the discussion periods; no record of their contribution appears in the minutes. The German work was limited to more careful definitions of a standard 'white point' using CIE colour coordinates, and the brightness of test surfaces.
49 The German delegate Dresler recommended a new standard 'illuminant E', representing sunlight, to add to the existing three illuminants. Other delegates criticized its poor approximation to sunlight, the adequacy of the existing 'illuminant C' for this purpose, and the desirability of *reducing*, rather than increasing, the number of standards.
50 Anon. 1930 'Illuminating engineering nomenclature and photometric standards', *Trans. Illum. Eng. Soc. (NY)* **25** 728–33.
51 MacAdam D L 1994 personal communication, 4 Feb, and Committee on Colorimetry, Optical Society of America 1953 *The Science of Colour* (Washington), Introduction.
52 See Judd D B and Kelly K L 1939 'Method of designating colors' *J. Res. NBS* **23** 355–85.
53 Nickerson D 1938 'The Inter-Society Color Council' *JOSA* **28** 357–9. The diversity of groups concerned with colour is illustrated by the council members, which included the American Association of Textile Chemists and Colorists, American Ceramic Society, American Psychological Association, American Society for Testing Materials, Illuminating Engineering Society, National Formulary, American Pharmaceutical Association, Optical Society of America, Technical Association of the Pulp and Paper Industry and the United States Pharmacopoeial Convention. In the UK, the British Colour Council was set up at about the same time, and published a set of silk colour swatches as colour references in 1934.
54 This American adoption of a proprietary colour system was not copied by other countries. The CIE and Munsell systems co-existed there, suggesting the decrease in internationalism through the decade.
55 Ives H E 1932 'Irwin Gillespie Priest' *JOSA* **22** 503–8.
56 The original committee had had five members, the two chief contributors being Priest and its chairman, Leonard Troland. The 23 members of the 1932 committee included 11 from industry, four from government, three from universities and five with unlisted affiliations, with roughly half espousing a psychological view.
57 Troland (1889–1932), gaining a PhD in psychology in 1915, worked for two years at the Nela laboratory, and was elected president of the OSA in 1922–3 at the age of 33. He became Research Director of the Technicolor Motion Picture Corporation in 1925, while holding an academic post at Harvard. See Southall J P C 1932 'Leonard Thompson Troland' *JOSA* **22** 509–11. Loyd Ancile Jones, an associate editor of *JOSA* for over 25 years and OSA Ives medallist for 1943, specialized in the physics of photography.

58 See Collins H M 1985 *Changing Order: Replication and Induction in Scientific Practice* (London) pp 142–5.
59 The data represented the mean measurements of ten observers measured by William Wright at Imperial College in 1929, and the seven measured by Guild from 1926 to 1928. See Guild J 1934 'The instrumental side of colorimetry', *J. Sci. Instr.* **11** 69–78.
60 Troland *op. cit.* note 34, p 565.
61 This assumption had been championed a half-century earlier by Helmholtz, but criticized as too 'physicalist' and simplistic by the proposer of an alternate system, Ewald Hering. Helmholtz's theory found stronger support among physicists, while Hering's was defended chiefly by physiologists and opthalmologists. See Turner R S 1994 *In the Eye's Mind: Vision and the Helmholtz–Hering Controversy* (Princeton).
62 James W 1892 *Psychology* (London) p 12.
63 Troland L T 1929 'Optics as seen by a psychologist' *JOSA* **18** 223–36.
64 Richardson L F 1929 'Quantitative mental estimates of light and colour' *Brit. J. Psychol.* **20** 27–37; quotation p 27.
65 Troland supported this approach when he noted 'the subjective study of color... in respect to those nuances which the German psychologists call... modes of appearance offers a fascinating field for investigation' [*op. cit.* note 34, p 233]. The Germans to whom he referred were David Katz (1884–1953), a Gestalt psychologist who specialized in colour perception, and Ewald Hering (1834–1918), a physiologist and psychologist. Katz's *The World of Colour*, espousing the psychological rather than the physiological or physical viewpoints, was first published in English in 1935, but was preceded by German editions in 1911 and 1930.
66 Katz D 1935 *The World of Colour* (London).
67 Turner *op. cit.* note 61, pp 238, 251.
68 Jackson M 1994 'A spectrum of belief: Goethe's "Republic" versus Newtonian "despotism"' *Soc. Stud. Sci.* **24** 673–701.
69 Collins *op. cit.* note 58, p 129.
70 Danziger *op. cit.* note 36, pp 136–55 discusses how psychologists embraced quantification as a means of simultaneously grounding, justifying and extending their subject.
71 Jones L A 1937 'Colorimetry: preliminary draft of a report on nomenclature and definitions', *JOSA* **27** 207–13.
72 Committee on Colorimetry 1953 *The Science of Color* (Washington) p 10.
73 Troland L T 1929 *Psychophysiology* Vol 2 (New York) p 57.
74 *Ibid.*, p 71.
75 Colorimetry Committee *op. cit.* note 72, p 10.
76 MacAdam was a research associate at Eastman Kodak from 1936, when he obtained his PhD. His association with the OSA began earlier, becoming a member of committees from the 1930s, Fellow in 1932, a director 1942–45 and President in 1962. He was later to trace the history of color metrics from an unproblematic 'internal' viewpoint [MacAdam D L 1970 *Sources of Color Science* (Cambridge, MA)].
77 Its author noted that his devised measuring units and definitions were strongly influenced by physicist Percy Bridgman's philosophy of *operationalism*, citing passages such as the following: 'Physics, when reduced to concepts [defined in terms of their properties], becomes as purely an abstract science and as far removed from reality as the abstract geometry of the mathematicians, built on postulates. It is the task for the experiment to discover whether concepts so defined correspond to anything in nature.' [Bridgman P 1927 *The Logic of Modern Physics* (London) pp 4–5]. This was reiterated by the ISCC committee: 'in the science of colorimetry a great many years

were spent deriving a precise operational concept of color which would represent a careful specification of operations performed' [Burnham R W, Hanes R M and Barleson C J 1963 *Color: A Guide to Basic Facts and Concepts* (New York) p 3].

78 Colorimetry Committee *op. cit.* note 72, p 13.

79 Pinch T J 1985 'Towards an analysis of scientific observation: the externality and evidential significance of observation reports in physics' *Soc. Stud. Sci.* **15** 3–16.

80 MacAdam D L 1994 personal communication, 4 February.

81 Colorimetry Committee *op. cit.* note 72, p 221.

82 See, for example, a special issue devoted to the Munsell Color System in *JOSA* 1940. As late as 1944, evidence seemed to show that heterochromatic photometry could not be made to give consistent results [Wright W D 1944 *The Measurement of Colour* (London)].

83 Anon. 1940 'Cooperation among color experts', *JOSA* **30** 573.

84 The publishing of *The Science of Color* in 1953 was contemporaneous with the adoption in America of the National Television System Committee (NTSC) standard for colour television. The earlier colorimetric research that informed the report was directly applied to the technical decisions taken by the television committee [Carnt P S and Townsend G B 1961 *Colour Television: N.T.S.C. System, Principles and Practice* (London)]. On the other hand, earlier colour television systems (e.g. J L Baird's system of 1928) implicitly drew upon the Maxwell–Helmholtz theory which formed the foundation of the CIE system of colour.

85 Jones L A 1943 'The historical background and evolution of the colorimetry report' *JOSA* **33** 534–43.

86 Murray H D 1952 *Colour in Theory and Practice* (London) p 264.

87 Judd 1940 'The Munsell color system', *JOSA* **30** 574.

88 Burnham *op. cit.* note 77. This positivistic ISCC catechism classified the definition of color as a 'basic fact', colorimetry as 'applied facts' and color vision theory as 'marginal facts' (p vi). Two of the OSA colorimetry committee members served on the ISCC committee, and four others, including Deane Judd, reviewed the report.

CHAPTER 8

MARKETING PHOTOMETRY

Victorian firms had sold gas photometers; their Edwardian counterparts supplied equipment for government and industrial laboratories. But light and colour measurement acquired a new lustre—an important commercial dimension—during the inter-war years. Quite suddenly, light measurement was everywhere.

Commercialization changed everything. Practitioners increasingly purchased ready-made equipment rather than constructed their own[1]. Technical communities were newly seeded and extended. Technological innovation opened new markets. Expertise in light measurement shifting from tedious protocols of visual observation to the design principles of electronic apparatus. The embodiment of techniques and standards into purchasable hardware was the culmination of a process that converted a human-centred activity into one manifested in instruments[2]. The spread of commercial instruments conferred a new legitimacy on the subject. To be *photoelectric* was to be up-to-date, precise and fast. A clear transition was in progress: the industry expanded; the technology evolved; the number and types of practitioners exploded.

Commercial development marshalled a complex interplay of influences. Writing of related domains, Davis Baird has described the period 1920–50 as a 'scientific revolution' in analytical chemistry because of the rise of instrumentation[3]. Contemporary chemists made the same observation; one, introducing a *Symposium on New Research Tools*, noted:

> it is particularly fitting that chemists and physicists should appear together . . . for the most remarkable aspect of the science of the past twenty years has been the way in which chemists and physicists have played into each other's hands. . . science and its tools develop together.[4]

Much of the change in analytical practice since the Great War can be correlated with the commercialization of light-measuring instruments, particularly colorimeters and spectrophotometers. The availability of ready-made instruments for light measurement neatly removed a class of problems—the construction of apparatus—from the user and at the same time opened the subject to communities of practitioners that previously had little contact with it. The new practitioners, in turn, influenced the course of light measurement. Robert Bud and Susan Cozzens have observed that 'new technologies can radically alter the access of a community of scientists to its phenomenon of study' and that

> people are an important element in spanning the institutional boundaries between the laboratory and the industrial firm. Scientists clearly do get involved in the development of instruments, in particular because of their ability to merge scientific and technical aims in the process of scientific work. Instrument makers, likewise, do interact with the laboratory as they develop and refine new products.[5]

But the process was more cohesive, more seamless, for light measurement. Practitioners, devices and techniques crossed disciplinary boundaries repeatedly. Relationships were promiscuous. The inter-relationship between the availability of technology and the evolution of practice was murky and changeable.

The discourse of light measurement had shifted from questioning the *need* for quantification to the *instrumental means* of achieving it. This dialogue also took place in new contexts: in advertisements, in the evaluations of designs to be found in scientific papers, and in the 'New Products' pages of scientific journals. The growth of industrial and commercial markets for photometric apparatus had, in turn, cultural, scientific and technological consequences. New communities of practitioners became associated with light measurement, including commercial designers, industrial chemists and production engineers. These groups extended light and colour measurement to new applications demanding the development of new kinds of measuring equipment. With this new apparatus, scientists having had no previous concern with light measurement were able to apply its methods to their particular problems. Particularly in industry, these early applications had mixed success. By the end of the decade, physical methods had almost entirely replaced visual observation, but the first flush of enthusiasm for the automated measurement of light in industry was fading.

8.1. BIRTH OF AN INDUSTRY

The fledgling photometric instrument industry grew out of a pre-existing scientific and precision instrument industry[6]. The commercial manufacture of light-measurement apparatus began on a small scale as soon as a market, in the form of professional photometric laboratories, became established[7]. Commercial photometers proliferated after the passing of gas testing legislation, and again upon the introduction of electric lighting.

The competition between gas and electric lighting systems caused a flurry of commercial development. There was a significant rise in photometric publications in the 1880s as a result of the commercial introduction of electric lighting. The appropriate *type* of photometric measurement was contentious: gas and electric lighting generally produced a different distribution of illumination on horizontal and vertical axes. Quantities such as 'mean horizontal candlepower' and 'mean spherical candlepower' were increasingly measured by purpose-built commercial instruments, and employed to argue for the superiority of their respective illuminants[8]. Photometric standards also promoted production runs of standard light sources and instrument designs.

By the First World War, the sale of photometric devices was a stable if small-scale enterprise. In America, the war triggered an upswing in the instrument industry. The heavy reliance on European instruments existing before the war was rapidly reversed. 'We now manufacture over 85 per cent of our industrial and scientific instruments and appliances', wrote the director of the NBS in 1924, 'where before the war over 80 per cent of these were imported'[9]. The instruments included light-measuring devices such as photometers, spectrophotometers and colorimetric apparatus. Far from being merely the adaptation of designs originated by academic or government scientists or the copying of European apparatus, this activity involved research, development and manufacture proceeding in parallel and often within a single company. As discussed earlier, commercial research laboratories played an important role in the development of light measurement during the 1920s. By the late 1930s, an American government survey listed at least four companies—Bausch & Lomb, General Electric, Westinghouse, and Weston—with dozens of staff members active in the research and development of light measuring instruments[10].

The war caused a similar expansion of the British precision instruments industry[11]. With the creation of the Ministry of Munitions in 1915, instrument firms were expanded, redirected or re-sited to meet the requirements of military instruments. When the war ended and government contracts were withdrawn, many companies found themselves overextended in production capacity compared to the available markets for their goods. To encourage research and cooperation between firms, the newly founded Department of Scientific and Industrial Research supported the formation of the British Scientific Instruments Research Association (BSIRA) in 1918[12]. Government initiatives played a minor role in the continued commercialization of light measurement.

The post-war expansion of the photometric instrument industry was a direct response to the needs of practitioners who were unable or unwilling to design and construct their own equipment. Several factors determined these user requirements: the development of research programmes, the increase in routine light measurement and a rise in appreciation for the benefits of quantitative light measurement.

This motive for the early expansion of the industry is at variance with conclusions drawn by Yakov Rabkin, who suggests that the integration of instruments into science 'occurs through vigorous supply of advanced instruments on the part of industry'[13]. The 'supply of advanced instruments' as an impetus to change was a feature of the early 1930s and beyond, but not of the preceding period. Indeed, the case of light measurement closely follows the four stages in the development of new instruments suggested by the National Academy of Sciences in America[14]:

(1) discovery of suitable means of observing some phenomenon,
(2) exploration of this phenomenon with special, home-made instruments or commercial prototypes,
(3) widespread use of commercial instruments and
(4) routine applications of the instrument to control industrial production as well as research.

That is, the spread of instrumentation was mediated as much by users as by manufacturers. Stage (1) and parts of stage (2) of this process have been discussed in previous chapters.

8.2. TECHNOLOGICAL INFLUENCES

A major impetus for the commercialization of light measurement was the development of reliable physical methods of detection. As discussed earlier, practitioners by the 1920s had refined the visual method of measurement, making evident its ultimate reliance on unfatigued and unbiased observers. Such a human-centred technology was not amenable to extensive commercialization. But the advent of reliable phototubes and electrical meters as commercially available components promised improvements of two types: first, lower costs by removing the need for numerous observers and second, more trustworthy results. This dual advantage led to numerous light-measurement devices for a host of applications.

There were two stages and two unrelated technologies behind the commercialization of photoelectric light measurement. First, detectors relying on the photoelectric effect were refined, particularly at research laboratories such as GEC's. Incorporating exotic materials in evacuated glass enclosures, and supplied with high voltage and monitored by sensitive electrometers (and, later, by galvanometers connected to valve amplifiers), these devices were suitable for some laboratory applications of photometry, but were considered by most to be too delicate for industrial use. Nevertheless, GEC in the UK and Westinghouse Electrical & Manufacturing Company in the USA targeted this market by constructing demonstration devices as diverse as photoelectric smoke recorders, newspaper bundle counters and automatic door openers[15]. By stripping away quantification and retaining merely the ability to *detect* light, these devices found a ready market. Thus, cultural needs translated this improbably fragile and high-precision technology into a reliable and attractive means of automation.

The second, and more financially significant, stage of commercialization was made with 'flat plate' photocells (figure 8.1)[16]. The first versions of these were simply variants of selenium, which practitioners had used sporadically since the 1880s. Relatively inexpensive and imprecise, these detectors were small and simple to operate. Quite suddenly, some five years after the commercial introduction of photoelectric tubes, instrument manufacturers began to market portable instruments employing improved variants of the selenium cell. Ironically, these relatively inaccurate sensors proved more successful than their predecessors in bringing quantification to industry[17]. The Weston Electrical Instrument Company in 1932 claimed to have introduced 'the first commercial dry disc

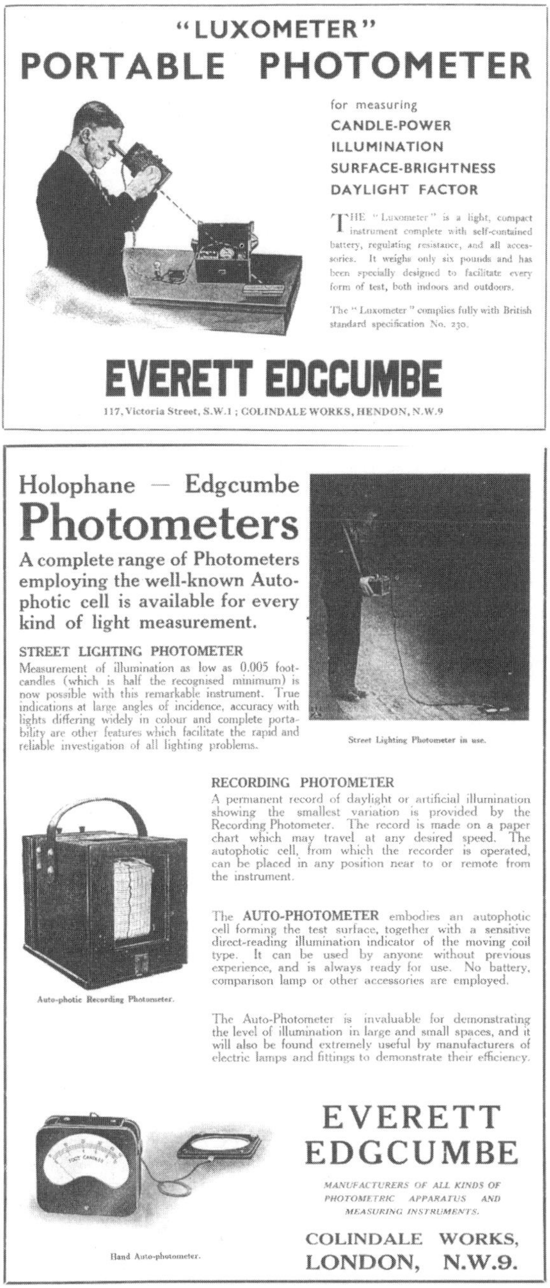

Figure 8.1. The shift of authority from eye to machine. Everett Edgcumbe advertisement for visual (top) and photoelectric (bottom) photometers. *Illum. Eng.* **24** (1931) xix and *Illum. Eng.* **28** (1935) 295.

Figure 8.2. Establishing status. Weston advertisement, *Illum. Eng.* **28** (1935) 26.

type' photocell under the trade name *Photronic*, and rapidly marketed a variety of portable meters based on it (figure 8.2)[18]. Such cells made practicable a variety of products owing to their small size and modest electrical requirements. Other manufacturers responded: Everett Edgcumbe & Co, for example, announced their *Autophotic* plate-type cell a year later[19]. Companies such as Salford Electrical Ltd used the same idea to produce a variety of instruments for light measurement. Commercial secrecy obscured the technical differences and relative advantages of these devices from the customer[20]. To differentiate their more elaborate and precise—and expensive—products from these flat-plate cells, manufacturers of the earlier devices dubbed them *phototubes*. Flat-plate photocells, unlike phototubes, were seldom sold as components because the flat-plate detectors comprised most of the cost of the simple photometers constructed from them. It was in the manufacturers' interest to exploit the technology by selling a complete product, which could have a considerably higher selling price than the detector alone. Moreover, the performance of such devices was not adequate for precise applications of the type performed in photometric laboratories; selling the components on their own would make their limitations more obvious to design engineers attempting to employ them. The commercial success of flat-plate photocells from the early 1930s is attributable as much to marketing as to technological advantages.

The technological benefits of the photoelectric detection of light were publicized on several fronts in Britain: by 1930, members of the NPL photometry department, gradually convinced of the practical superiority of such detectors to the eye, cautiously endorsed their use; their collaborators at the GEC Research Laboratory were demonstrating prototypes of commercial instruments and small firms were introducing portable photometers. As noted by one reviewer for *Nature*, 'the introduction of various forms of rectifier photoelectric cell has certainly simplified many problems in the use of instruments such as colorimeters (chemical type), densitometers and the like'[21]. In 1933, the Science Museum recognized this technical and commercial wave by mounting a three-month exhibition of photoelectric equipment[22].

8.3. LINKING COMMUNITIES

Who were the groups responsible for supporting this commercial growth of light measurement? The links between the communities of designers, producers and users of commercial light-measuring instruments were closely intermeshed, particularly in the early years. These communities interacted in ways that have received relatively little attention in the historiography of instruments or of modern science. While connecting a scientific revolution with the availability of commercial instruments, Baird does not clearly indicate how such inter-dependency operated. Similarly, Rabkin scarcely touches on the subject when he writes:

> The advent of serial, mass-produced scientific instrumentation increased the ease of exploitation. This led to certain alienation of the scientist from the actual design of the instrument, particularly in the 20th century.... However, even in earlier centuries the production of instruments, mainly for astronomy and physics, was often affected by non-researchers, popularizers of science or instrument collectors. This phenomenon may not be quite so recent.[23]

Historians have broached the interaction of technical communities, however, for other forms of instrument developed almost contemporaneously with photometers. Christine Blondel, for example, discussing the adoption of the D'Arsonval galvanometer in the latter decades of the 19th century, writes:

> At the beginning of the 1880s the scientific and technical territory of industrial electricity is not yet defined. There results, in fact, three intermingled paths, each making its interests felt: that of the inventor, the man of machines; that of the savant, man of the laboratory; and finally that of the manufacturer, subjected to the market and to competition, and who left his name only on the plates of his apparatus.[24]

Brian Gee has also explored the relationship between the scientific research worker and the instrument manufacturer, seeing it as fixed and determined by separate career paths: 'instrument makers descend from and are tied to their trade in the practical arts by the genealogy of master and apprentice'[25].

In the case of photometry, and perhaps generally for peripheral sciences like it, the relationship was instead a complex and changing one. The design and production of light-measuring instruments did not involve simply a one-way wresting of control from the hands of scientists to manufacturers. At least four types of relationship between the designer, the manufacturer and the user can be discerned:

(i) a scientific instrument maker constructing custom-made apparatus according to the user's specification;
(ii) an instrument company manufacturing apparatus developed by or for one user or community of users but made available to other practitioners;
(iii) a company marketing a device originally developed for its own use and
(iv) a firm developing and manufacturing equipment specifically for a perceived market.

Although there was a gradual development from relationships (i) to (iv), examples of each type can be found over the period covered, and indeed up to the present day[26]. Moreover, the definition of the terms 'manufacturer', 'designer' and 'user' varied in each case, although stabilizing considerably in the decade before the Second World War. Each term could refer, in specific instances, to a scientist, engineer, industrialist or lay-person, this interchangeability of commercial roles indicating from another perspective the seamless structure of the subject of light measurement. Some brief examples will illustrate the taxonomy of commercial relationships and introduce the firms active in the field.

Custom manufacturing

In Britain, scientific instrument makers had a long history of custom manufacturing devices based on the designs of scientists[27]. These instrument makers employed the technologies of their day and mastered new technologies as they arose. Continuing this tradition, some produced photometric apparatus. Among the earliest commissions of the Cambridge Scientific Instrument Co, for example, were 'colour mixers' and photographic light meters for William Abney[28].

Manufacturing designs in collaboration with designers

Popular photometer designs could be licensed by the original scientist-designer for sale to others, thus converting him from customer to profit-sharer, when instrument manufacturers perceived a wider market for a custom-made device. The arrival of gas regulation in the 1860s provided just such a market: the firm of William Sugg & Co manufactured photometers initially for the Metropolitan Board of Works, and the Harcourt pentane standard lamp was designed by one of the Gas Referees[29]. This apparatus was subsequently sold in a variety of forms to gas supply companies, the Board of Trade, and for export to customers as far afield as the Canadian government[30].

By the turn of the century, the manufacture of licensed photometric apparatus was an active, if limited, business. In collaboration with the PTR in Germany, for example, Schmidt & Haensch manufactured the highly successful Lummer–Brodhun photometer from 1892; Foote, Pierson & Co of New York manufactured the Ulbricht sphere integrating photometer under licence from its German designer and Kipp & Zonen in Holland manufactured photoelectric microphotometers and galvanometers according to the designs of W J H Moll. In Britain, Alexander Wright & Co manufactured photometric benches of a type originally supplied for the NPL, and themselves based on PTR models. They also supplied standard Harcourt pentane lamps which the NPL and British industry had adopted as an intensity standard, and even carried out the chemical refining necessary for the purified pentane itself[31].

Commercial adaptation generally began by seeking new *markets* for an existing design, rather than by modifying the design itself. Thus a 'lustre meter' designed for the Linen Industry Research Association was later marketed unchanged by the Cambridge Instrument Co to measure the surface gloss of any surface[32]. In the more complex or potentially more versatile designs, however, the manufacturer re-engineered the instrument for commercial production and new applications. The GE recording spectrophotometer of 1935, for example, was the commercial successor to prototypes constructed by A C Hardy at MIT from the late 1920s[33]. Contemporary publications document well the history of this product, indicating its unique status and enthusiastic reception[34].

Collaborations between the scientist–inventor and instrument manufacturer could benefit both, since the scientist obtained wide recognition for the design, the manufacturer extended his product range and markets, and both generally made money. The association with a prominent scientist could confer status as well as improved sales on the manufacturer. Just as importantly, recognition as a designer could be as important as conventional scientific publications in raising the esteem of scientists in this peripheral subject area. Both W J H Moll and A C Hardy, for example, were widely acclaimed by their peers as both innovators in instrumentation and as research scientists, roles that they cultivated by publishing several papers on their instrument designs[35].

Commercialization of an in-house development

Other products were brought out by companies that had developed them for internal use. An example of this form of commercialization is the Kodak Research Laboratory photoelectric colorimeter, designed to evaluate the characteristics of colour films[36]. The device proved useful to film processors and users as well as to manufacturers. This form of commercialization was restrained, though, for at least two reasons: manufacturers had little incentive to make available apparatus that could benefit their competitors, and such apparatus usually fell outside the product lines of the company.

Manufacturing for a perceived market

In the last decades of the 19th century, when enthusiastic amateurs were still able to make significant contributions, some devices were designed and then directly marketed by their inventors. The 'Tintometer' of Joseph Lovibond is an example of one such device that has seen continuous development for nearly a century[37]. A similar case is the colour books and instruments arising from the Munsell colour system[38].

The successful products of such lone inventors formed the basis of small firms. More frequently, however, an existing manufacturer developed light-measurement apparatus when it had mastered a technology and perceived a commercial need. A particularly early example of this is the Siemens & Halske selenium photometer introduced in 1875. The Hefner lamp, developed by the same company as a proposed standard for German photometry, had been preceded by earlier, less successful light sources. Photometric products were a small but nurtured sideline for this dominant electrotechnical company.

8.3.1. Extension of commercial expertise

As in the national laboratories before the war, two technological traditions became involved in commercial light measurement in the 1920s. The first was supported by optical instrument companies that previously had produced spectrometers and visual photometers, and the second by companies with expertise in electrical instrumentation.

Photometry via optics

In Britain, several optical firms entered the field of light measurement. Most of these came to manufacture photoelectric devices after having previously marketed versions relying on either visual or photographic technology. Adam Hilger & Co, for example, 'manufacturers of scientific instruments adapted chiefly for astronomy, mathematics and optics' since 1875, was producing microphotometers by 1906 to measure the optical density of spectrographic plates[39]. The photographic recording of spectra was now a routine operation in a variety of laboratory contexts, but practitioners required a means of reducing the data to a graph for quantitative analysis or for publication. Scanning photometers of a variety of designs—nearly all for photographic use—were offered by Kipp & Zonen, Cambridge Instruments Ltd, C F Casella & Co and Holophane, among others[40]. Some optical designs were manufactured long after more precise alternatives were available. Casella, for example, manufactured a visual 'extinction meter' for meteorological use after the Second World War[41]. The German optical company Carl Zeiss drew upon its experience as a manufacturer of microscopes and accessories to sell photometers. In a series of advertisements in 1922, they promoted their Pulfrich (visual) photometer for use as a colorimeter, nephelometer, glossimeter and photometer, claiming that it 'meets the requirements of the chemical, physiological, textile, paint and other industrial laboratories'[42].

Photometry via electronics

The second technical tradition becoming involved with photometry—that of electrical measurement—was supported by electrical equipment manufacturers.

Weston, an American company, and the British firms Salford Electronics and Edgcumbe & Co, had specialized exclusively in electrical equipment through the 1920s, but photoelectric photometry became a major interest by the early 1930s. Each benefited from prior experience in electrical measurement or from links with other sources of funding or technical expertise. Weston had a longstanding reputation for electrical standards; Salford Electronics was a subsidiary of GEC Ltd; and Everett, Edgcumbe & Co had links with photometry through co-founder Kenelm Edgcumbe's membership on the British Illuminating Committee and the Commission Internationale de l'Éclairage.

Among companies from the electrical tradition, the General Electric Company, both in America and England, was the most influential player in the inter-war period. Opening research laboratories in 1919, the British firm with that name, GEC Ltd, initially concentrated on lighting and photoelectric tubes. The American operations of General Electric Inc. delved into similar areas of measurement, although concentrating on photometric instruments and applications rather than components[43].

8.3.2. New practitioners

Besides the re-definition and consolidation of existing communities of manufacturers and users, commercialization caused wholly new groups to take up light measurement. These newly involved communities comprised designers, chemists and industrial engineers.

Instrument designers

The merging of optical and electrical traditions in instrument companies was embodied in individual scientists and engineers, with some designers becoming adept in a new subject that could be termed photoelectric engineering (as with the study of light measurement itself, the design of instruments did not have a cogent label, both subjects tending towards conjunctive prefixes such as 'electrotechnical', 'opto-electrical' and 'electro-optical'). This demanded an intimate knowledge of both electrical and optical sciences.

New publications in the early 1930s signalled the appearance of a self-recognized community of designers. The staff of the GEC Research Laboratory, attempting to convince engineers of the reliability of the photoelectric components that they had developed, and to encourage their use, wrote articles and books aimed at engineers and technically competent practitioners. At least one of these was aimed squarely at the nascent photometric engineering community: *Illuminating Engineering Equipment: its Theory and Design* promoted the use of photoelectric methods in a new generation of commercial products[44]. Such documentation extended the influence of the instrument makers to a second phalanx of practitioners, loosely binding these peripheral communities which still lacked the unity provided by courses and standards of training.

Chemists

Since the late 19th century, chemists had accumulated a growing body of knowledge concerning the measurement of chemical concentrations by colour changes. Nevertheless, as late as the First World War the term *quantitative chemistry* generally referred to 'wet' techniques such as gravimetric (weighing) and volumetric (measuring) methods[45]. *Indicator* methods relied upon noting the colour change of a solution to detect a change of—for example—acidity, and were inherently non-quantitative[46]. More general quantitative colorimetric analysis demanded standardized methods and benefited from instruments to ease the task of colour comparison[47]. Unlike photometers, visual colorimeters proved to be technologically undemanding and to have a large market. By 1942 'the number of colorimetric instruments on the market [was] unusually large'[48].

Production engineers

As manufacturers knew well, a convenient method of verifying the uniformity and suitability of many products is to observe their visual appearance. Discoloration of paper, mismatching of fabric colours and inadequate brightness of electric lamps had all been monitored by human observers since the turn of the century. Such visual verification was awkward to carry out on the industrial scale, as discussed in chapter 3, and engineers sought means of supplementing or replacing human observers by physical methods. The culture of industrial production could support this transition. Photoelectric measuring instruments may have been accepted in some factories and plants because of the earlier acceptance of cruder photoelectric sensing devices. For the industrial engineer, the knowledge required to operate and maintain a photoelectric paper-bale counter was little different from that needed for a paper-whiteness monitor. The employment of the new technology, and the staff to support it, could be self-perpetuating. By the mid-1930s one engineer reported that such usages were commonplace, and indeed that 'many miles of street lighting' were controlled by light-actuated switches, and that 'most of the large power stations' employed photoelectric smoke detectors[49].

By stepping back from the problematic physical quantification of light, the crude but simple applications of photoelectric detectors vied with the high-precision applications for the attention of industry

8.4. MAKING MODERNITY

The evolution of commercial photometry portrayed here suggests a technology-driven advance. But the commercial advance of photometry, radiometry and colorimetry was also fuelled by genuine industrial needs.

Probably the first major application of light measurement in industry was the measurement of temperature. The first non-contact method to become commercially important was radiation pyrometry. In this technique, a thermocouple or thermopile generates a voltage when illuminated by light from a hot object such as a steel furnace or pottery kiln. When coupled to a direct-reading indicator or chart recorder, the signal could directly indicate temperature. For materials hot enough to emit visible light instead of radiant heat or 'infrared', the industrial engineer could use optical pyrometry. In this technique the intensity of the sample is equated to that of the filament of a small electric lamp superimposed on the field of view. The current supplying the filament is calibrated in terms of source temperature. An alternative technique was colour-temperature measurement, in which the colour of the glowing body was either compared with a standard by eye or else monitored at two wavelengths by a physical detector. Optical, radiation and colour pyrometers and temperature recorders, researched at the national laboratories before the First World War, came into common use in chemical plants through the 1920s[50].

Some manufacturers saw the industrial application of colorimetry for verifying product colours as 'a matter of very great importance'[51]. From its early customers working in academic or government laboratories, the small photometry industry began to turn in the 1920s increasingly towards industrial laboratories and plants. By the 1930s, the measurement of light spanned applications from pure research to quality control in factories. Over 600 American companies manufactured industrial instrumentation, particularly temperature- and pressure-measuring devices. The fraction of instrument sales relative to all machinery increased even during the American depression[52]. Methods that had been used solely in the academic laboratory were applied to industrial problems. Chemists saw spectroscopy, in particular, as a new tool for the quantification of mixtures[53]. Transforming the method from a research technique used by academic physicists to chemists measuring the trace components of steel in a works laboratory demanded standardization and simplification. Practitioners combined photographic methods of recording with reliable, automated scanning densitometers to yield a viable industrial technique. By 1930, such visible spectroscopy was being supplemented by growing interest in infrared analysis. Chemists at large industrial research laboratories began to adopt infrared spectroscopy in the decade before the Second World War, a trend that accelerated rapidly during the war[54]. University research into the development of visible and infrared recording spectrometers expanded.

Photometry and colorimetry also began to diffuse from the research laboratories to industry. The new availability of what managers regarded as reliable and objective instrumentation led to wide-scale interest in applying quantitative light measurement to industrial problems. All applications calling for the evaluation or standardization of colour were affected. The textile industry, for example, began to employ colorimeters for matching the colours of dyed fabrics[55]; and paint manufacturers tested new formulations and the uniformity of production[56].

The adoption of light measurement by industry fed back into the technology itself. The requirements of industrial apparatus were different from their laboratory counterparts. For routine applications, equipment had to be robust, simple and reliable. Reliability demanded devices to be insensitive to environmental factors and to be stable over weeks or months. This, in turn, required that the optical detectors, electronic and mechanical components did not degrade with time—an impracticable goal, given existing phototube and thermionic valve designs. To overcome hardware limitations, designers used the strategy of correcting for imbalances, drifts and fluctuations. The need for 'self-compensation' of imperfections and the desire for automatic recording were rapidly combined into self-registering photometric instruments almost as soon as photoelectric methods of measurement became available[57]. As John Walsh had predicted, the greater precision of photoelectric photometry also allowed more rapid measurements, opening new directions of research[58].

8.5. BACKLASH TO COMMERCIALIZATION

Portions of the process industry, where analysts were trained, if at all, in more traditional wet chemistry techniques, received light measurement coolly. Indeed, the new photometric and colorimetric instruments appeared almost *too* easy to use by unskilled personnel, endangering existing jobs for chemists at industrial plants. One trade editorialist felt it necessary to calm concern by emphasizing the skill needed for photometric techniques:

> It may be mentioned that the fear of certain chemists that the introduction of a spectrograph into their laboratories might tend to prejudice their position and prospects is entirely without foundation. It is obvious that only a worker trained in the use and theory of scientific instruments could hope to control successfully the more delicate operations involved, and while unskilled workers can, and do, operate a kind of spectroscope in the sorting sheds of many steel works, it needs scientific training of no mean order to operate a logarithmic wedge sector and interpret the results correctly.[59]

While rejecting the idea that chemists should have to behave like physicists, the editorial called for both elementary and advanced training in optical methods for industrial application, noting that 'when the importance of applied optics generally is remembered, it is a matter of surprise that such has not already been done'[60].

The conservatism of users and their lack of training for industrial application of the techniques were not the only difficulties, because the ease of use was deceptive. Commercial light measurement proved to have associated technical problems. The instrument firms had marketed automated photometry and colorimetry as a straightforward method of increasing efficiency and reducing overheads in industrial applications. Like the scientists in the standards laboratories, however, workers in industry began to recognize unanticipated complexities in the new techniques.

Quantification did not always provide solutions. Discussing the automatic detection and recording of smoke levels from factories, one engineer noted:

> it is often considered—and with justification—that a qualitative record which merely shows 'smoke' or 'no smoke' is preferable to the quantitative record which indicates degrees of smoke density. Not only is it difficult to establish a calibration for all thicknesses of smoke strata, but any such device which is operated by the valve anode current depends for its accuracy on the constancy of that current which cannot be guaranteed throughout the whole of its working life.[61]

Moreover, physical photometers, just like the eye itself, were subject to errors that were not always obvious. Observing that 'photoelectric cells are good when used very cautiously, but are apt to lie "without blushing"', one designer vaunted the more faithful spectral, angular and linear characteristics of his own device[62]. The complexities of photoelectric devices were as mistrusted as visual methods had been three decades earlier.

The quantification offered by the manufacturers was increasingly seen as incomplete or misleading. Research into light and colour, particularly when related to real industrial situations, had enlarged the number of visual characteristics to be quantified. Besides the hue, saturation and brilliance of coloured light, the surfaces of real materials had optical attributes such as lustre, sparkle, luminosity and gloss. Discussing these problems, the chairman of the American Committee on Colorimetry wrote:

> [The modes of colour] are strictly phenomenal or experiential attributes, not reducible to physical terms, and demonstrable only by introspection. However... the conditions for their presence in consciousness can be specified objectively, if we assume the response system to be normal in its other stages.[63]

Separating the subjective and physical characteristics of light and colour was no longer just a problem for scientific committees: it was being faced daily and directly on the factory floor. Writing of his mixed experiences with colorimetric instruments, a representative of the Printing and Allied Trades Research Association (London) observed:

> Unfortunately, the spectrophotometer is a costly instrument and requires skilled operation: as a result, many so-called reflectometers, whiteness- and brightness-meters have made their appearance. In the commonest of these, light from the sample is received by a photocell, and readings are taken with red, green and blue filters in front of the cell; such instruments are inexpensive and simple to operate. It is not generally realized, however, that papers are not necessarily a good match even when the 'red', 'green' and 'blue' readings are the same; conversely, papers may be a good visual match and yet give different readings...it is not commonly appreciated in the trade that colour is 'three-dimensional', and that consequently no single instrument reading can define a colour.[64]

Contrasting earlier pronouncements, even the Head of Colorimetry at the NBS cautioned that physical methods were not a panacea:

> in spite of claims made by manufacturers and others using photoelectric cells the eye is often a better instrument than the photoelectric cell.... For certain portions of the spectrum they are much better than the eye, but in others, and in many problems in photometry, the chief advantage is speed.[65]

The measurement of light and colour was proving to be unexpectedly recalcitrant in converging towards a technological solution. Colour was a subjective sensation difficult to quantify and accord between different observers, let alone 'physical' instruments. The 1931 CIE specification of the 'standard observer' made possible the numerical expression of colours, but did not make colour matching any easier. Nor did it encompass the properties of surfaces. Two options were available: either to use human observers and visual photometers—i.e. to revert to conventional but tedious colour matching—or to employ physical photometers. The adoption of physical instruments could ensure more repeatable measurements, but at the expense of generality: their numbers were not necessarily related closely to the visual perception of appearance. The demand for rapid and reliable testing of products during the 1930s argued for physical methods, just as the testing of incandescent electric lamps had done in the national laboratories a decade earlier. Practitioners once again made the shift from physiological to physical methods. Their pragmatic solution was to develop specialized instruments to measure more of the awkward visual characteristics.

8.6. NEW INSTRUMENTS AND NEW MEASUREMENTS

The discussion of new communities of practitioners and technologies cannot be separated from that of new types of measurement. The new communities, in some cases, attempted new forms of quantitative light measurement, to which the firms in light measurement responded by selling instruments. In other cases, new technology made possible a measurement that proved widely useful to practitioners. The spectrometer manufacturer, Hilger, exemplified the latter

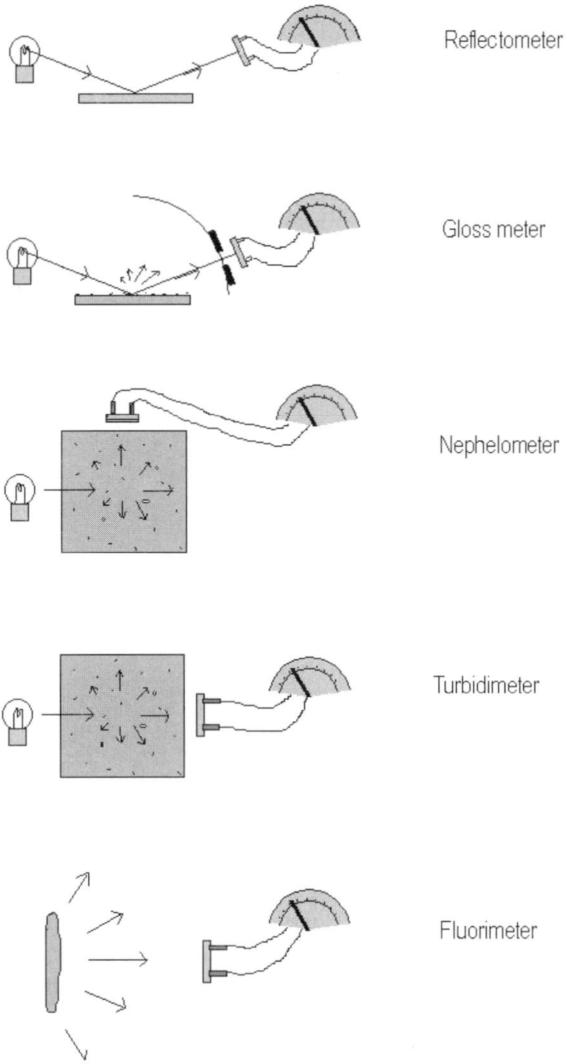

Figure 8.3. New types of photometric instrument commercialized between the wars. Rearrangements of light source, sample, photocell and meter generated new forms of measurement.

case, publicizing the technique of absorption spectrophotometry by publishing bibliographies of papers on the subject[66].

Photoelectric technology made practicable a variety of measurements that previously had been laborious or inaccurate (figure 8.3). But the measurement process had to be diversified. With a carefully designed instrument, the reflection of light from surfaces could now be quantified straightforwardly[67].

For surfaces that did not have a mirror finish, the surface texture caused light scattering. 'Gloss', this diffuse/shiny characteristic of surfaces, was important in the porcelain, cloth, ceramic and metals industries, and was measured by an instrument bearing the ungainly name *roughometer* in America and *glossmeter* in Britain[68]. From the early 1930s, Adam Hilger & Co manufactured the *blancometer*, a photoelectric instrument design to match nearly white surfaces of similar texture[69]. In it, light from an incandescent source was reflected into a photocell, either from a white magnesium oxide reference or from the sample under investigation. Adjustable wedges of graded transparency could be positioned to yield the same reading from both materials on an electrometer connected to the photocell. To determine the colour of the sample surface, coloured filters could be interposed in the light path to pass red, green and blue light. In another instrument, turbidity, a measure of the light transmitted by a liquid or gas containing particles, was employed to infer the size of dust particles[70]. The same principle was used in the closely related *nephelometer*, which measured the light *scattered* from liquids containing particles. This version proved popular in measuring the purity of water supplies. Other characteristics that had previously been estimated by eye gained dedicated photoelectric instrumentation, e.g. *fluorimeters* to measure the fluorescence from materials[71] and polarimeters to measure the polarization of light reflected from surfaces.

For most users, though, photoelectric methods remained a two-step process. The majority still employed photometric instruments principally for measuring the density of photographic plates. Scanning photometers for analysing photographically recorded spectra were the most common type of instrument developed in the decade before the Second World War[72].

8.7. PHOTOMETRY FOR THE MILLIONS

Spencer Weart has observed that 'the 1920s were a golden age of scientific faith, not only among scientists and industrialists but also for the public at large'[73]. The public, while able to marvel at the demonstrations of photoelectric devices, could not participate in this aspect of the golden age until inexpensive and simple devices became available[74]. Moreover, the entities measured had little relevance for the general public. But the disc-type photocells introduced in the early 1930s caused photoelectric technology to diffuse widely, multiplying the number of devices and users. Two products based on disc-type photocells proved immediately popular and were produced in numerous variants: *illumination meters*, used to measure the lighting level in buildings or on streets, and *exposure meters* for photography. Illumination meters were frequently calibrated in terms of the 'daylight factor', i.e. the fraction of illumination compared to unobstructed daylight[75]. Holophane, a major supplier of prismatic light fittings, also became the chief British source for light measuring instruments in the 1920s. In 1930 the company introduced a 'sill ratio meter' specifically to measure the daylight factor. Their promotional literature emphasized the legal importance of such a measurement, noting that the Prescription Act of 1832 endowed windows that had enjoyed free access of light uninterruptedly for 20 years with certain rights of light. Since 1865, attempts had been made to consider the questions involved in such cases in a quantitative manner. The shadows of tall buildings, increasingly common from the turn of the century, caused property depreciation and brought photometry to building law[76]. Holophane's solution was to compare the intensity of a uniformly bright or dull sky with that of the room by means of a sill-mounted visual photometer[77].

As discussed in chapter 2, early photographers had made little use of light measurement devices. Commercial 'exposure meters' had not had much success until the end of the 1870s, when gelatine plates manufactured with a predictable and sensitive response to light became widely available. A range of exposure devices trickled onto the market after that time, relying on a variety of technologies[78]. But the range of commercial exposure devices remained broad and static until the early 1930s. Yet when the photoelectric version first became available, it found a ready market in the growing hobby of amateur photography[79]. Physical light measurement entered the popular domain with the electrical 'exposure meter' having a dial calibrated in terms of film sensitivity and camera apertures. Photographers—a larger fraction of them enthusiastic but inexperienced amateurs now—began to recognize technical arguments for using such meters. Photographic films were much 'faster' than 50 years earlier, and camera shutter speeds covered a broader range: both factors increased the likelihood of over- or under-exposure. Errors in exposure could no longer be compensated easily by adjusting the development time of film, because photographers increasingly relegated this task to commercial laboratories. Films were, in any case, now too sensitive to view while the latent image appeared. There was an element of art as well. Inter-war photography had moved beyond merely candid reproduction; it was now inspired and extended by photographic artists such as Edward Weston, Man Ray and Alfred Steiglitz. Photographers now sought a richly graduated range of monochrome tones from deep black to palest white, which demanded close attention to exposure. But beyond all this, owning and using an exposure meter became a mark of status for the careful, modern (and affluent) amateur photographer. The success of such devices owed as much to consumer fashion as to technical benefit[80].

By the mid-1930s, simple physical photometers of this type were popular among engineers and photographers alike. A Swiss lighting engineer commented:

> The development of the inexpensive, fairly reliable and fairly accurate photovoltaic cell photometer was itself an item of major importance to the development of better lighting. For the first time, the travelling agent, the consulting engineer, the student of lighting, every person interested in establishing a record of an intensity of lighting was given the means to do so. The instrument is so much simpler than those previously used that these have been completely superseded for demonstration purposes.[81]

[Copyright.

MUSIC TRANSMITTED BY A LIGHT BEAM.

The wireless set on the left controls the brightness of a neon lamp connected up instead of a loudspeaker, which then gives light variations in place of music. The light is concentrated by a reflector into a thin beam and shines on a photo-electric cell in the small black box. The changes in light on this cell are amplified and reproduced on the loudspeaker. If the demonstrator places his hand in the light beam, the music stops. This apparatus in the picture will work up to a distance of 50 feet; with more powerful amplifiers greater distances can be covered.

[Copyright.

PHOTO-ELECTRIC PHOTOMETER.

In this elaborate machine the efficiency and light of electric lamps is measured by a photo-electric cell. The lamp is placed in the centre of the large sphere and a portion of the light falls on the cell in the smaller sphere. In this way a measurement of the total luminous output of the lamp is obtained.

Figure 8.4. Defining modernity. From 'The electric eye—the photoelectric cell', *The Wonder Book of Electricity* (circa 1932) pp 45 and 54.

Nor were photoelectric detectors confined solely to photometry. Many practising engineers found that 'the simplest applications of photocells are frequently the most useful ones'[82]. Inventors realized that the simple photocell could be integrated into ever more complex products produced in larger volume and with higher profit (figure 8.4). Even Albert Einstein co-patented an automatic exposure system for a camera[83].

8.8. A BETTER IMAGE THROUGH ADVERTISING

The advertisement of commercial light-measuring products had a significant influence on the status of the technology and its perception by the scientific and engineering communities. At the close of the First World War, photometry had been relatively stagnant; publications had fallen, and visual observing techniques had been taken close to their practical limits. The introduction of photoelectric technology to a wider community in the early 1920s was initially slow, as it appeared unreliable and complex. But, as Brian Gee has noted, for both contemporary scientists and historians 'the first appearance of an item in a trade catalogue often signals that research and development [has] reached the point of commercial viability'[84]. Advertising and commercial demonstrations not only documented this faltering subject, but transformed its image into one of modernity and control.

The earliest print advertisements, appearing in trade journals, simply publicized the availability of a type of apparatus. Established firms such as The Tintometer Co and Hilger & Co, for example, advertised in *The Journal of Scientific Instruments*. Advertisements for photometers by Alexander Wright & Son and Holophane appeared in *The Illuminating Engineer*. As competition for customers rose and new customers unfamiliar with the technology sought instruments, however, advertisements assumed a more didactic and propagandistic theme[85]. Ready-made apparatus for the neophyte began to appear. The Holophane company presented the *Lumeter* as the solution to the problem of measurement of the illumination from light sources, although no description was given of its principle of operation or method of use[86]. Instead, advertisements curtly provided the company address, the product name, and a brief description of the size, weight and intended use of the instrument[87]. Such advertising strategies not only literally 'black-boxed' the instrument, but attempted to 'black-box' the not inconsiderable operating complexity as well. Through the 1920s, the Lumeter was the only regularly advertised photometer in Britain. Its commercial success in a changing market is implied by frequent design updates. Such remodelling of designs was novel in a field that only a few years earlier had been commercially dormant, and soon caused it to rival the automotive industry in innovation. An advertisement claiming the Lumeter to be 'entirely redesigned, and a number of improvements made'[88], was followed a few months later by another announcing that 'the 1926 Model is now available conforming with all requirements of the new British Engineering Standards Association Specification No 230, 1925'[89]. Despite its commercial dominance the Lumeter, based on the visual comparison of an internally and externally illuminated screen, lost its privileged status the following year when inexpensive photoelectric meters began to appear. These newer devices stressed versatility for a variety of uses. The *Luxometer* of Everett, Edgcumbe & Co, for example, was advertised 'for measuring candle-power, illumination, surface brightness and daylight factor', making it capable of performing all the tasks required by practitioners of light measurement[90].

As quickly as manufacturers marketed the new instruments for physical photometry, their purchasers deployed them to convince the next tier of customers of their modern practices. An advertisement by Regants Lamps Ltd, for example, was aimed at optical manufacturers, and emphasized the scientific basis of their own production:

> The Regants glass is the only glass of its kind on the British market...come and see it in our laboratory. Test it out on our spectrometer. Get its spectral wave lengths. In your search for the better, GET THE BEST.[91]

The ability to *measure* and *illustrate* the transparency of glass became a selling

By courtesy of] [*Daily Herald, Manchester.*

COUNTING BY ELECTRICITY.

As each bundle of newspapers passes a photo-electric cell a beam of light is broken; this affects the sensitive plate in the cell and a record is made by means of an electric impulse.

Figure 8.5. Making light count. From 'The electric eye—the photoelectric cell', *The Wonder Book of Electricity* (circa 1932) p 21.

point. Light measurement was being co-opted to demonstrate the quality of other products, transferring its own heightened status to them. A similar theme is apparent in a 1932 advertisement that announced 'photoelectric cells from the "His Master's Voice" laboratories for efficiency and reliability'[92]. Such cells had had, even five years earlier, a reputation for precisely the opposite characteristics: irregular performance, poor uniformity and instability.

Demonstrations, more than print, served as an effective advertising medium. General Electric and Westinghouse devoted considerable engineering time to designing demonstration apparatus as well as to publicizing their products in advertisements, magazines and books. GEC demonstrated phototube technology with relatively undemanding exhibits. Typically, a beam of light shining on the phototube, when interrupted, would trip a relay to operate a motor or other device. These so-called 'electric eyes' found commercial application in the following decade as automatic door-openers. Other common applications included the counting of objects on conveyor belts (figure 8.5), and the detection of web fractures on paper-making and printing machines[93]. The Osram subsidiary of GEC also used photoelectric cells to advertise its products, producing several demonstration novelties to encourage the use of its cells by other companies[94]. In one such gimmick, a customer's hand picking up leaflets from a distribution box interrupted the light beam to ring a bell. In another, the demonstrator could use an electric hand torch to steer a model motor car by directing the beam onto one of two phototubes connected to corresponding thermionic valves and relays controlling a steering motor. These 'magic' demonstrations emphasized the qualities of automated seeing, effortless manipulation and action at a distance. Indeed, 'magic eye' became a popular and enduring euphemism[95]. In this way the phototube's potential for detection and control were brought home to a receptive public. As a direct result of such exhibits and portrayals, the trend to physical photometry grew during the following decade, and was virtually complete by the Second World War.

The commercialization of light measurement—that is, trade in instruments themselves—was thus one of the last and most powerful factors to shape its social presence. This economic dimension, fuelled by advances in technology, supported the most rapid evolution that the subject had yet undergone. For the first time, the measurement of light was convincingly portrayed and almost universally perceived as a useful and accurate technique for scientist and layman alike.

Yet the increased public profile and commercial success of light measurement was not solely, or even predominantly, a technology-driven affair. Indeed, the cultural invention of a *need*—that of industrial matching and testing—pre-dated reliable photoelectric detectors. Nor did the consensus regarding quantification alone impel its acceptance: the first commercial inroads were made by devices that merely *sensed* rather than *measured* light. Other, cultural, factors played an important role, particularly in the placing of an increased value on automation and standardization. By 1939, the term *photometer* was almost universally preceded by the adjective *photoelectric* in the titles appearing in instrument journals[96]. Photoelectric methods recreated light measurement as the very image of stability, accuracy and modernity.

NOTES

1 The commercialization of light measurement involved primarily goods rather than services. Although the national laboratories of Britain, America and Germany provided calibration and testing services, these were on a relatively small commercial scale and did not significantly influence the marketing of photometry. At the NBS, for example, assuming the full gamut of standardizing, candlepower and lifetime tests, the calibration of 1000 incandescent lamps brought in no more than $8000 annually. For the companies and commercial laboratories using such services, photometric testing represented a small fraction of their operating costs. This chapter therefore concentrates on the commercialization of hardware.

2 An echo of Gaston Bachelard's discussion of instruments as 'reified theories' [Bachelard 1933 *Les Intuitions Atomistiques* (Paris) p 140].

3 Baird D 1993 'Analytical chemistry and the 'big' scientific instrument revolution', *Ann. Sci.* **50** 267–90.

4 Anon. 1931 'Editorial', *J. Indus. & Eng. Chem.* **23** 1223.

5 Bud R and Cozzens S E 1992 *Invisible Connections: Instruments, Institutions and Science* (Bellingham) pp xii–xiii.

6 The term 'scientific instrument', following a working definition by James Clerk Maxwell and widely accepted in Britain, specifically referred to a piece of apparatus designed for scientific experimentation. This excluded identical instruments made for commercial or utilitarian purposes such as photometers for gas inspectors. See Warner D J 1990 'What is a scientific instrument, when did it become one, and why?' *BJHS* **23** 83–93.

7 Such growth is notoriously difficult to document. Reliable figures for the numbers of products available, quantities sold and prices have not been amassed. In the absence of such data, growth has been inferred from references in contemporary publications.

8 By 1925, with the dominance of electric lighting established, only mean spherical candlepower was much used, mean horizontal candlepower 'now recognized as having little or no meaning' [Anon. 1925 'Cube photometer' *J. Sci. Instr.* **2** 201].

9 G K Burgess, quoted in Cochrane 1966 *Measures for Progress: a History of the National Bureau of Standards* (Washington) p 269.

10 Bausch and Lomb in Rochester, NY, researched photometers and spectrophotometers with a total of 46 staff; the General Electric Incandescent Lamp Laboratory at Nela Park, Cleveland, employed 47 engineers and scientists and 59 support staff in the engineering and lighting research laboratories, where research included 'spectrophotometry, photometry, physical, biological, physiological, photochemical and psychological aspects of light utilization; the science of seeing, and many phases of color'; the Westinghouse Electric and Manufacturing Co (East Pittsburgh, PA) Lamp Division in Bloomfield, NJ, had an Engineering Department employing 108 staff including 34 engineers studying photometry and physical measurements, and its Research Department employed 15 for research including photoelectricity and spectroscopy; and the Weston Electrical Instrument Corporation employed 30 staff to 'develop instruments for measuring electrical... means for measuring light... and any quantity which can be made a function of an electrical quantity'. See Hull C 1938 'Industrial Research Laboratories of the United States, 6th edition' *National Research Council Bulletin No 102* (Washington, DC) pp 33, 90, 222 and 223. This survey undoubtedly underestimated the amount of research being performed, asking the companies themselves to judge whether their work was research or merely 'the improvement and development of products'. The efficiency of data collection is also uncertain: some 454 of the 1769 companies 'for various reasons did not find their way into' the 1933 edition.

11 Williams M E W 1989 'Crisis or complacency? The precision instrument industry in Britain and France, 1900–1920' in Blondel C, Parot F, Turner A and Williams M (eds) 1989 *Studies in the History of Scientific Instruments* (London) pp 273–81 (my translation).

12 This initiative attracted member firms specializing in either optical, electrical or x-ray instrumentation and had limited success. The organization continued with government support (owing to its identification as a 'key' industry) through the Second World War. While becoming peripherally involved in the design of photometric instruments, the association was of little importance to the commercial development of the subject in Britain. For details of the activities of BSIRA, see Williams *op. cit.* note 11, pp 85–9 and 123–36.

13 Rabkin, 'Rediscovering the instrument: research, industry, and education' in Bud and Cozzens *op. cit.* note 5, p 66.

14 National Academy of Sciences 1965 *Chemistry: Opportunity and Needs* (Washington, DC) p 65, quoted in Rabkin *op. cit.* note 13, p 66.

15 See The Physical Society and Optical Society 1932 *22nd Annual Exhibition of Scientific Instruments and Apparatus* (London) p 136, and Lance T M C 1932 'The electric eye—the photoelectric cell' in *The Wonder Book of Electricity* (London).

16 The financial success is inferred from the number of companies manufacturing or incorporating photocells rather than phototubes into products. Much of the commercial importance of phototubes centred not on the measurement of light intensity for scientific purposes, but rather for applications such as sound reproduction in talking pictures and the scanning of photographs for phototelegraphy.

17 Principally because of the simpler electronics and procedures needed to obtain 'a reading'.

18 The new cells were publicized in advertisements and in scientific articles which, however, revealed more concerning the cells' performance than their design. See,

for example, Romain B P 1933 'Notes on the Weston Photronic photoelectric cell' *RSI* **4** 83–5, and Shook G A and Scrivener B J 1932 'The Weston Photronic cell in optical measurements' *RSI* **3** 553–5. The name *photronic* found brief use as a generic term, thus reinforcing Weston's priority claim and helping to consolidate their market. The lack of constructional details, however, led practitioners increasingly to prefer descriptive terms and other manufacturers' detectors.

19 E I Everett, having served his apprenticeship at the Cambridge Scientific Instrument Co, left in 1884 and 12 years later founded Everett & Co. In 1898 he was joined by Kenelm Edgcumbe, with the new company specializing in electrical engineering instruments; see Cattermole M J G and Wolfe A F 1987 *Horace Darwin's Shop: a History of the Cambridge Scientific Instrument Company 1878 to 1968* (Bristol) pp 23–4. In 1934, the company collaborated with Holophane Ltd to produce 'Autophotometers' employing their Autophotic cells.

20 Besides the 'photronic' design, newly-marketed photovoltaic and photoconductive materials for cells in the early 1930s included cuprous oxide and lead sulphide. The photovoltaic cells generally comprised a metal disc coated on one side with selenium or cuprous oxide whose surface was covered in turn by transparent layers of metal and protected by lacquer.

21 Anon. 1936 'Clarity tester for gelatine' *Nature* **137** 861.

22 Anon. 1933 'Exhibition of photoelectric equipment' *Illum. Eng.* **26** 97. This included displays of the major types of photocell and their principles, and industrial examples such as package counters, burglar alarms, street lamp switching and daylight brightness meters.

23 Rabkin *op. cit.* note 13, p 59.

24 Blondel *op. cit.* note 11, pp 179–91 (my translation).

25 Gee B 1990 'On attending to the instrument maker in physics history', in Roche J (ed) 1990 *Physicists Look Back* (Bristol) pp 205–25; quotation p 217.

26 Mari Williams, in case studies of early 20th century instrumentation firms, has noted that no simple pattern of commercial innovation can be discerned. See Williams 1988 'Technical innovation: examples from the scientific instrument industry' in Liebenau H 1988 *The Challenge of New Technology: Innovation in British Business Since 1850* (Aldershot).

27 For the instrument-making trade prior to the 19th century, see Daumas M 1953 *Les Instruments Scientifiques aux XVIIe et XVIIIe Siècles* (Paris). For surveys of products and manufacturers of the following century, see Turner G L'E 1983 *Nineteenth Century Scientific Instruments* (London); Clerq P R (ed) 1985 *Nineteenth Century Scientific Instruments and Their Makers* (Amsterdam); and Payen J 1986 'Les constructeurs d'instruments scientifiques en France au XIXe siècle' *Arch. Int. Hist. Sci.* **36** 84–161.

28 Cattermole and Wolfe *op. cit.* note 19.

29 Dibdin W J 1889 *Practical Photometry* (London).

30 *Ibid.*, p 30.

31 Abady J 1902 *Gas Analyst's Manual* (London) lists Alexander Wright & Co as being able to furnish 'all the apparatus for testing gas and materials used in gas works'.

32 Anon. 1931 *J. Sci. Instr.* **8** 356–8. The company, founded in 1881, was the source of new instrument companies as well as instruments. Some of its former apprentices and managers formed W G Pye & Co (1895), Everett & Co (1896), the Foster Instrument Co (1910) and Unicam Instruments (1934). See Cattermole and Wolfe *op. cit.* note 19.

33 Hardy, Professor of Optics and Photography at MIT, was prominent in the field of colour research and spectrophotometry from the 1920s to 50s. He was a key member

of the Colorimetry Committee of the Optical Society of America which debated the nature of colour in the 1930s, as discussed in the previous chapter. His recording spectrophotometer and subsequent *Handbook of Colorimetry* were cited as playing 'pre-eminent roles in establishing the industrial use of colorimetry' [MacAdam D L 1981 'The Hardy recording spectrophotometer and the *MIT Handbook of Colorimetry*' in CIE *Golden Jubilee of Colour in the CIE* (Bradford) pp 19–22]. The voluminous data of the *Handbook*, like the earlier stellar magnitude catalogues of Pickering, persuaded practitioners of the reliability and applicability of the new method.

34 See Hardy A C 1938 'History of the design of the recording spectrophotometer', *JOSA* **28** 360–4; Michaelson J L 1938 'Construction of the General Electric recording spectrophotometer', *JOSA* **28** 365–71; and Gibson K S and Keegan H J 1938 'Calibration and operation of the General Electric recording spectrophotometer of the National Bureau of Standards' *JOSA* **28** 372–85. This instrument was quickly followed by other commercial efforts, including a compact instrument designed by the spectroscopist R W Wood for the Coleman Electric Company, and instruments by Beckman Ltd and Adam Hilger & Co.

35 For example Moll W J H 1921 'A new registering microphotometer' *Proc. Phys. Soc.* **33** 207–16; Moll W J H and Burger H C 1935 'Set of instruments for measuring spectral absorption' *J. Sci. Instr.* **12** 148–52; Hardy A C 1929 'A recording photoelectric color analyser' *JOSA & RSI* **18** 96–117.

36 Anon. 1933 *J. Sci. Instr.* **10** 116–18.

37 The Tintometer Co, founded in 1884, continues to sell photoelectric colorimeters at the beginning of the 21st century.

38 Upon the death of Albert Munsell in 1918, his son and wife extended the products of the Munsell Color Company to include a range of educational and measuring materials.

39 For more on Hilger, see Chaldecott J A 1989 'Printed ephemera of some 19th-century instrument makers', in Blondel *op. cit.* note 24, pp 159–68; A F 1897 'Adam Hilger' *Nature* **56** 34; and Cattermole and Wolfe *op. cit.* note 19, pp 141–3. On densitometers, see, e.g. Anon. 1936 'Photoelectric absorptiometer', *J. Sci. Instr.* **13** 268–9, manufactured by Hilger, and Toy F C 1930 'Improved form of photographic density meter' *J. Sci. Instr.* **7** 253–6. Various terms were used to describe essentially the same device: densitometer, photographic photometer or absorptiometer, with the prefix *micro-* implying an examining region smaller than about one millimetre. For a general discussion of microphotometers, see Walker R C and Lance T M C 1933 *Photoelectric Cell Applications* (London) ch 9.

40 For Casella, see Williams *op. cit.* note 11, pp 13–14.

41 C F Casella & Co 1948 'Gold visibility meter' *Meteorological and Scientific Instruments, Cat. No. 684* (London) p 16. The 'recycling' or retention of outmoded designs to satisfy a conservative market can oppose technological innovation, however. See P Brenni, 'The illustrated catalogues of scientific instrument makers', in Blondel *op. cit.* note 24 169–78.

42 Carl Zeiss advertisements 1922 *J. Indus. & Eng. Chem.* **14** 100, 142, 188.

43 For histories of GE relating to light measurement, see Wise G R 1985 *Willis R Whitney, General Electric, and the Origins of US Industrial Research* (New York) and Reich L S 1985 *The Making of American Industrial Research: Science and Business at GE and Bell, 1876–1926* (Cambridge).

44 Jolley L B W, Waldram J M and Wilson G H 1930 (London), and 1930 advertisement *Illum. Eng.* **23** 64b.

45 See, for example, Gooch F A 1916 *Representative Procedures in Quantitative*

Chemical Analysis (New York), and Szabadváry F 1966 *History of Analytical Chemistry* (Oxford).

46 Prideaux E B R 1917 *The Theory and Use of Indicators* (London).

47 See, for example, Snell F D 1922 *Colorimetric Analysis* (New York) and Strafford N 1933 *The Detection and Determination of Small Amounts of Inorganic Substances by Colorimetric Methods* (London).

48 Gibb T R P 1942 *Optical Methods of Chemical Analysis* (New York) p xiii.

49 Dobell C H 1936 *Trans. Illum. Eng. Soc.* **1** 143.

50 New product announcements and advertisements appeared, for example, in 'A compact form of optical pyrometer' *Chem. Eng. Works Chemist* **12** (1922) 167–8; **14** (1924) 183–4; **14** (1924) 208–9. See also Sosman R B 1922 'New tools for high-temperature research' *J. Indus. & Eng. Chem.* **14** 1369–74.

51 Barry H 1928 'Investigation of colour problems' *Chem. Age* **18** 319. For applications, see, for example, Draves C Z 1931 'Color measurements in the dyestuffs industry' *JOSA* **21** 336–46, and van Arsdel W B 1931 'Color measurement in the paper industry' *JOSA* **21** 347–57.

52 From 0.4% in 1919 to 1.4% in 1935. See Bennett S 1993 *A History of Control Engineering 1930–1955* (London) p 70.

53 Nitchie C G 1929 'Quantitative analysis with the spectrograph', *Ind. & Eng. Chem.* **1** 1–18.

54 Rabkin Y M 1987 'The adoption of infrared spectroscopy by chemists', *Isis* **78** 31–54, and Johnston S F 1991 *Fourier Transform Infrared: a Constantly Evolving Technology* (Chichester).

55 Nutting R D 1934 'The detection of small color differences in dyed textiles' *JOSA* **24** 135.

56 Benford F 1934 'A reflectometer for all types of surfaces' *JOSA* **24** 165.

57 For example, Randall H M and Strong J 1931 'A self recording spectrometer' *RSI* **2** 585–99, and Brackett F S and McAlister E D 1930 'The automatic recording of the infrared at high resolution' *RSI* **1** 181.

58 One new direction was the study of very short time scales in photometry made possible by the rapid response of phototubes. See, for example, McDermott L H and Cuckow F W 1935 'The time lag in the attainment of constant luminous output from tungsten filament electric lamps' *J. Sci. Instr.* **12** 323–7.

59 For a detailed description of the use of log-sector discs for determining the intensities of spectral lines (and thereby quantifying chemical constituents), see Gibb *op. cit.* note 48, pp 49–52.

60 Anon. 1935 'Industrial spectrum analysis' *Chem. Age* **33** 1.

61 Walker O J 1932 *Recent Applications of Absorption Spectrophotometry* (London) pp 132–3.

62 English S 1935 'Some properties of the cells used in Holophane–Edgcumbe Autophotometers' *Illum. Eng.* **28** 94–6.

63 Troland L T 1929 *Psychophysiology* (New York) Vol 1 p 254.

64 Harrison V G W 1941 'Physics in the printing and paper-making industries' *J. Sci. Instr.* **18** 103–9.

65 Gibson K 1930 'Progress in illumination' *Illum. Eng.* **21** 265–272; quotation p 271.

66 Walker *op. cit.* note 61 and Walker O J 1939 *Absorption Spectrophotometry and its Applications: Bibliography and Abstracts 1932 to 1938* (London).

67 Bergmann L 1933 'A practical photoelectric reflection meter' *Z. f. Tech. Phys.* **14** 157–8.

68 Salford Instruments Ltd 1937 'Comparative gloss meter' *J. Sci. Instr.* **14** 32–3. Other alternatives were glossimeter or reflectometer.

69 Anon. 1934 'New instruments' *J. Sci. Instr.* **11** 62.

70 Richardson E G 1936 'A photoelectric apparatus for delineating the size–frequency curve of clays or dusts', *J. Sci. Instr.* **13** 229–33. The technique came to the attention of many chemists through the paper by Tolman R C, Reyerson L H, Vliet E B, Gerke R H and Brooke A P 1919 'The relation between the intensity of Tyndall beam and concentration of suspensions and smokes', *J. Am. Chem. Soc.* **41** 300–3, which coined the alternative term tyndallmeter.

71 The fluorescence from radium intended for instrument dials had been the subject of an investigation at the NPL during the First World War, and employed visual methods.

72 For example Lees J H 1931 'A recording microphotometer' *J. Sci. Instr.* **8** 272–9 and Lance 1932 *op. cit.* note 15, pp 45–54.

73 Weart S R 1976 'The rise of 'prostituted' physics', *Nature* **262** 13–17; quotation p 14.

74 For example Lance 1932 *op. cit.* note 15

75 For example Barnard G P 1936 'Portable photoelectric daylight factor meter' *J. Sci. Instr.* **3** 392–403. The 'daylight factor' had been suggested by Alexander Trotter in 1895, and popularized by the NPL/DSIR studies by P J Waldram of Building Illumination from 1923. Room illumination 1% as bright as outdoors was deemed good, but $< 0.4\%$ poor.

76 Swarbrick J 1929 *Easements of Light: the Depreciation in Value of Property Due to High Buildings* (Manchester); Holophane Ltd 1930 *Illum. Eng.* **23** 19.

77 Anon. 1930 'The Holophane sill-ratio meter' *Illum. Eng.* **23** 278.

78 The devices in one important collection have been classified by their curator as either (i) exposure tables or calculators; (ii) tintometers, relying on the darkening of a standard photographic paper; (iii) extinction meters, employing apertures or absorbing filters to restrict the light reaching the eye to the threshold of detection or (iv) photoelectric meters. See Thomas D B 1969 *The Science Museum Photography Collection* (London) pp 37–44.

79 One of the first of these was the Weston 617 Universal Exposure Meter of 1931, which combined two selenium cells and a micro-ammeter. [Thomas *op. cit.* note 78 cat. no. 271] and The Physical Society and Optical Society 1935 *25th Annual Exhibition of Scientific Instruments and Apparatus* (London).

80 For contemporary descriptions of the new technology, see Harrison G B 1934 'Photoelectric exposure meters' *Photog. J.* **74** 169–77, and Nähring E 1938 'Photoelectric exposure meters' *Photog. Indus.* **36** 1358–62 and 1384–6.

81 Atherton C A 1935 'Comité d'études sur la pratique de l'éclairage' *Compte Rendu CIE* (London) p 653.

82 Walker R C 1936 'Some applications of light-sensitive cells' *Trans. Illum. Eng. Soc.* **1** 129–34; quotation p 132.

83 Einstein and Gustav Bucky, a radiologist, obtained US patent 2,058,562 in May, 1936 [Pais A 1982 *'Subtle is the Lord... ': The Science and Life of Albert Einstein* (London) p 495]. A cine camera marketed in Austria in 1935, the Eumig C-2, was the first to incorporate a photoelectric meter coupled to a lens aperture. Kodak sold a still-camera version from 1937 for the luxury market.

84 Gee *op. cit.* note 25, p 223.

85 A similar observation has been made about other types of industrial instrument in the inter-war period: 'Companies saw themselves as consultants and educators as well as suppliers of instruments' [Bennett *op. cit.* note 52, p 72].

86 The first version of the Lumeter was invented by J S Dow (a long-time officer of the Illuminating Engineering Society of London) and V H MacKinney in 1910. See Walsh J W T 1951 'The early years of illuminating engineering in Great Britain' *Trans. Illum. Eng. Soc.* **16** 49–60.

87 For example Holophane Ltd 1929 'The Holophane Lumeter' *Illum. Eng.* **22** 156.

88 Holophane Ltd 1926 'The Holophane Lumeter' *Illum. Eng.* **19** 30.

89 Holophane Ltd advertisement 1926 *Illum. Eng.* **19** 804.

90 Everett, Edgcumbe & Co advertisement 1931 *Illum. Eng.* **24** 226a.

91 Regants Lamp Ltd advertisement 1929 *Illum. Eng.* **22** 48.

92 The Gramophone Company 1932 *The Physical Society and Optical Society 22nd Annual Exhibition of Scientific Instruments and Apparatus* (London) p iv. Such cells were used in both sound films and experimental television systems from the late 1920s. See, for example, Burns R W 1991 'The contribution of the Bell Telephone Laboratories to the early development of television', *Hist. Technol.* **13** 181–213.

93 Walker *op. cit.* note 61. For a critical evaluation of 'electric eyes' by a firm specializing in visual photometers, see Fawcett A J 1951 *Electric Eyes: a Concise and Elementary Description of the Photoelectric Cell, for the Non-Technical Reader; its Uses in Industry, and its Uses and Short-Comings in Colorimetry* (Salisbury).

94 Walker and Lance *op. cit.* note 39, pp 81–3.

95 For example 'Eleven pairs of "magic eyes" have counted approximately 7,000,000 motor vehicles during the last year' [*Baltimore Sun* 22 February 1938 p 20].

96 A standard for flat-plate photoelectric cells was written during this period: *British Standard Specification for Photoelectric Cells No 586-1935*. Descriptions dealt with properties such as working voltage, colour temperature, ageing process, minimum sensitivity, maximum change of sensitivity, maximum slope, maximum dark current, frequency response and light flux incident on cell.

CHAPTER 9

MILITARIZING RADIOMETRY

Through the late summer of 1953, light was being measured on the bright Mojave desert of China Lake, California. The source was no longer a gas lamp, or incandescent bulb, or glowing pool of molten metal, or even the sun: it was a military jet, repeatedly approaching, banking and sweeping past[1].

There had been a side-step in this relocation. The quantification of intensity no longer seemed quite so important, but *detection* now mattered critically. And a shift in sponsors brought a shift in wavelength. By the end of the Second World War, photometry had largely stabilized in terms of standards, technology, institutional management and social specialization. Colorimetry, too, had attained several of the attributes of a stable subject. But the third specialism of this newly identified triumvirate—radiometry—was expanding disproportionately. Light measurement had broached a military dimension.

9.1. THE MYSTIQUE OF THE INVISIBLE

Until the early 20th century, radiometry had been the facet of light measurement least tarnished by the mundane, and the most imbued with an aura of exciting scientific discovery and mystery. This was due in no small part to the invisibility of the radiations detected. As discussed in chapter 2, the study of radiant heat had distinct historical origins and was, for some time, devoid of any compelling application. Nor was a connection between this elusive entity, affecting thermometers and other heat-measuring instruments, commonly connected with visible light[2]. Such factors tended to isolate the subject from the workaday concerns of photometry and colorimetry during the 19th century. Blondlot's investigations of n-rays, impelled by the turn-of-the-century scientific excitement at the discoveries of new and exotic radiations, were unusual in bringing photometric techniques to bear. But the vacillating methodology of Blondlot and his co-workers suggests just how tentatively invisible radiation was labelled as 'optical'.

Most investigators maintained a clear distinction between their research on invisible radiations and those of photometrists. Boundaries of several types existed: *occupational*, because radiometry developed in the exclusive domain of the physicist whereas photometry and colorimetry, as we have seen, had mixed parentage[3]; *workplace related*, because radiometric research was to be centred for a time at universities, the home of academic physicists; *application oriented*, because, as outlined earlier, it was divorced from practical utility; and in technical practice. Foremost in maintaining such distinctions of practice was the implicit rejection of direct human-centred observations. The complicated responses of the human eye were never an issue for radiometrists. Instead, they focused on investigating and developing physical detectors of radiation, and applying them either to discover more about the radiations themselves or in devising instruments to exploit the radiations. By unproblematically avoiding this perennial difficulty at the centre of photometry and colorimetry, specialists in radiometry had no difficulty in associating themselves with mainstream physical science. Nevertheless, the late 19th-century distinctions that set radiometry apart were to be reconstructed with the appearance of new sponsors and technologies.

9.2. MILITARY CONNECTIONS

The device-centred nature of this alluring research was eventually responsible for attracting an attentive sponsor: the military. The new sponsored research was, from the beginning, decidedly application oriented and new uses multiplied rapidly. The applications were bound up with the covert and clandestine—which unavoidably produces a patchy and unevenly weighted historiography. Military interest centred initially on the generation and detection of invisible radiation for signalling.

During the First World War, Theodore W Case in America found that sulphide salts were photoconductive (that is, altering in electrical conductivity according to the intensity of light falling upon them), and developed thallous sulphide (Tl_2S) cells. Their sensitivity, in fact, was principally to infrared rather than to visible radiation. Supported by the US Army between 1917 and 1918, Case adapted these relatively unreliable detectors for use as sensors in an infrared signalling device (and eventually patenting his 'Thalofide' cells in 1919). The prototype signalling system, consisting of a 60 inch diameter searchlight as the source of radiation (which would be alternately blocked and uncovered to send messages, akin to smoke signals or early optical telegraphs) and a thallous sulphide detector at the focus of a 24 inch diameter paraboloid mirror, sent messages 18 miles through what was described as 'smoky atmosphere' in 1917. The smokiness was not merely a passing observation: it was a strong selling point. A longstanding belief—largely unsubstantiated—about communicating and imaging with infrared radiation was that it was little affected by cloud, fog and smoke. This notion, widely repeated to and by the military for decades, promoted the technology's acceptance[4].

Nevertheless, Case's apparatus was not a success: the detectors were too irregular in performance to support even such a non-quantitative application. Their electrical response to radiation varied from cell to cell, and was proportional neither to the intensity of radiation nor to the applied voltage. Like visible light, invisible radiation was difficult to quantify. Work was discontinued in 1918; communication by the detection of infrared radiation appeared distinctly unpromising.

9.2.1. British research

Unlike their American counterparts, the interest of the British military centred on the detection of small aircraft by the heat they emitted[5]. Such an idea had been proposed by F A Lindemann (later Lord Cherwell) as early as 1916 but not taken up, and an investigation in 1926 by A B Wood of the Admiralty also looked unpromising. In 1935, R V Jones was developing infrared detectors at the Clarendon laboratory at Oxford under Lindemann's guidance. He was occasionally diverted from this work—intended for observations of the sun—to produce detectors for a retired American Navy inventor, Commander Paul H MacNeil, who was promoting his own version of an infrared detector of aircraft. While the MacNeil device was also unsuccessful, it reinforced interest at the Air Ministry, which in January of that year had set up a Committee for the Scientific Survey of Air Defence. Jones and an NPL scientist, J S Anderson, performed their own trials late in 1935, again with poor results. Detecting the radiation from hot engine surfaces appeared difficult.

The Committee nevertheless asked Jones to continue with full-time development, even if it was recognized to be a peripheral line of investigation. Unlike the concurrent radar research, which 'had a large research team... devoted to it', Jones 'for much of the time, had only [him]self'[6]. He devised equipment based on infrared detectors coupled to a small telescope, with signals amplified by a four-stage valve amplifier and indicated on a galvanometer—an arrangement employed tentatively in spectroscopy laboratories since the 1920s[7]. With various versions of the system developed over two years, it proved possible to detect single-engined aircraft from the ground at a distance of up to two miles, or about a half-mile air-to-air. Compared to radar, though, this radiometric equipment was incapable of detecting the range (distance) of aircraft. And experience belied myth: the detection was not effective through clouds. In March 1938, the small project was ended in favour of radar. A year later, however, Jones briefly joined a new Infra-Red Group, 'Group E', at the Admiralty Research Laboratory (ARL). E G Hill, its head, had earlier explored infrared signalling at HM Signal School in Portsmouth, and the group focused on such applications[8]. Infrared 'light' could be detected in some circumstances, to be sure—especially when emitted by cooperative targets—but appeared too weak to be measured for the planned military applications.

9.2.2. American developments during the Second World War

Aircraft detection initially excited little interest in America but, early in 1940, Theodore Case's idea was revived for a ground-based signalling system to be used during times of radio blackout. Several such projects were sponsored by the American government, which organized directed research for military applications. Through the war, this military sponsorship became wide-ranging and pervasive. In 1940, the National Defense Research Committee (NDRC) was formed to coordinate the funding of research for military purposes. One of its original four Divisions was 'Detection, Controls and Instruments'; within it, two of the four sections were 'Instruments' and 'Infrared Devices'[9]. The following year, after complaints that the NDRC was responsible only for researching and prototyping, and not for developing instruments to a manufacturable stage, the wider-ranging Office of Scientific Research and Development (OSRD) was set up, making NDRC a sub-section within it[10]. An Optics Division and Physics Division were formed with George R Harrison, formerly director of the Instruments Section, as Chief and Deputy Chief, respectively.

The NDRC drew upon some of the most prominent American optical scientists for its membership, bringing together physicists, colour scientists and psychologists for some projects. It also intermixed these specialisms and imposed a military impetus unseen in the previous war. Harrison led a 13-man team for 17 months as Chairman of the Instruments Section of the original NDRC. Of the three other Sections in the Detection, Controls and Instruments Division at that time, a Section alternately labelled 'Infrared Devices' or 'Heat Radiation' Section was chaired by A C Bemis and included infrared spectroscopist J D Strong with five other members and consultants. From December 1942 until the war's end, Harrison was Chief of the Optics and Camouflage Division, and Deputy Chief of the closely associated Physics Division. Optics included a five-member Infrared Section, a 13-member Illumination & Vision Section which included W E Forsythe and spectroscopists A H Pfund and H E White, and a Camouflage Section which included colour scientists and psychologists such as A C Hardy, L A Jones and E G Boring[11]. Significantly for the cognitive unity and management of these sections, Harrison had made his name in the inter-war years for his refinement of photographic photometry[12].

Even so, radiometry seemed to be a technology in search of an application. While the NDRC was the central organization in America responsible for wartime military-directed research, the Optics and Physics Divisions tackled miscellaneous problems, 'which had no particular relationship to each other and defied ready classification'. Indeed, reports the official history,

> except for a few instances, their work was almost entirely lacking in continuity... their primary goal was to create or improve any physical instrument which was needed by the Army or Navy which did not fall into one of the specialized, major fields of investigation.[13]

Nevertheless, among the wide range of sponsored wartime projects, the collective radiometry research and development were significant—'a program which ranks with radar as a prime example of the application of theoretical science to the practical problems of war'—and were to influence American post-war developments[14]. The work of Robert J Cashman, a physicist at Northwestern University in Illinois, was a fertile seed. Cashman had been extending Case's work on photoconductive thallous sulphide detectors since 1935. His efforts to develop a stable detector with reproducible characteristics in production were spurred by the knowledge that Case had seen such detectors in Germany in the early 1930s[15]. Cashman consequently received one of the first contracts from the NDRC—from among some 126 granted in December 1940—to make a systematic study of Thalofide cells, and his work was supported throughout the war[16]. The NDRC organized American military research on a new model which was to become the template for post-war funding: rather than requiring scientists and technologists to accept a military commission and to work at a government facility under military command procedures, as had been the practice in the previous war, the NDRC preferred to channel money via short-term contracts to existing groups at large institutions[17]. Through them, Cashman's research led to reliable production procedures for detectors. Further fundamental research on the cells started at MIT in 1943, and in 1944 the NDRC contracted General Electric at West Lynn, Massachusetts, to manufacture the cells. Within 11 months some 6800 had been produced with a reported 90% yield[18]. Other successful programmes included an infrared-guided bomb which used a bolometer as sensor, and heat-sensitive phosphors for sniperscopes and scanning systems used for the detection of heat-radiating targets. Several NDRC contracts directly benefited from, and publicized, such detector and infrared systems research, which led to 'nearly a score of infrared systems for a variety of highly specialized military applications'[19]. Few of these projects entered full-scale production during the war, but there was a hint that perhaps radiometry could be as applicable as photometry after all.

9.2.3. German experiences

Despite post-war claims by the NDRC that 'American scientists won by a wide margin in their race to be the first to make practical use of infrared light'[20], German work clearly surpassed it in pursuing new technical directions and concepts. A novel variety of infrared detector, the lead sulphide (PbS) photoconductive detector, had been developed in Germany from 1932 when Edgar W Kutzscher at the University of Berlin began to study them[21]. Like his British and American contemporaries, within a year he obtained military sponsorship—from the German Army, in this case. Kutzscher was Director of Infrared Research and Development of the Electracustic Company in Kiel during the war, where he and his teams developed the new detectors and infrared systems based on them[22]. Compared to British and American work, German infrared research was wide-ranging, theoretically based and innovative. Indeed, according to a 1944 report to the German Air Ministry, infrared homing devices were a more promising technology for missile guidance, owing to simplicity and technical advancement, than either radar or acoustic methods[23].

The breadth of development is suggested by the variety of wartime actors involved. A decade after the war, Kutzscher listed seven collaborators within his own company with whom he had studied the basic physics of detectors, materials and the atmosphere, as well as production techniques and applied systems engineering for infrared detection. Other techniques of fabricating infrared detectors were also developed by Bernhard Gudden[24] from the 1920s, Dr Gorlich at the laboratories of Zeiss-Ikon in Dresden, and others. The most successful wartime development was the 'Kiel IV', an airborne infrared system that—unlike Jones' English prototypes—had excellent range, and which was produced at Carl Zeiss in Jena under Werner K Weihe[25].

Developing such detection systems demanded a mixture of optics, electronics and materials science. German advances were made in materials that transmit infrared radiation (as glass transmits visible light). 'KRS5', a mixture of thallium iodide and thallium bromide, was developed by the Zeiss firm; infrared-transmitting 'Duran' glass was fabricated by Schott Glassworks. Other aspects of infrared systems were developed at German firms such as AEG, Kepka and Rheinmetall-Borsig. Yet Kutzscher stated that the design of efficient systems mated to their most important recognized potential application, the guidance of missiles, 'was not accomplished at the end of the war'[26]. Like the other combatants, the German military deployed only limited production runs of some infrared devices during the war, for example using the radiation reflected from targets such as tanks to direct guns and the *lichtsprecher* or optical telephone[27].

9.2.4. *Post-war perspectives*

These extensive German developments remained largely unknown to the Allies until after the war. While identified as a useful and potentially fertile wartime expedient, radiometry never received American funding remotely comparable to the technologies of radar, the proximity fuse, solid-fuel rockets or the atomic bomb; in Britain, its limited funding was a pre-war casualty. The 'night scopes' employed by US riflemen were credited with being responsible for 30% of the Japanese casualties in the early stages of the battle for Okinawa[28], but the technology of infrared measurement in both countries remained both technically and organizationally marginalized.

The sponsorship of this American wartime research appeared equally ephemeral. Its chief architect, Vannevar Bush, saw the NDRC as a temporary organization purely to deal with the requirements of the wartime emergency; it was already being dismantled in the final year of the war. In mid-1945, the Office of Scientific Research and Development was effectively replaced by the creation of a new body, the Research Board for National Security (RBNS). This was a joint board consisting of Army, Navy and civilian representatives to organize post-war research for military purposes. At about the same time, an Office of Naval Research (ONR) was formed by the Navy to provide continuity to maintain the wartime research impetus while other organizations still awaited approval, and to gain a central role in military research and development[29]. The ONR proved to be a liberal source of funding for civilian science, and became the principal contractor for fundamental research at universities in the post-war years.

The end of the war rapidly brought new information but just as rapidly closed off certain avenues for unclassified research. Cashman extended his studies in the early post-war years and discovered other lead salts that showed promise as detectors of infrared radiation[30]. But the wide-ranging German successes, newly uncovered by the Allies, did more than flesh out these findings: they redirected the thrust of research and development. The German trajectory of research was essentially the direction continued in the USA and Britain under military sponsorship after the war. From 1946, detector technology was rapidly disseminated to firms such as Mullard Ltd in Southampton, UK, as part of war reparations, and sometimes was accompanied by the valuable tacit knowledge of technical experts. E W Kutzscher, for example, was flown to Britain from Kiel after the war, and subsequently had an important influence on American developments when he joined Lockheed Aircraft Co in Burbank, California as Research Scientist[31].

Some aspects of this information were recognized as having considerable post-war potential and were classified. Where information about 'Metascopes', or night-vision devices based on infrared phosphors, was widely publicized as an example of American wartime ingenuity[32], information about 'heat detectors' became as invisible as the radiation itself. The 1948 history of the NDRC Optics Division reports tersely that 'the details of the actual adaptation of heat-detection principles to military needs are still locked in the files of the War Department'[33]. At a NATO conference discussing wartime German infrared homing devices a decade later, Kutzscher—now representing the Americans—spoke in intentionally vague terms of the physics of detection and deflected detailed questions with the statement that 'results of recent measurements are classified'[34].

Although infrared devices had seen only limited deployment by the Germans and Americans during the war, they appeared to show promise. How was a strong post-war development programme, supported almost entirely by military funding, justified? Four factors were prominent. First, the new military aircraft and missiles developed at the end of the war proved ideal targets for infrared sensors. Kutzscher's teams had studied infrared detection of reciprocating engines in aircraft, for which the hot exposed exhaust pipes were the principal source of infrared radiation. As the British had long realized, such heat sources could easily be shielded by engine shrouds[35]. The NDRC history of the American developments, in fact, fatalistically omitted any mention of such targets, describing its infrared systems as being 'instruments that could guide missiles toward the hot smokestacks of ships and factories', and reported that post-war investigations had found a similar Japanese heat-seeking bomb intended for 'the hot innards of ships in the invasion fleet'[36]. But the new jet aircraft and rocket motors, by contrast, produced more concentrated and hotter plumes of exhaust gases that were radiometrically bright and hence much more easily detectable by infrared detectors. Nevertheless, development programmes strongly encouraged fundamental research because improvements in the sensitivity of detectors translated directly into longer-range detection of these much faster-moving targets.

The second factor in favour of infrared technology was its ability to be used 'passively', i.e. by measuring the radiation emitted by warm bodies, rather than having to illuminate the targets with another source. This made infrared detection more covert than radar. Third, there was the German (and limited American) evidence that sensitive infrared detectors could be produced in volume and employed successfully in military contexts. And fourth, theoretical research was suggesting that considerable improvements in such detectors should be possible. Thus a newly ripe application, combined with manufacturing confidence and theoretical potential, created a new military market. Nevertheless, a fifth factor outweighed the others: these technical factors merely facilitated the general military pressure for tactical post-war advantage. The extension of radiometry was fuelled primarily by the political context of Cold War.

9.2.5. New research: beyond the n-ray

The military engagement with radiometry bore striking parallels with Blondlot's study of n-rays a half-century earlier. The very properties of their radiations were unclear, intriguing and communicated by hearsay. Both were concerned with *detection* rather than quantification. The radiation they sought was perpetually at the limit of detectability. Infrared detection systems, like Blondlot's laboratory assistants, merely signalled 'presence' or 'absence' of the elusive signal. And military designers shared Blondlot's philosophy of observation. There was little need to measure the size of the signal; what was important was to extend the threshold of detectability as far as possible.

An awareness of a greater potential for the technology emerged from spectroscopy. The spectroscopy community was eager to extend observations to ever-longer wavelengths (and correspondingly weak energy sources) of infrared radiation and to more difficult (e.g. thicker, more absorbing or more scattering specimens)[37]. But unlike the military, spectroscopists had a more central need to quantify their measurements. The co-evolution of commercial infrared spectroscopy for applications such as organic and analytical chemistry was another active research area immediately after the war, and was responsible for most of the published research at that time[38]. Research focused as much on instruments as on experiment, including several new types of infrared detector and studies of the ultimate sensitivity of such detectors[39]. While this work placed limits on the feasibility of infrared detection, it also demonstrated the gulf between practical systems and their theoretical potential.

9.2.6. New technology

Into the early 1950s, detectors developed in Germany included the thallous sulphide and lead sulphide (PbS); Americans added the lead selenide (PbSe), lead telluride (PbTe) and indium antimonide (InSb) detectors. British workers introduced mercury–cadmium–telluride (HgCdTe) infrared detectors. These developments were largely a product of military funding, but were available (if often expensive) to academic spectroscopists.

These devices rapidly found military applications. A guided aircraft rocket (the GAR-2) was in production from 1956; the similar infrared-guided Sidewinder missile was first used militarily against Chinese aircraft in 1958 (figure 9.1).

Figure 9.1. GAR-2A infrared Guided Aircraft Rocket (left) developed by US Air Force, beside James J O'Reilly, engineering test pilot for the Hughes Aircraft Company. In production from 1956, some 16 000 of the original design were produced. Reproduced with permission of HRL Laboratories, Malibu, CA.

By the early 1960s, the American military had missile guidance systems, fire control systems, bomber-defence devices, thermal reconnaissance equipment and others, all employing infrared measurement devices[40]. Despite such apparently rapid deployments and funding on a scale hitherto unknown by the scientific community, infrared research and development remained a rather secondary technology for the American military in the first post-war decade. As one early compendium on the technology reported,

> Infrared engineering, like radar engineering, has evolved under cover of military security. Many current applications are still highly classified, and details cannot be divulged... unlike radar, which received a monumental development effort during World War II, operational infrared has evolved rather slowly, on a limited-budget

> basis. With the advancement of military strategy into environments which are more favorable to the infrared technique, such as high altitude and space, infrared devices are receiving more serious attention.[41]

Indeed the American space programme, and particularly military projects for communications systems and the remote sensing of information by infrared radiation, maintained the momentum throughout the 1960s and 70s. In an environment free of atmospheric absorption and unhindered by earlier restrictions in project budgets, infrared radiometry research attained unearthly levels.

9.3. NEW CENTRES

Given the relatively large scale of American funding compared to its pre-war levels, it is not surprising that new loci of expertise in radiometry sprang up in the post-war years, mainly at military contractors. The government and private laboratories of the first decades of the century were joined by something different in scale and practice. The new laboratories operated by research and development contracts, and proliferated in proportion to military expenditure. For the writing of the 1965 text *Handbook of Military Infrared Technology*, some of the institutions and companies providing technical information were the Raytheon Co; Minneapolis-Honeywell Regulator Co; Westinghouse Electric Corp; Garrett Corp; Naval Ordnance Test Station; Barnes Engineering Co; Servo Corporation of America; Eastman Kodak Co; Air Force Cambridge Research Laboratories; Malakar Laboratories, Librascope Division; General Precision Co; A D Little Co; The RAND Corp; Texas Instruments Inc; Leesona Moos Corp; Infrared Detector Department of Radiation Electronics Inc; Engelhard Industries, Inc; National Bureau of Standards; Fish Schurman Corp[42]. Firms providing entire chapters for the text included Sylvania Electronic Systems; Infrared Industries Inc; Itek Corporation; Grumman Aircraft Engineering Corp and Mithras Inc. Many of these firms were located near the institutions that had benefited from wartime NDRC contracts such as MIT, and contributed to a growing belt of technology firms in the north-east USA.

In Britain, the principal government-directed research centre was the Radar Research Establishment (RRE) at Malvern (later the Royal Signals and Radar Establishment)[43]. Several British firms had research and development departments devoted to infrared work from the early 1950s, including de Havilland Propellers and EMI. Owing to the Official Secrets Act and government policy, their work was kept substantially separate.

Yet government-sponsored bodies organized interactions. Replacing the former word-of-mouth communication between academic physicists were new, more formal, structures. Organizations such as IRIS (Infrared Information Symposia) and IRIA (Infrared Information and Analysis Center) existed by 1961 to collate information from the large number of development projects. The following year the US Department of Defense further coordinated efforts by establishing the Joint Services Infrared Sensitive Element Testing Program

(JSIRSETP) at the Naval Ordnance Laboratory in Corona, California (later moved to the Naval Electronics Laboratory Center, San Diego). A sifting out of the firms participating in the original research projects eventually occurred. The major detector firm by the late 1960s was the Santa Barbara Research Center (SBRC), a subsidiary of the Hughes Aircraft Company[44]. Thus, apart from the University of Michigan, itself a major beneficiary of military contracts, the bulk of radiometric research was being undertaken by private firms. Previously centred in universities, radiometry had been redirected by the war to join photometry as a shadowy specialism outside the mainstream of academic science.

9.4. NEW COMMUNITIES

As the discipline was translated, so were its specialists. They increased in number, and the centre of mass was displaced from physicists to a new breed of appropriating specialist.

From a small group of researchers in the early 1950s, infrared meetings drew 500 to 1000 participants by 1965[45]. The collective biographies of these communities mutated as they expanded. The special status of physicists in the American and British military began to be eroded by the mid 1950s. By that time, although they were still valued for the development of novel instruments, their role as generalists—juggling information of markets, engineering, production expertise and strategy—had been grasped by electrical engineers[46]. The new catch-all subject of 'electro-optics' was becoming a more useful description. The *Handbook of Military Infrared Technology* mirrored this new concoction, acknowledging publications mainly of the IEEE (*Proc. IEEE, Proc. Inst of Radio Engineers*), the OSA (*Applied Optics, JOSA*), and, in Britain, the Institute of Physics (*J. Sci. Instr., Physics in Technology*). The editors categorized infrared detectors as a sub-category of 'modern optics' entwined 'intimately with the contemporary field of solid-state physics'.

Physicists continued to lose ground within this new specialism. The Society of Photo-Optical Instrumentation Engineers (SPIE, and renamed 'The International Society for Optical Engineering' in the 1980s), a small organization bringing together technologists primarily in the photographic and motion-picture industries in the post-war years, was transformed by an influx of researchers benefiting from military contracts. The initial connection was for specialized cameras and tracking devices to monitor missile launches. Gradually, however, these new 'electro-optical engineers', versed in mechanical, optical and electronic design to varying degrees, began to work with radiometric systems. The military component was so significant that some SPIE meetings were restricted to American citizens during the 1970s and 1980s.

Thus, unlike photometry and colorimetry, radiometry by the 1960s arguably *did* succeed in attracting its own appropriating specialist community—the electro-optical engineers. While sub-fields became concerned with the specialism, optical engineers had the strongest claim to control it—from theory, to design, installation and operation of its technology. That electro-optical engineers took over this role can be attributed to the dominance of bountiful sponsors and controlling applications—governments funding military usages.

9.5. NEW UNITS, NEW STANDARDS

The specialism of radiometry adapted to its new sponsorship not only by shifting its occupational locus, but by altering its language and technical guideposts: its measurement units and standards.

To some—who were developing an infrared version of what had previously been optical technology, such as optical telegraphs and optical telephones—a connection with photometry had seemed natural, if implicit. The US Navy specified the sensitivity threshold of Metascopes in terms of specified sensitivity in terms of ungainly 'nautical-mile-candles'. During Cashman's wartime work on thallous sulphide cells, infrared sources were calibrated in terms of *visual* response, sometimes in Hefners or foot-candles. As one chronicler states, an NDRC contractor 'chose to adopt a system of *photometry* for the infrared', constructing 'analogies to photometric concepts... such as the "holocandle" and "infrawatt"'. By the late 1960s such quantities were derided as 'cumbersome concepts' long discarded in favour of direct, energy-related units[47].

The very notion of a reference standard was also problematic. As turn-of-the-century photometrists at national laboratories had found, a good standard of brightness had to be very similar to what was being evaluated. Gas lamps had to be compared with flames; electric light bulbs needed to be compared with other glowing metal filaments of similar temperature. The distribution of radiation also generated its own 'standard units': gas lamps were amenably described by 'horizontal candlepower', while incandescent electric lamps were more suited to 'spherical candlepower'. So it was with military aircraft. But the nature of aircraft as sources of light is complex. The leading surfaces of a jet aeroplane or missile are heated by aerodynamic friction, and emit infrared light something like a blackbody source. Jet and rocket nozzles are much hotter. And the exhaust gases themselves are often a combination of blackbody radiation and 'emission' lines (strong radiation of isolated wavelengths due to chemical species in the burning fuel). Indeed, the spectral distribution of radiation could serve as an accurate 'signature' of the airborne body unique to it. In such circumstances, the inter-comparison of instruments was difficult. 'Traceability of instrument performance to the National Bureau of Standards is more and more a real question', noted William Wolfe, editor of the *Handbook of Military Infrared Technology*[48]. Calibration of the detection equipment was therefore a fraught process involving a combination of crude laboratory comparisons, theoretical estimates and expensive field trials.

The very form of the units also changed to suit new circumstances. The new light sources of interest did not remain at rest on a laboratory optical bench; aircraft and rockets, soldiers and tanks changed distance, angle, orientation and apparent shape. Consequently the units of radiometry ceased to be adequate. Why should investigators be concerned with the *total* power (the 'radiant flux', in watts) emitted by a light source or the power emitted from its *surface* ('the radiant emittance' or 'exitance', W/m^2), when its size and even distance might be unknown? When 'sources' became uncooperative 'targets', new measurement philosophies and units gained relevance. All were based on what could be measured by the detector rather than on how the light source could be manipulated. The power falling on the detector ('irradiance', W/m^2), the power radiated into a solid angle ('radiant intensity', W/sr) and, given the luxury of knowledge of the target size, the power radiated into a solid angle per area of the source ('radiance', W/sr m^2) became the new values of interest[49]. This shifting of consideration from source to detector has parallels with illuminating engineering, which had moved from the characterization of sources to that of reflective surfaces (roads, walls and windows) some 50 years earlier.

9.6. COMMERCIALIZATION OF CONFIDENTIAL EXPERTISE

9.6.1. New public knowledge

By the early 1960s, the large number of firms and technologists connected with infrared technology demanded a wide distribution of information. Civilian applications were also sufficiently widespread to promote popular articles and texts. The major source of information, however, was the *Handbook of Military Infrared Technology* sponsored by the US Office of Naval Research, and contracted by the Advanced Research Projects Agency (ARPA) of the American military. ARPA had contracted the University of Michigan to supervise the writing of the book[50]. Given the military background to this work, it is unsurprising that many of the sources of information were connected with the analysis of targets. Among the sources of information and acronyms were: BAMIRAC, the Ballistic Missile Radiation Analysis Center; TABSAC, the Target and Backgrounds Signature Analysis Center; BAC, the Background Analysis Centre [all at the Institute of Science and Technology, University of Michigan]; RACIC, the Remote Areas Conflict Information Center, the Battelle Memorial Institute in Columbus, Ohio; CINFAC, the Counterinsurgency Information Analysis Center, American University, Washington, DC. Radiometry, the central subject of the book, was extended to the meteorology of clouds, properties of the atmosphere, vegetation and ground covers, tracking system design, linear systems engineering, thermal coatings and optical materials.

This compendium was updated as the ostensibly civilian *Infrared Handbook* in 1978[51]. In it, military connections with radiometry were distinctly downplayed. Chapters on 'Targets' and 'Backgrounds' were subsumed into 'Artificial Sources' and 'Natural Sources'. The technology was recast as less aggressive: descriptions of 'Control Systems' gave way to 'Warning Systems'. The sponsor remained, however, the Infrared Information and Analysis Center—a 'Defense Logistics Agency administered Department of Defense Information Analysis Center' and supported by American defence contracts. Similar research and development programmes were instituted in the Soviet Union, and produced similar technical compendia, both overtly and covertly military in origin[52].

Only from the 1990s, with the end of the Cold War and the search for new markets, did firms transfer their energies frankly to civilian applications of radiometry.

9.7. A NEW BALANCE: RADIOMETRY AS THE 'SENIOR' SPECIALISM

While having distinct origins from those of both photometry and colorimetry, radiometry began to subsume the other two specialisms as it mushroomed after the war. The sources of this coalescence were threefold:

(1) the general acceptance of visible and invisible radiation as electromagnetic, and analysable by conventional physics in terms of energy and wavelength;
(2) the strong unifying effect of measurement standards and
(3) the existence of an integrating sponsor.

The combination of a cognitive viewpoint with government-directed applications was a common feature of post-war science. Having an affluent sponsor moulded the measurement of light and colour. It promoted the majority of research and applications for two decades, supported the integration of research at disparate companies and institutions and controlled the communication and publication of such research. Strong ties were irresistible. Government sponsorship transcended boundaries: it broke down the occupational boundaries by mixing specialists; removed workplace-related boundaries by encouraging new research environments in well funded private laboratories; promoted novel applications and equally new technological collaborations and lowered technical boundaries by supporting novel solutions.

Thus the story of radiometry between 1930 and 1970 can be summarized as being impelled by military funding and actioned by a plethora of firms in Germany, America, Britain and elsewhere. The post-war subject was based on the theoretical trajectory launched by German wartime studies and the NDRC organizational/funding model. As much as late 19th century photometry and early 20th century colorimetry, radiometry from mid-century was the product of formal organizations acting in a particular social and cultural context.

NOTES

1 The China Lake Naval Ordnance Test Station, used for tests of the US Navy's Sidewinder air-to-air missile, was one of several sites used for post-war radiometric observations. Other important locations were the White Sands Proving Ground in New Mexico, the Redstone Arsenal Complex (US Army) in Alabama and the salt flats of Utah.

2 I will avoid the term 'electromagnetic radiation' (a connection first mooted by James Clerk Maxwell's work), which suggests an anachronistic identification between visible light and invisible radiations that was seldom pressed by non-physicists before the First World War.

3 The close connection between 'pure' and 'applied' physics for combined photometric and radiometric research at the PTR is a national and temporal exception to this occupational separation.

4 More careful atmospheric research later showed the limitations, as for other forms of radiation, caused by absorption and scattering by atmospheric molecules.

5 Jones R V 1961 'Infrared detection in British air defence, 1935–38' *Infr. Phys.* **1** 153–62; Jones R V 1979 *Most Secret War: British Scientific Intelligence 1939–45* (London) especially ch 4.

6 *Ibid.*, p 161.

7 In America, the technique of infrared spectroscopy spread substantially from two centres: the National Bureau of Standards at Washington, DC, and The Johns Hopkins University some 50 miles away. William Coblentz at the NBS had been measuring infrared absorption spectra of materials since the turn of the century. At Johns Hopkins, the research group of Harrison Randall concentrated on developing instrumentation and extending measurements to ever-longer wavelengths. The group also devoted considerable effort to improving methods of detecting radiation. The thermocouples they used were conceptually the same as those used in the previous century. Randall's group developed schemes for discounting the effects of changing temperature (which caused the thermocouple voltage to drift). This perturbation from outside disturbances was the major limitation in measuring infrared intensity. Just as importantly for acceptance of the techniques, Randall's collaborators developed recording spectrometers. These early systems had to be proven to give results as accurate and repeatable as manual measurements. For an account of this crucial American work, see Randall H M 1954 'Infrared spectroscopy at the University of Michigan' *JOSA* **44** 97–103.

8 Jones *op. cit.* note 5, pp 46–50.

9 Stephenson H K and Jones E L with Harrison G R (ed) 1948 'OPTICS: a History of Divisions 16 and 17, NDRC' in *Science in World War II: Applied Physics* (Boston).

10 Leslie S W 1993 *The Cold War and American Science: The Military–Industrial–Academic Complex at MIT and Stanford* (New York).

11 Stephenson *et al op. cit.* note 9.

12 See chapter 6 note 37.

13 Stephenson *et al op. cit.* note 9, p 196.

14 *Ibid.*, p 199.

15 Lovell D J 1971 'Cashman thallous sulfide cell' *Appl. Opt.* **10** 1003–8.

16 Kevles D J 1978 *The Physicists: a History of a Scientific Community in Modern America* (New York). On the NDRC, see pp 297–303.

17 Zachary G P 1999 *Endless Frontier: Vannevar Bush, Engineer of the American Century* (Chicago).

18 Stephenson *et al op. cit.* note 9, pp 232–5.

19 For example H E White at the University of California under NDRC contract developed a portable infrared optical telephone; other projects were carried out at RCA Indianapolis; Baird Associates, Cambridge, Massachusetts; and a Navy speech/communications system was developed at Northwestern University. Lovell *op. cit.* note 15; Stephenson *et al op. cit.* note 9.

20 Stephenson *et al op. cit.* note 9, p 227.

21 Significantly, British research neglected photoconductive detectors. Jones [*op. cit.* note 5, p 160] claims that this was a consequence of long research at a 'Government Establishment' which showed them to have poor sensitivity for hot targets, and because the detectors would have required cooling to be effective.

22 Kutzscher E W 1956 'The physical and technical development of infrared homing devices', in Benecke T and Quicke A W (eds) *History of German Guided Missiles*

Development (Brunswick, Germany) pp 202–17.

23 *Ibid.*, p 201. Acoustic methods relied on detecting aircraft range and direction from the sounds of their engines.

24 For his pre-war work, see Gudden B 1928 *Lichtelektrische Erscheinungen* (Berlin).

25 Kruse P W, McGlauchlin L D and McQuistan R B 1962 *Elements of Infrared Technology: Generation, Transmission and Detection* (New York) pp 6–7.

26 Kutzscher *op. cit.* note 22. This may have been disingenuous, given Kutzscher's post-war employment by an American military contractor with an interest in continued development, and the predominance of Electroacustic's technological solutions in American infrared compedia through the 1960s.

27 Jamieson J A, McFee R H, Plass G N, Grube R H and Richards R G 1963 *Infrared Physics and Engineering* (New York).

28 Stephenson *et al op. cit.* note 9, p 243.

29 Schweber S S 1988 'The mutual embrace of science and the military: ONR and the growth of physics in the United States after World War II', in E Mendelsohn, M R Smith and P Weingart (eds) *Science, Technology and the Military* (Dordrecht) pp 3–45; Sapolsky H M 1990 *Science and the Navy: the History of the Office of Naval Research* (Princeton).

30 Cashman R J 1946 'New photoconductive cells' *JOSA* **36** 356.

31 Bower T 1987 *The Paperclip Conspiracy: the Battle for Spoils and Secrets of Nazi Germany* (London) p 149; Kutzscher *op. cit.* note 22. On the same subject see also Judt M and Ciesla B (eds) 1996 *Technology Transfer Out of Germany After 1945* (Amsterdam).

32 The Metascope was, in fact, a development of an ultraviolet-radiation imaging device developed by university physicists, and not an innovation of wartime research.

33 Stephenson *et al op. cit.* note 9, p 245.

34 Kutzscher *op. cit.* note 22, p 217.

35 Jones *op. cit.* note 5, p 154.

36 Stephenson *et al op. cit.* note 9, p 200.

37 See, for example, Perfect D S 1924 'Some instruments for detecting infrared radiation' *J. Sci. Instr.* **1** 312–29, 353–3. Randall H M 1954 'Infrared spectroscopy at the University of Michigan' *J. Opt. Sci. Am.* **44** 97–103. Palik E D 1977 'History of far-infrared research I. The Rubens era' *J. Opt. Sci. Am.* **67** 857–64; Ginsburg N 1977 'History of far-infrared research II. The grating era, 1925–1960' *J. Opt. Sci. Am.* **67** 865–71.

38 Ballard S S 1951 'Spectrophotometry in the United States', in *Proc. London Conference on Optical Instruments* (London) ch 13; Randall *op. cit.* note 37; Edlen B 1966 'Frontiers in spectroscopy' *JOSA* **56** 1285.

39 See, in particular, Golay M E 1947 'A pneumatic infra-red detector' *RSI* **18** 357; Golay 1949 'The theoretical and practical sensitivity of the pneumatic infra-red detector' *RSI* **20** 816; Hornig D F and O'Keefe B J 1947 'The design of fast thermopiles and the ultimate sensitivity of thermal detectors' *JOSA* **37** 474–82; Jones R C 1947 'The ultimate sensitivity of radiation detectors' *JOSA* **37** 879–90; Fellgett P B 1949 'On the ultimate sensitivity and practical performance of radiation detectors' *JOSA* **39** 970–6; Smith R E, Jones F E and Chasmar R P 1957 *Detection and Measurement of Infrared Radiation* (Oxford).

40 Jamieson *op. cit.* note 27, pp 4–5.

41 *Ibid.*, p 1.

42 Wolfe W L 1965 *Handbook of Military Infrared Technology* (Washington).

43 Gummett P 1988 'The government of military R&D in Britain' in Mendelsohn *et al* *op. cit.* note 29, pp 481–506.
44 Hudson R D Jr and Hudson J W (eds) 1975 *Infrared Detectors* (Stroudsburg).
45 Wolfe *op. cit.* note 42.
46 Schweber *op. cit.* note 29, pp 5, 35–6.
47 Lovell *op. cit.* note 15.
48 Wolfe *op. cit.* note 42, p v.
49 American National Standard Institute 1986 *Nomenclature and Definitions for Illumination Engineering* (ANSI Report); Rea M S (ed) 1993 *The Illumination Engineering Society Lighting Handbook* (New York, 8th edn).
50 Wolfe *op. cit.* note 42. The idea of publishing unclassified information was conceived in 1961, and information was collated between 1962 and 1963.
51 Wolfe W L and Zissis G J (eds) 1978 *The Infrared Handbook* (Washington, DC). The introduction [pp vi–vii] hints at the cultural differences between communities: 'Nomenclature uniformity was...difficult to obtain. Our first rule, of course, was to define the terms as they are used. The most troublesome technical word was "intensity". Most astronomers use "intensity" or "specific intensity" as a term referring to the distribution of flux (or radiant power) with respect to area and solid angle. We use "radiance" for this. Workers in the fields of electromagnetic theory often use "intensity" when they refer to the distribution of flux with respect to area alone. We use "irradiance" or "exitance" for this. We use "intensity" only for referring to the distribution of flux with respect to solid angle.' The 'we' behind the revised *Handbook* continued to be American electro-optical technologists supported by US military contracts.
52 For example Kriksunov L Z and Usoltsev I F 1963 *Infrakrasnyye Ustroystva Samonavedeniya Upravlyayemykh Snaryadov [Infrared Equipment for Missile Homing]* (Moscow); Margolin I A and Rumyanstev N P 1957 *Fundamentals of Infrared Technology* (Moscow, 2nd edn); Bramson M A 1968 *Infrared Radiation: A Handbook for Applications*, transl. by B Rodman (Plenum). The latter two texts are devoid of any mention of military applications.

CHAPTER 10

AN 'UNDISCIPLINED SCIENCE'

When George Biddell Airy called for 'some notion or measure of the degree of darkness' during the eclipse of 1858, he had a variety of techniques in mind. His immediate contemporaries, though, were little motivated to mathematize light and colour. Not until a quarter-century later did a strong pulse of interest develop for quantitative light measurement[1]. As previous chapters have shown, the dilatory transition from qualitative 'notions' to quantitative 'measures' of intensity developed into an 'undisciplined' science: a subject without widely recognized professional underpinnings or intellectual coherency. But was it as atypical a science as it seems? This chapter argues that the episodic evolution of the subject illuminates quite common, but under-represented, features of science in the professional period.

10.1. EVOLUTION OF PRACTICE AND TECHNIQUE

The history of light measurement cannot be told neatly in terms of intellectual challenges or experimental discovery. It involved relatively few academic scientists and laboratories. Nor can it convincingly be told as a Whig history—a tale of steady progress towards comprehensive and sophisticated understandings. But the story *is* intimately bound up with the growth of institutions and technical professions, and with shifting scientific cultures.

Consider the technical 'problems': accounting for the disappointingly fickle response of the human eye, oft conceived as the final arbiter of brightness; overcoming the confusion of units of measure; employing contentious 'standards' of intensity which could be maintained only to relatively poor physical tolerances; replacing the eye by seemingly more promising physical detectors which introduced new complexities of their own to the measurement process. Three important technical transitions were promoted more by faith than by substantiated advantage:

(a) the widespread identification of quantification as a desirable goal around the turn of the 20th century;
(b) the supplanting of visual by physical methods from the late 1920s;

(c) a convergence of the techniques used for measuring light, colour and invisible radiation by the Second World War.

Even in outline form, these problems hint at a strong social component.

For colour measurement these hints are more explicit. Practitioners seeking utilitarian application of colour metrics faced one key problem: to standardize a meaningful description of colour despite the vagaries of the human eye. To do so, they consciously limited the boundaries of their subject. Replacing the substantial complexities of human colour perception by a nominal 'standard observer', they were able to construct a framework within which quantitative analysis was possible. But colour measurement, even after the 1931 standardization, remained contentious: the approximations misrepresented and limited the description of more complex colour properties. The standardization was unsatisfactory for psychologists, for whom the utilitarian advantages were of little consequence and avoided the deeper issues of colour perception that they and philosophers wished to address. The quantification of colour was, then, seen by the Second World War as a pragmatic accomplishment—a convenient makeshift suited to the dominant technical sub-culture.

The evolution of these intellectual features of light measurement can be viewed as a gradual convergence, selection and stabilization. From a collection of isolated communities (including astronomers, gas inspectors and photographic researchers), the practitioners moved towards a shared viewpoint favourable to quantification and to the physical methods of measurement that facilitated it. There was a *convergence* of ideas regarding how light and colour should be described and treated. A greater number of scientific communities became familiar with light measurement as the technology developed, and began to accept the goal of quantitative measurement of light intensity and colour[2]. But this trend towards quantification cannot be seen as a natural progression; rather, the desire for measurement is a consequence of particular cultural goals emphasizing the comparison and standardization of goods and services[3]. The general acceptance of quantification implicitly involved *selection* of concepts deemed important. Thus the assurance of uniform manufactured goods and demonstrably adequate lighting was widely perceived as being more worthy of attention than, for example, a poetic, aesthetic or psychologically meaningful vocabulary of light and colour[4]. Such self-limiting standards *stabilized* the subject and aided consensus.

A second factor in the convergence of practice was the underpinning of the new conceptual objectives by technological development. Investigation of the photoelectric effect allowed the realization of physical photometry. Practitioners (mainly engineers and physicists) deemed the modelling and ultimate replacement of human visual characteristics by physical analogues—even averaged and highly simplified models—as important in enabling applications of light and colour measurement. Hence the ready acceptance that the photocurrent produced by illuminating a phototube was a measure much like human vision—even a superior measure, in that it was unaffected by other human characteristics such as fatigue.

There was a clear shift in authority from eye to machine. The consensus of the practitioners in all communities on this point is indicated by the rapid transition from visual to photoelectric methods, which occupied a period of scarcely 15 years. Within a portion of the career of a practising scientist or engineer, then, the measurement of light was transformed from a human-centred to an instrument-centred activity. Even so, widespread acceptance of such detectors hinged not on their ability to quantify but rather on their facility to automate.

A third determinant in the convergence of practice was the portrayal of light as a particular manifestation of electromagnetic radiation. Through the 1930s the subjects of photometry, colorimetry and radiometry were increasingly being lumped together[5]. For example, the opening pages of W E Barrows' (1938) *Light, Photometry and Illuminating Engineering* detail respectively the electromagnetic spectrum, spectral energy distribution curves of light sources and the spectral sensitivity of the eye. This format became *de rigeur* for books on colour by the Second World War. Colorimetry—now described as mapping the effect of particular wavelengths of radiation on visual perception—came to be viewed as a sub-set of photometry (defining and measuring the intensity of 'white', or eye-averaged, radiation) which was in turn seen as a particular case of the more general practices of radiometry (measuring the intensity of radiations of any wavelength). Such a hierarchical linking carried implications about what constituted valid methods of observation and analysis. Interpreting the human eye as merely one form of energy detector strongly supported the argument for physical methods. Wolfe, the editor of *The Handbook of Military Infrared Technology* (1965), reiterated the point for radiometry:

> The chapters of this Handbook are arranged in a sequence that is now almost traditional, and it is logical. The radiators come first, then the medium of propagation, the receiver system, the transducers and electronics, and finally a number of special applications....[6]

The seeming 'common sense' of this categorization is a reflection of the dominance of physics in the hierarchy of 20th century science.

These intellectual changes to the subject were implicitly social in motivation. The other deciding factors in the subject's evolution were overtly social and cultural in origin:

(a) adoption of photometry for illuminating gas inspection circa 1860, with an emphasis on uniformity of practice;
(b) a shift in interest towards electrotechnical uses after 1880, when electric and gas lighting systems began to compete, and promoting higher precision;
(c) rise of the illuminating engineering movement circa 1900, having the 'scientisation' of photometry as a major goal;
(d) research at government laboratories from circa 1900, and at industrial laboratories a decade later, tasked with the standardization of intensity to promote national industries;
(e) efforts at regulation and definition of the light and colour by delegations during the inter-war period;

(f) commercialization and industrialization of photoelectric instruments after 1930 and
(g) a second wave of commercialization based on military radiometry from 1950 to 1970.

10.2. THE SOCIAL FOUNDATIONS OF LIGHT

The social changes in the practice of light measurement during the early 20th century can be characterized as a transition towards an increasingly cooperative enterprise involving progressively larger groups of practitioners. This emergence of collective activity did not represent merely a rising popularity for increasingly standardized techniques, but rather the growing organization of separate communities. The growth of organization among academic scientists has been discussed, for example, by Donald Cardwell, who attributes the British case to 'a highly successful take-over bid for science and scholarship generally' by universities, converting the subject from the domain of amateurs to career educators and researchers[7]. This interpretation neglects the utilitarian concerns that motivated the development of light measurement. More pertinent illustrations concentrating on the case of American and British electrotechnics have been given, for example, by David Noble, Thomas Hughes and Graeme Gooday[8].

The most convincing successes of the subject were *social* successes: light and colour measurement provided a means of standardizing discussion. Astronomers could compare observations; inspectors could pass or fail lighting installations; industrialists could match and specify tints. Light measurement promoted scientific communication and unity by facilitating such common bases. On the other hand, the main thrust of the quantitative method—its numerical specification and arithmetic manipulation of intensity values—can be seen as having been less encompassing and fruitful. Practitioners repeatedly voiced concern about the ability and desirability of replacing the unreliable human eye by an unrepresentative physical measurement, and this was paralleled by the discovery of imperfections of the physical methods themselves. Human vision remained inextricably part of the process of light measurement, whether manifested in a human observer or as a disembodied table of average visual response.

Light measurement was a subject shaped by socially mediated processes. This is perhaps unsurprising for a study which, at heart, relies upon the relationship between the practitioner and human sources of data[9]. But it is also a specialism located outside universities. The most widely accepted models of scientific development still accepted by most scientists, however, neglect the role of peripheral subjects such as photometry and colorimetry, denying their place in the taxonomy of science altogether.

Karl Popper, for example, emphasizes the intellectual interplay between hypothesis and its experimental refutation in scientific change[10]. While observing that 'the growth of scientific knowledge may be said to be the growth of ordinary human knowledge *writ large*', he downplays the social factors in the creation of scientific knowledge. From this perspective, applied science and technology are merely applications of hard-won facts. Issues central to the field of light measurement—the roles of communities of practitioners, technological innovation and cultural pressures—receive scant attention. Indeed, light measurement can be assimilated only with difficulty into the Popperian view of science.

The second popular picture originates with Thomas Kuhn, who sees science as a series of 'normal' periods interspersed with revolutions in scientific orthodoxy[11]. 'Normal' science, a cumulative process of accreting new facts onto an existing theoretical framework, is interrupted when the scientific community decides collectively that new facts can no longer be incorporated. At this point, a new framework is established that replaces, either in whole or in part, the old one. The change in world view may redefine which 'facts' are important and make the previous views incomprehensible. The importance of the social component in this scientific development is evident. Indeed, Kuhn stresses that

> scientific knowledge, like language, is intrinsically the common property of a group or else nothing at all. To understand it we shall need to know the special characteristics of the groups that create and use it.[12]

His analysis nevertheless centres on theory rather than experiment and practice. For Kuhn, experimental science is an adjunct rather than a central component of scientific advance. His history of the blackbody laws, for example, stresses the development of theories to the almost complete exclusion of experiment—a case which David Cahan has convincingly shown to have been motivated by utilitarian concerns[13]. More particularly, Kuhn's views of quantification relegate it to a secondary role in the development of science. In normal science, he argues, measurements reveal 'no novelty in nature', but merely make explicit 'a *previously implicit* agreement between theory and the world'[14]. This view neglects the role of quantification in making possible a discourse—in providing a language of description and comparison. Light measurement in Kuhnian terms is distinctly peripheral in scientific importance, fulfilling at best a verificatory role[15].

The history of light measurement shows the centrality of cultural factors in determining the choice of scientific topics studied, the methods employed and the investigators who study them, and thus the selection of which facts, from the pool of 'natural' knowledge, are pursued. Indeed, some of the cases argue that the resulting knowledge is itself culturally moulded—that beliefs, in the words of John Law, 'might have been otherwise'[16]. The significance of this social shaping is seen most clearly in the case of colour, in which the complexities of human perception were progressively simplified and normalized to make them amenable to quantification, a goal having particular value in 20th-century consumer society. Similarly, physical photometry was socially transformed from a complex technology dubiously related to visual perception into a powerful means of automating industrial processes. Some examples of artificiality are obvious: light measurement seems to have attracted progressive 're-mappings' of observation into highly abstract and clearly 'constructed' quantities, e.g. the CIE chromaticity coordinates, or astronomers' Hertzsprung–Russell diagrams. This 'seduction of simplifications and conventions' may be a more ubiquitous feature of knowledge-production than generally acknowledged[17].

The social perspective can be extended further for fresh insights. Bruno Latour and Michel Callon, for example, describe the development of science and technology by an 'actor–network' theory. In the language of Callon all factors influencing the practice and development of a science are actors that interact through networks[18]. These actors and networks operate at many levels: for the subject of light measurement some of the principal actors can be identified as the CIE, the human eye, incandescent lamps, Alexander Trotter and photometers. The networks comprise interactions of varying importance between humans, instititutions, instruments and the scientific subjects. The inclusion of non-human factors as protagonists in a story couched in terms of battles of control is what distinguishes the Latourian perspective from social constructivism *per se*[19]. Indeed, to limit the analysis to human actors—to the social dimension—is as misleading as restricting it to a discussion of mere technology, suggests Latour.

Perhaps Latour's most fertile theme is his claim that historians often mistake the *direction* and *complexity* of cause-and-effect relationships[20]. Thus the monitoring of gas supplies for illuminants and the changing emphases in astronomy influenced the technologies adopted for comparing light intensities rather than vice versa. That is, photometry during this period was impelled by the cultural invention of problems—the 'need' for stable gas supplies and for reliable catalogues of stellar magnitudes, respectively—rather than by the availability of new technology. Similarly, the creation of photometric standards made possible the growth of new scientific communities, rather than being a consequence of cooperating, pre-existing communities. And instead of the properties of human perception solely defining the single, 'correct' science of colorimetry, the subject was fashioned by social, technological and historical factors. Overturning our expectations, colorimetry defined which aspects of human colour perception were deemed significant and which should be ignored.

Latour's emphasis on the importance of the laboratory as a key feature of scientific development has some relevance here. He has argued, for example, that Pasteur was able to convince his critics of his microbial research by converting cow fields into laboratories, where experimental variables could be strictly controlled[21]. In the case of light measurement, the marshalling of laboratory techniques by workers of the late 19th and early 20th century had more ambivalent effects: on the one hand, observational methods were refined there; on the other, a raft of new 'problems' and nonlinear effects were identified. The primary point of contention for colorimetry was not the production of *facts* but the production of a coherent *subject*. Rather than disputing the reliability and meaning of experimental evidence—the products of laboratory work—the historical actors differed in their opinions regarding the range of evidence to incorporate into their subject (i.e. defining the scope and borders of colorimetry). Physicists frequently judged psychologists' 'facts' and organizing principles to be irrelevant to constructing the subject and vice versa. Moreover, the criteria defining good measurement were reshaped by different communities. Thus, issues of competition are curiously off-centred in this peripheral subject. Points of contention, such as a recognition of a *need* to quantify light, and the utility of human versus physical measurement, were played out over decades during which the scientific communities changed as much as the questions they posed did.

In discussing how technoscience is shared between large and small actors, Latour further suggests that the trend is inevitably towards agglomeration and the eventual control of a subject by players that can marshal the greatest resources; small countries, for example, lack autonomy[22]. Replacing the word *country* by *astronomical community* or *illuminating engineering fraternity*, however, it is clear that this trend is not universal. Sub-cultures need not merge or even grow into internally sufficient entities to control a subject. In the case of light and colour measurement, they merely mutated the subject to suit their own ends—ends such as the pragmatic and particular scale of magnitude adopted by astronomers or the colour charts employed by bird fanciers or automobile manufacturers. These communities experienced no pressure to converge as long as their goals of quantification were expressed in particular and local terms. Light and colour measurement consistently failed to achieve autonomy.

10.3. A PERIPHERAL SCIENCE?

The *immiscibility* of these communities is an enduring feature of the subject. As noted in the last chapter, boundaries related to occupation, workplace, application and technical practice kept them separate. From the late 19th century onwards these communities fitted imperfectly into the disciplinary map. Neither scientists nor engineers claimed the subject (or subjects) as their own. What qualities relegated the subjects to the margins of scientific discourse? In what ways was light measurement different?

10.3.1. On being at the edge

Photometry and colorimetry were, over the period covered in this work, 'on the side-lines', and 'on the borderline of interest' rather than 'at the frontier of knowledge'. That is, they occupied a region between recognized disciplinary sciences (e.g. physical chemistry or hydrodynamics) and something else, identified by its practitioners alternately as a technique, a technology or an applied science.

Sciences have commonly been described as 'peripheral' in a geographical sense[23] or in circumstances of inadequate funding or resources[24]. Some definitions of 'marginal' science have been proposed having resonances with 'peripheral'. For Thomas Gieryn and Richard Hirsch, a scientist is 'marginal' if young or if recently migrated from another field[25]. They cite an earlier definition of a marginal scientist as one who is 'a cultural hybrid... living and sharing intimately in the cultural life and traditions of two distinct people'[26]. Jonathan Cole and Harriet Zuckerman have explored this definition, distinguishing between those subjects that are consistent with a 'central discipline', such as molecular biology or sociobiology, and those that are 'cultural hybrids' spanning science departments. They suggest that the hybrid type encounters more initial resistance from practitioners than the 'centrally based' type[27]. Nevertheless, their case studies show that the hybridisation invariably is transitory; the fields inevitably coalesce to form self-contained disciplines. Similarly, David Edge and Michael Mulkay cite three forms of marginality in the early history of radio astronomy, a field recognized as a discipline within two decades of its emergence[28].

These characterizations are inadequate for discussing light measurement. The equating of peripheral science as 'new science' is inappropriate, because photometry arguably remains a 'science on the side-lines' even today. Nor was it either geographically or economically marginalized.

The failure to achieve autonomy was a central characteristic of the subject of light measurement and one that sets it apart from disciplinary sciences. Previous sociological studies of scientific disciplines reveal the particularities of this case study. To paraphrase G Lemaine *et al*, disciplines during early stages loosely define the research problems, and results are open to widely differing interpretations. With specialization, agreement tends to increase, consensus grows, publications occur in more specialized journals, the proportion of references by authors not centrally engaged in research declines markedly and a small number among the many early papers come to be viewed as paradigmatic and get cited regularly. Research areas develop in response to major innovations as well as from government support and university expansion programmes. The rate, direction and intellectual content of development depend on such social factors[29]. This list of attributes accords only weakly with the history of light measurement, which corresponds only to the first of the preceding stages. At best, it appears as a discipline suffering arrested growth.

As for the case of radio astronomy, it has been common to postulate a connection between discipline formation and the maturity of a subject. According to this model, 'specialties' eventually and inevitably evolve into disciplines. John Law, for example, identifies three types of specialty and distinguishes between 'mature' and 'immature' specialties. A 'method-based' specialty such as x-ray crystallography is defined 'on the basis of shared scientific gadgetry'; 'theory-based' specialties have a shared formalism and 'subject-based' specialties have members working on a particular subject matter[30]. Law suggests that the first two of these are later stages in development than the third. Such an evolutionary path is inappropriate for peripheral science. While the subject of light measurement arguably could be labelled as a subject-based specialty, it cannot be said to have achieved 'maturity on a basis of shared methods' or 'on a basis of shared theories'[31]. Despite the shared subject matter, and the eventual practical consensus on photoelectric techniques, light measurement has remained a tenuously defined 'specialty'—but it does not follow that this makes it immature. In the same vein, Nicholas Mullins denotes Law's former two cases as being at the 'cluster' stage, and the latter as at the 'network' stage, with specialties seen as growing from nuclei of researchers bound by communications, colleagueship and co-authorship[32]. Having successfully traversed these stages, he says, a subject becomes a specialty, 'an institutionalized cluster which has developed regular processes for training and recruitment into roles which are institutionally defined as belonging to that specialty'[33]. These prior studies have all stressed the importance of an academic nucleus, if not in the early emergence of a new phenomenon, then in its development into a coherent discipline[34]. The emphasis on clustering highlights the insufficiency of Mullins' model for a peripheral science: it is the lack of a single centre that distinguishes light measurement from the case studies that these authors cite.

10.3.2. Technique, technology or applied science?

If a peripheral science lacks the central attributes of an academic science, is it, then, merely technology? I have used the term in previous chapters to describe aspects of the subject, but it is inadequate to characterize it fully. Previous attempts to distinguish science from technology, e.g. by Derek de Solla-Price, have been unconvincing, and this is particularly so for light measurement[35]. In distinction to his definition of technology, the field of light measurement was arguably a 'papyrocentric' activity and one closely associated with astronomy and spectroscopy, although lacking both discipline and an active network of co-citation. Barry Barnes has argued that, in any case, science and technology cannot easily be separated, and that neither is subordinate nor wholly reliant upon the other[36]. The subject of photometry also lacks some of the characteristics commonly associated with technology such as developing primarily in response to market forces. Light measurement cannot be relegated to mere technology or tool-making because only in the latter part of the period studied (after 1920) was some photometric research funded solely and directly for commercial ends (e.g. GEC phototube research); several aspects of the subject had little commercial or industrial motive, for instance photographic photometry[37]. Furthermore, unlike pure technologies, peripheral science does not develop a coterie of professionals. For example, light measurement could not be described as engineering, because the training and licensing of practitioners remained sporadic and uninfluential in its development. Of course, the definition of a 'technology' can be widened to include most of the learned and skilled activities of human life, but this merely dilutes the term to the point of meaninglessness. For the same reasons, the term 'technoscience' popularized by Bruno Latour is not sufficiently specific[38].

To a few practitioners, light measurement was merely a *technique* to be *applied* to problems. This definition is ultimately unsatisfactory because of the breadth of methods employed, the range of problems studied and the variety of investigators who used them. It minimizes the scope of the subject and neglects its pretentions for the status of a science[39]. This was clearly the case for colorimetry, which until the 1930s had little reliance on elaborate observing techniques or apparatus. Rather than being centred on a particular technique or apparatus, colorimetry was defined by its goal.

Is a peripheral science, finally, just another term for applied science? The primary difficulty with the term *applied science* is its implicit assumption of a direction of development, i.e. scientific discovery followed by practical application. Such a categorization also frequently implies an inadequate or unsuccessful science. D S L Cardwell is dismissive in his description of many early 20th-century career practitioners as members of a hitherto non-existent 'rank and file', with applied scientists often 'of the second and third rank'. He tempers this, however, with the statement that

> researches of the applied scientist are guided not by purely scientific considerations, but by the requirements of industry...this does not mean that the applied scientist and technologist are...truncated scientists.[40]

I suggest that peripheral science is not merely technology or applied science, nor a subject of lower intellectual stature. Instead, it is a qualitatively different enterprise; much of technology is peripheral to science and vice versa. Rather than being invariably linked with technology or applied science, peripheral science is a distinct and persistent category that shares some of their attributes, but evincing distinct developmental features. This perspective is supported by other recent work in the history and sociology of science.

Terry Shinn, for example, has characterized subjects such as magnet science as 'research-technologies', a fertile classification having much in common with this notion of peripheral science[41]. Shinn sees research-technology as embracing a set of practices, devices and institutional arrangements, often centred on instrumentation. He distinguishes these activities from experimentation and scientific theorizing, as well as from hands-on engineering. These fluid practices are performed by communities connected to both science and industry but, to some extent, separate from each[42].

Peter Galison, who has focused on the history of the instrument-making tradition, argues that it has been central to the evolution of modern physics[43]. Instead of the conventional hierarchy of theory, experiment and application, he reverses the perspective to place scientific instruments, not theories, centre-stage. Machines are not merely convenient tools, he claims: they draw together disparate scientific cultures, seed the nuclei of new working practices and even determine how their users visualize the world. As this study argues for the measurement of light, so too Galison mistrusts dichotomies in particle physics. Understandable neither as a struggle between theory and experiment, nor merely as intellectual rule-making versus social interests, physics is 'a complicated patchwork of highly structured pieces'[44]. Nor was his collection of instrument makers, experimenters, theorists and their associated social resources immutable. The nature of experiments and the experimenter have changed dramatically over the century.

Shinn's case studies of 'multi-lateral professional and institutional association' in France and Germany have much in common with the technical groups that came to measure light and colour. Tracing the roots of this approach to late 19th-century Germany, he suggests that these interstitial communities really became established in the mid 20th century. The communities of light measurement suggest an even earlier chronology. I would argue for a more naturally integrated co-evolution of professional science and 'peripheral science': these research-technology communities have been occupying gaps between disciplines and engineering specialties for as long as there have been disciplines and specialties. Combined with a re-evaluation of other case studies investigated as research-technologies and instrument-based subcultures, the experiences of the measurers of light suggest general features for such groups.

10.3.3. Attributes of peripheral science

Some of the identifiable characteristics that place a peripheral science outside the traditional views of both scientific disciplines and engineering specialties are:

(1) a lack of autonomy and authority over the subject by any one group of practitioners;
(2) a persistent straddling of disciplinary boundaries;
(3) a lack of professionalization among the subject's practitioners and
(4) a continuous and fluid interplay between technology, applied science and fundamental research.

These points are inter-related and follow from one key feature: the sharing of the subject between distinct scientific and technological sub-cultures.

Lack of autonomy and authority by any one group of practitioners

The absence of 'ownership' by a single community deprived light measurement of a clear definition and purpose. Without focus, it was both shared and unclaimed, constraining its standardization.

Case studies displaying the sharing of control between communities have, in previous historical analyses, evoked dichotomies: technology versus science, internal versus external influences or theory versus experiment. For example, the idea of two communities—e.g. 'practical engineers' versus 'academic engineers' and scientists—has been proposed for the situation of the subjects of refrigeration/thermodynamics in Germany and British chemistry at the turn of the 20th century[45]. Such neat dichotomies, while evidentally satisfactory for some historical episodes, are of limited usefulness for describing light measurement. There, such two-way splits of influences could be postulated only for restricted time periods or subject areas, if at all (e.g. Victorian gas inspectors versus astronomers; visual versus physical methods of photometry circa 1900–20; optical versus electrical engineering traditions in photometry; industrial versus governmental laboratories circa 1910–30; physicists versus psychologists in colorimetry between the wars). Far from being determined by a playing-off of rival influences, the subject depended on sporadic attention from several communities.

Persistent straddling of disciplinary boundaries

A *discipline* can be defined briefly as a subject based on systematic knowledge and uniting its practitioners in a self-regulating system of training and intellectual approbation. The key elements are *self-definition* by the practitioners and *external recognition* by non-practitioners. Lacking both these features, photometry and colorimetry certainly never developed into disciplines[46]. Its practitioners did not adopt any specific term for the field which found itself practised in such diverse contexts—individual departments of electrotechnics, gas engineering and optics. Borrowing elements from one another and shifting definition, these peripheral subjects have defied classification by both practitioner and historian. This lack of cohesion is a characteristic that persists for these subjects to the present day. The difference between 'disciplinary' and 'undisciplined' science has been discussed previously.

Lack of professionalization

The distinctions between an *occupation* and a *profession* have been discussed in the earlier context of illuminating engineers. These practitioners did not attempt to define themselves either as professional engineers or as scientists of a distinct specialty[47]. The discussions of this point at the early Illuminating Engineering Societies reveal that their members' aversion to such labels stemmed from a lack of confidence in their body of knowledge as a coherent subject and from their disparate backgrounds. The new members voiced both their wish to encourage research and communication and the concern that their differing vocations would impede this goal. A profession, involving career and societal characteristics in addition to the intellectual features of a discipline, is unlikely to develop where a discipline does not. The lack of professionalization may thus be a consequence of the disciplinary straddling of a peripheral science.

Changing interplay between technology, applied and pure science

A seamless web of influences is appropriate to describe peripheral science. Occupying a nexus between more easily identified subjects, it borrows from each—its position on the science/technology divide both drifting with time and dependent on the perspective of the observer. The social networks are transient, 'coalescing briefly around single theoretical and technical problems they share for brief periods, as passing aspects of longer term goals'[48]. In a subject not driven by theoretical impetus, social factors play a decisive role.

10.4. EPILOGUE: DECLINING FORTUNES

These traits suggest a consequence: a subject unnurtured by a long-lived and active scientific community inevitably languishes; a technique of limited or unappreciated utility is abandoned or under-utilized. This was the case for light and colour measurement. By the end of the 1930s the consolidation of practice was nearly complete: although Germany had long resisted change in standards of light intensity, it adopted a platinum-based standard along with France, America and Britain in the early months of the Second World War, on New Year's Day, 1940[49]. The subject's status as an active research area fell once the central concerns were satisfied and techniques were rendered routine.

Previous chapters have chronicled the progressive organization of light measurement by technical societies, research laboratories and appointed delegations. While these collective efforts encouraged a convergence of practitioners, the increased attention devoted to photometry and colorimetry by committees and industry was not sustained. The inter-war period saw both the ascent and decline of light measurement as a collective enterprise.

By the early 1930s the practice of illuminating engineering had become gradually less concerned with light measurement than with the design of lighting. Where texts before the First World War carried titles such as *Illumination and Photometry*, *Illumination: its Distribution and Measurement* and *Electrical Photometry and Illumination*, the subject of photometry was later relegated to single chapters in *Modern Illuminants and Illuminating Engineering*, *The Scientific Basis of Illuminating Engineering* and *Illuminating Engineering*[50]. According to the President of the Illuminating Engineering Society of New York some two decades after its foundation, this was a natural consequence of the maturity of the subject. Sciences, he claimed, pass through three stages: (1) the observation of elementary phenomena, (2), the measurement and deduction of laws and (3) the application of knowledge. The early years of the Society, he argued, had concentrated on stage (2) and 'it was natural that the first ten years of the illuminating engineering movement should be occupied mainly in developing methods of measuring light'[51]. The evidence presented in this book refutes his simple sequence; indeed, 'elementary phenomena', 'measurement' and 'application' continued to mingle in photometric practice. Nevertheless, the measurement of light ceased to be of direct concern to the illuminating engineering community.

A similar *de*volution can be seen in the Society that provided the initial impetus for standardizing light measurement: the Illuminating Engineering Society of London merged with the Chartered Institution of Building Services Engineers as recently as 1980. The subject, once it had been rendered routine, failed to retain the interest of the originally high proportion of scientists, and was instead sustained by a coterie of career engineers. The shift of interest is signalled by the subtitle of its periodical, which changed in the 1920s from *The Journal of Scientific Illumination* to *The Journal of Good Lighting*.

The inter-war period was the most active for research into heterochromatic photometry and colorimetry. With the contentious issues settled by delegations, attention devoted to these subjects declined considerably during and after the Second World War. An indication of its faltering status is given by the reduced emphasis at the National Bureau of Standards, where responsibility for colour research was reorganized seven times between 1948 and 1974, eventually devolving to become a part of the Sensory Environment Section of Building Research.

Similarly, the Commission Internationale de l'Éclairage continued to study colour standardization after the Second World War but limited this to relatively minor iterations of its 1931 work[52]. A loss of vitality in the CIE is suggested by the 50th anniversary meeting (Vienna, 1963) which reported the deaths of several past delegates including John Walsh, who had been associated with the Commission continuously from its origin[53].

Despite the relative prominence given to light measurement in the inter-war period and its faltering fortunes thereafter, the subject continued to exist, if not flourish. The decisive changes of the inter-war period had stabilized it to produce a generally recognized and definable subject. Light measurement was now based on physical measurement, and linked to vision by agreed conventions concerning 'average' humans. Subsequent work at research laboratories centred on refining measurement technologies and psychophysical definitions, and in exploring further the visual characteristics that fell outside the prescribed areas. The expansion of post-war radiometry and optical engineering, fuelled for a time by ballooning military budgets, consolidated these definitions.

These disparate contexts illustrate the patchwork that has characterized light and colour measurement; its threads are stitched from distinct technical sub-cultures and diverse intellectual components. Just as this peripheral subject was woven from the disciplinary fabrics of physics, technology, psychology and physiology, so too did its practitioners decide that the properties of light and colour were necessarily shared between the eye, instruments and energy.

NOTES

1. There was a significant rise in publications on photometry between 1880 and 1905, and a similar rise in publications on photoelectricity between 1931 and 1936. *Royal Society Catalogue of Scientific Papers 1800–1900*, Subject Index Vol III, Physics, Part I (Cambridge, 1912), Category 3010 ('Photometry, Units of Light'); *International Catalogue of Scientific Literature*: Physics, 1901–1914, Category 3010 ('Photometry, Units of Light, Brightness'); *Physics Abstracts* **1–41** (1898–1939): *Photometry and Photoelectricity*.
2. Exceptions to this are few indeed. For light measurement, at least, there appear to have been few proponents of a non-quantitative treatment of light after the First World War. Interest in light measurement was by then restricted to 'scientific' applications (in the broadest sense, and as opposed to metaphysical or artistic appeal) and 'scientific' methods, which by the inter-war period were firmly equated with quantification. On the other hand the subject of colour, engaging the interest of artists and philosophers, was never convincingly constrained by the desire for quantification. Examples of metaphysical and philosophical enlargements of the concept, and influence, of colour include: Matthaei R and Aach H (eds) 1971 *Goethe's Colour Theory* (New York); Westphal J 1987 *Colour: a Philosophical Introduction* (London) and Hilbert D R 1987 *Colour and Perception: a Study in Anthropocentric Realism* (Stanford). Such dimensions fall outside the scope of this work, which traces the progressive *narrowing* of the notion of colour by physical scientists to suit their objective of quantification.
3. On the cultural motives for quantification, and its limited penetration into everyday life, see Lave J 1986 'The values of quantification', in J Law (ed), *Power, Action and Belief: a New Sociology of Knowledge?* (London) pp 88–111.
4. A few scientists could wax poetic about the beauty of light. Albert Michelson, for example, using rhetoric typical of turn-of-the-century popular scientific works, lamented his inability to describe light and colour as clearly as could a poet or artist: 'I hope that the day may be near when a Ruskin will be found equal to the description

of the beauties of coloring, the exquisite gradations of light and shade...which are encountered at every turn' [Michelson A A 1901 *Light Waves and Their Uses* (Chicago) pp 1–2]. Even he devoted his energies, when not popularizing his work for the general public, to quantifying light, however. For an overview of the changing mental models of light, see Zajonc A 1993 *Catching the Light* (New York).

5 Forsythe W E (ed) 1937 *Measurement of Radiant Energy* (New York) and Moon P 1936 *The Scientific Basis of Illuminating Engineering* (New York). Forsythe, working at the Incandescent Lamp Department of GE at Nela Park, brought together scientists specializing in radiometry, photometry and colorimetry for his book. This can be seen as the product of a 'culture of unification' which had been nurtured at Nela Park since its foundation, owing to the research policies of its first directors. Similarly Moon, an illuminating engineer and relative outsider to the scientific community, attempted to broach the separation by allying illuminating engineering with scientific principles.

6 Wolfe W 1965 *The Handbook of Military Infrared Technology* (Washington) p 1.

7 Until the turn of the 20th century, British photometry in particular, and British science in general, was nearly devoid of organization and government support. Cardwell refers to a '*fin de siècle* lassitude' in British science, which he ascribes to the diversion of interest from science and technology during the 'age of imperialism'; strangulation of scientific enthusiasm by an oppressively time-consuming examination system; and, excessive specialization with little attention paid to applied problems [Cardwell D S L 1972 *The Organization of Science in England* (London) p 191].

8 Noble D F 1979 *America by Design: Science, Technology and the Rise of Corporate Capitalism* (New York), Hughes T P 1983 *Networks of Power: Electrification in Western Society 1880–1930* (Baltimore) and Gooday G J N 1991 'Teaching telegraphy and electrotechnics in the physics laboratory: William Ayrton and the creation of an academic space for electrical engineering in Britain 1873–1884', *Hist. Technol.* **13** 73–111. Noble discusses how 'during the closing decades of the 19th century, the new institutions of science-based industry, scientific technical education, and professional engineering had gradually coalesced to form an integrated social matrix (composed of the corporations, the schools, the professional societies)' [p 50]. Hughes' 'systems approach' emphasizes the interplay of interests beyond those of academic scientists. Gooday documents the transition of electrotechnics from an engineering craft to academic subject.

9 A feature shared with the related subject of psychology; see Danziger K 1994 *Constructing the Subject: Historical Origins of Psychological Research* (New York) pp 8–10.

10 Popper K 1972 *Conjectures and Refutations* (London, 4th edn) p vii.

11 Kuhn T S 1970 *The Structure of Scientific Revolutions* (Chicago, 2nd edn).

12 *Ibid.*, p 210.

13 Kuhn T S 1978 *Blackbody Theory and the Quantum Discontinuity* (Oxford) and Cahan D 1989 *An Institute for an Empire: the Physikalisch-Technische Reichsanstalt 1871–1918* (Cambridge) ch 4.

14 Kuhn T S 1961 'The function of measurement in modern physical science' in H Woolf (ed) *Quantification* (Indianapolis) pp 31–63; quotation p 41 (author's italics).

15 Colorimetry sits awkwardly in a Kuhnian analysis for two reasons. First, Kuhn's 'preparadigm' and 'revolutionary' periods are difficult to identify for colour measurement, and arguably telescope into a brief period during the 1930s. Second, the 'incommensurability' is across disciplines rather than time periods.

16 Law J (ed) 1991 *A Sociology of Monsters: Essays on Power, Technology and*

Domination (London) pp 1–23. Law suggests that a sociology of special cases, or 'monsters', is required to deal with the myriad differences between heterogeneous case studies.

17 For the case of the construction of valid tests of water quality, for example (and involving one of the prominent Victorian photometrists, Joseph Dibdin), see Hamelin C 1990 *A Science of Impurity: Water Analysis in Nineteenth Century Britain* (Berkeley, CA); quotation p 40.

18 For example Callon M, Law J and Rip A 1986 'Glossary' and 'How to study the force of science' in Callon *et al Mapping the Dynamics of Science and Technology: Sociology and Science in the Real World* (London) pp xvi–xvii and 3–18.

19 More restrained accounts of social constructivism are espoused, for example, in the works of Trevor Pinch and Harry Collins. Collins' empirical programme of relativism is particularly relevant to describe the negotiated consensus in 1930s colorimetry [Collins H M 1981 'Knowledge and controversy: studies of modern natural science' *Soc. Stud. Sci.* **11** 1–3].

20 See Latour B 1987 *Science in Action* (Cambridge, MA) pp 7–14.

21 Latour B 1988 *The Pasteurization of France* (Cambridge, MA).

22 Latour *op. cit.* note 20, p 167.

23 For example for 'peripheral or newly civilised countries' [de Candolle A 1885 *Histoire des Sciences et des Savants Depuis Deux Siècles* (Geneva)], or 'division of the world of science into centre (or centres) and periphery' [Crawford E 1992 *Nationalism and Internationalism in Science, 1880–1939* (Cambridge) pp 18–23] or French 'provincial' science [Nye M J 1975 'The scientific periphery in France: the Faculty of Sciences at Toulouse (1880–1930)' *Minerva* **13** 374–403].

24 Schott T 1988 'International influence in science: beyond center and periphery', *Soc. Sci. Res.* **17** 219–38.

25 Gieryn T F and Hirsch R T 1983, 'Marginality and innovation in science', *Soc. Stud. Sci.* **13** 87–106.

26 Robert Park, quoted in Gieryn and Hirsch *op. cit.* note 25.

27 Cole J R and Zuckerman H 1975 'The emergence of a scientific specialty: the self-exemplifying case of the sociology of science' in Coser L A (ed) *The Idea of Social Structure* (New York) pp 139–74.

28 Edge D O and Mulkay M J 1976 *Astronomy Transformed: the Emergence of Radio Astronomy in Britain* (New York) pp 362–3. The marginal characteristics include: (i) initial discovery by an 'applied' scientist indirectly linked to the 'basic' research networks; (ii) wartime discoveries of academic scientists that then seeded academic research and (iii) the introduction of new astronomical techniques by researchers trained as physicists, studying problems not initially identified as astronomical.

29 Lemaine G, McLeod R, Mulkay M and Weingart P (eds) 1976 *Perspectives on the Emergence of Scientific Disciplines* (The Hague) p 6.

30 Law J 1973 'The development of specialties in science: the case of X-ray protein crystallography' *Sci. Stud.* **3** 275–303.

31 *Ibid.*, p 303.

32 Mullins N C 1972 'The development of a scientific specialty: the phage group and the origins of molecular biology' *Minerva* **10** 51–82, and Mullins N C 1973 'The development of specialties in social science: the case of ethnomethodology' *Sci. Stud.* **3** 245–74.

33 *Ibid.*, p 274.

34 Edge and Mulkay [*op. cit.* note 28, pp 356–7] describe the early history of

radio astronomy in terms of several cooperating academic research groups which differentiated the scientific problems selected.

35 De Solla-Price cites technology as having features including (1) little or no discipline, i.e. lacking professionals trained in universities by other 'experts', dedicated journals, literature dominated by a close-knit group of co-citators and neglect of archival literature; (2) literature centred on catalogues, handbooks etc and (3) little influence on mainstream science [de Solla-Price D J 1965 'Is technology historically independent of science? A study in statistical historiography' *Technol. Culture* **6** 553–68.

36 Barnes B 1982 'The science–technology relationship: a model and a query' *Soc. Stud. Sci.* **12** 166–72.

37 Commercial products such as microdensitometers were introduced in response to market demand.

38 See Latour *op. cit.* note 20, pp 157–9, 174–5. Latour uses *technoscience* as an all-encompassing term to include not just technology and science, but the networks that make them possible.

39 For example by J Walsh, who as a Division leader of the NPL perhaps not surprisingly referred to photometry as an *applied* science and a branch of technical physics. Edward Hyde, first director of the Nela laboratory, denoted it one of the 'great middle fields of science' (see ch 5 note 102).

40 Cardwell *op. cit.* note 7, pp 229, 235.

41 Shinn T 1997 'Crossing boundaries: the emergence of research-technology communities', in H Etzkowitz and L A Leydesdorff (eds) 1997 *Universities and the Global Knowledge Economy : A Triple Helix of University–Industry–Government Relations* (London) pp 85–96; Shinn T 1993 'The Bellevue grand électroaimant, 1900–1940: birth of a research-technology community *HSPS* **24** 157–87.

42 Joerges B and Shinn T (eds) 2001 *Instrumentation: Between Science, State and Industry* (Dordrecht).

43 Galison P L 1987 *How Experiments End* (Chicago) and Galison P L 1997 *Image and Logic: a Material Culture of Microphysics* (Chicago).

44 Galison 1997 *op. cit.* note 43, p xx.

45 See Dienel H-L 1993 'Industrial refrigeration in Germany 1870–1930: interactions between two engineering subcultures' *Conference on Technological Change* (Oxford). University researchers approached refrigeration from the point of view of thermodynamic theory, and spent considerable time in consultancy work, acting as 'science notaries' to validate practical research. The working engineers employed empirical methods to select the best form of refrigeration technology. For a comparable case of the negotiation between emergent communities in academic and industrial chemistry, see Donnelly J F 1986 'Representations of applied science: academics and the chemical industry in late 19th-century England' *Soc. Stud. Sci.* **16** 195–234.

46 The situation of international colorimetry in the early 20th century was reminiscent of that in German research into colour *perception* during the late 19th century. As R S Turner 1987 has noted ['Paradigms and productivity: the case of physiological optics, 1840–94' *Soc. Stud. Sci.* **17** 35–68; quotation p 43], 'it never constituted a true disciplinary grouping. Vision studies *per se* (as opposed to medical applications) never achieved institutional recognition in the European universities, never possessed a journal addressed exclusively to its concerns, and never generated arguments for its methodological or philosophical autonomy *vis-à-vis* other branches of science. Likewise, virtually none of its practitioners pursued vision research to the exclusion of other problems. Instead, researchers from several legitimate disciplines contributed to

the study of vision.' Thus peripheral sciences may spawn others, as colour perception, colour measurement and photometry shared similar features.

47 Illuminating engineers and photometrists were often on the outskirts of the developing hierarchies of science and of industry. R Torstendahl argues ['Engineers in industry, 1850–1910: professional men and new bureaucrats. A comparative approach' in Bernhard C G, Crawford E and Sörbom P (eds) 1982 *Science, Technology and Society in the Time of Alfred Nobel* (Oxford) pp 253–70] that the professionalization and career differentiation of groups of employees, such as the electrotechnicians at Siemens & Halske, was contingent on their firms devoting resources to research and development. Only a handful of illuminating engineers thus found career definition through this industry- and government-sponsored bureaucratization.

48 Edge and Mulkay *op. cit.* note 28, p 127.

49 This was essentially the long-sought Violle standard, first proposed in 1881 and actively pursued by the PTR, NPL and others from the 1890s. Formal international ratification was, however, delayed by the war and did not occur until 1948. See Walsh J W T 1940 'The new standard of light' *Trans. Illum. Eng. Soc.* **5** 89–92, and Jones O C and Preston J S 1969 *Photometric Standards and the Unit of Light* (London).

50 Wickenden W E 1910 (New York); Trotter A P 1911 (London); Bohle H 1912 (London); Gaster L and Dow J S 1920 (London); Moon P 1936 (New York) and Boast W B 1942 (New York), respectively.

51 Dow J S 1930 'Illuminating engineering: what it is and what it may become' *Illum. Eng. (NY)* **23** 295–8.

52 The '1931 standard observer' was revised and augmented in 1960, 1964, 1971 and 1976, notably to include a wider field of view (10° instead of the original 2°).

53 CIE 1963 *Compte Rendu CIE* 12–13.

BIBLIOGRAPHY

ABBREVIATIONS

The following abbreviations have been used in the notes and bibliography.

Periodicals

Am. J. Sci.	American Journal of Science
Am. J. Phys.	American Journal of Physics
Ann. Harvard Coll. Obs.	Annals of the Harvard College Observatory
Ann. Physik	Annalen der Physik
Ann. Sci.	Annals of Science
Appl. Opt.	Applied Optics
Arch. Hist. Exact Sci.	Archive for the History of the Exact Sciences
Arch. Int. Hist. Sci.	Archives Internationales d'Histoire des Sciences
Astron. & Astrophys.	Astronomy & Astrophysics
Astrophys. J.	Astrophysical Journal
Biog. Mem. Nat. Acad. Sci.	Biographical Memoirs of the National Academy of Sciences of the USA
BJHS	British Journal for the History of Science
Brit. J. Psychol.	British Journal of Psychology
Bull. Bur. Standards	Bulletin of the Bureau of Standards
Chem. Age	The Chemical Age
Chem. Eng. Works Chemist	Chemical Engineering and the Works Chemist
Coll. Res. NPL	Collected Researches of the National Physical Laboratory
Comptes Rendus	Comptes Rendus Hebdomadaires des Séances de l'Académie des Sciences
Compte Rendu CIE	Recueil des Travaux et Compte Rendu des Séances de la Commission Internationale de l'Éclairage
Daedalus	Daedalus
DNB	Dictionary of National Biography
DSB	Dictionary of Scientific Biography
Elec. Perspectives	Electrical Perspectives
Electrician	The Electrician

GEC Rev.	GEC Review
GEC J.	GEC Journal
Hist. Sci.	History of Science
Hist. Stud. Phys. Sci.	Historical Studies in the Physical Sciences
Hist. Stud. Phys. Biol. Sci.	Historical Studies in the Physical and Biological Sciences
Hist. Technol.	History of Technology
Ind. & Eng. Chem.	Industrial and Engineering Chemistry
Indus. Chemist	The Industrial Chemist
Illum. Engineering	Illuminating Engineering
Illum. Eng.	The Illuminating Engineer (London)
Illum. Eng. (NY)	The Illuminating Engineer (New York)
Infr. Phys.	Infrared Physics
Isis	Isis
J. Am. Chem. Soc.	Journal of the American Chemical Society
J. de Phys.	Journal de Physique
J. Franklin Inst.	Journal of the Franklin Institute
J. Gas Lighting	Journal of Gas Lighting
J. Hist. Astron.	Journal of the History of Astronomy
J. Indus. & Eng. Chem.	Journal of Industrial and Engineering Chemistry
J. Inst. Radio Engrs.	Journal of the Institute of Radio Engineers
J. IEE	Journal of the Institute of Electrical Engineers
J. Res. NBS	Journal of Research of the National Bureau of Standards
J. Sci. Instr.	Journal of Scientific Instruments
JOSA	Journal of the Optical Society of America
JOSA & RSI	Journal of the Optical Society of America and Review of Scientific Instruments
J. Vac. Sci. Tech.	Journal of Vacuum Science & Technology
Lum. Élec.	La Lumière Électrique
Minerva	Minerva
Mon. Not. Roy. Astron. Soc.	Monthly Notices of the Royal Astronomical Society
Mém. Acad. R. des Sci. Paris	Mémoires de l'Académie Royale des Sciences de Paris
Mind	Mind
Nat. Acad. Sci. Proc.	National Academy of Science Proceedings
NPL Report	National Physical Laboratory Report for the Year
Nature	Nature
Obit. Not. Roy. Soc.	Obituary Notices of Fellows of the Royal Society of London
Opt. & Phot. News	Optics and Photonics News
Opt. Eng.	Optical Engineering
Osiris	Osiris

Phil. Mag.	Philosophical Magazine
Phil. Trans. Roy. Soc.	Philosophical Transactions of the Royal Society of London
Photog. Indus.	Photographic Industry
Photog. J.	Photographic Journal
Photog. News	Photographic News
Phys. Rev.	Physical Review
Phys. Today	Physics Today
Proc. Am. Acad. Arts. Sci.	Proceedings of the American Academy of Arts and Sciences
Proc. IEE	Proceedings of the Institute of Electrical Engineers
Proc. Opt. Convention	Proceedings of the Optical Convention
Proc. Phys. Soc.	Proceedings of the Physical Society of London
Proc. Roy. Astron. Soc.	Proceedings of the Royal Astronomical Society
Proc. Roy. Soc.	Proceedings of the Royal Society of London
Proc. Roy. Soc. Edin.	Proceedings of the Royal Society of Edinburgh
Rev. Opt.	Revue d'Optique
Rev. Sci. Instr.	Review of Scientific Instruments
Sci. Context	Science in Context
Sci. Stud.	Science Studies
Soc. Sci. Res.	Social Science Research
Soc. Stud. Sci.	Social Studies in Science
Technol. Culture	Technology and Culture
Trans. Illum. Eng. Soc.	Transactions of the Illuminating Engineering Society of London
Trans. Illum. Eng. Soc. (NY)	Transactions of the Illuminating Engineering Society of New York
Trans. Opt. Soc.	Transactions of the Optical Society

Organizations

BCC	British Colour Council
BEAMA	British Electrical and Applied Manufacturers Association
BESA	British Engineering Standards Association
BSIRA	British Scientific Instruments Research Association
CIE	Commission Internationale de l'Éclairage
CIP	Commission Internationale de Photométrie
DSIR	Department of Scientific and Industrial Research
ELMA	Electric Light Manufacturers Association
GEC	General Electric Company (UK)
IRC	International Research Council
ISCC	Inter-Society Color Council (USA)
NBS	National Bureau of Standards (USA)
NDRC	National Defense Research Committee

Nela	National Electric Lamp Association (USA)
NPL	National Physical Laboratory (UK)
OSA	Optical Society of America
OSRD	Office of Scientific Research and Development
PTR	Physikalisch-Technische Reichsanstalt
RRE	Radar Research Establishment
RSRE	Royal Signals and Radar Establishment
SBRC	Santa Barbara Research Center
SPIE	Society of Photo-Optical Instrumentation Engineers

Other

J	Energy in joules
sr	Solid angle in steradians
W	Power in watts

SOURCES

The primary sources for this work have been principally contemporary papers, articles, reports and books. As light measurement was frequently perceived as a technique—a means to an end rather than the end in itself—it was often confined to specialist and trade journals. Nevertheless, the subject was highly fragmented, and the published sources were diverse. The most important of these were journals dealing with applied science, engineering and instrumentation. *The Journal of the Optical Society of America and Review of Scientific Instruments* (published together between 1921 and 1929, and separately thereafter) and *Journal of Scientific Instruments*, a British journal founded in 1924, proved to be useful primary sources. The relatively small number of contributors to the subject of light measurement over the period studied made the exhaustive study of some sources practicable. A reasonable longitudinal survey of the subject was obtained by surveying a number of English language journals. Laboratory reports were also fairly frequent sources of information on light measurement. *NPL Report for the Year*, *Collected Researches of the NPL*, *Bureau of Standards Journal of Research* (later renamed *Journal of Research of the NBS*) and *GEC Review* contained the research products of these laboratories. Another major source was the *Compte Rendu des séances de la Commission Internationale de l'Éclairage*, the international body responsible for lighting standards. This account, generally published at four-year intervals, included the resolutions, minutes of meetings and lists of attendees at the CIE sessions.

Apart from journals self-described as 'scientific', trade magazines and popular accounts have also provided useful information. The practice of light measurement involved several independent communities of workers, but the self-styled 'illuminating engineers' made the strongest efforts to define the subject. *The Illuminating Engineer* (London) and *Transactions of the Illuminating Engineering Society of New York*, both founded in the early years of the 20th century and responsible for much of the early enthusiasm for light measurement, provided considerable detail regarding the social evolution of the subject. These and similar publications such as the *Journal of the Franklin Institute* covered, among other things, work at government laboratories, commercial developments and international legal standards. Moreover, the informal tone they presented through editorials, varied articles and occasionally opinionated news items provided clues that the scientific journals omitted. The *New Products* sections of such publications helped trace the contemporary firms and technologies, as did patent records. The variety of groups concerned with light measurement, and responsible for its peripheral character, are reflected by the diversity of sources in which their activities were recorded.

Last among primary published sources, books gave a reasonably clear account of the contemporary state of the art. In most cases, such books were survey texts intended for practitioners in the field. Such texts generally provided a broad survey of the subject of intensity standards, photometric apparatus, recent references and photometric data for engineers or students of physics. Even for such seemingly 'objective' sources, the sub-text has considerable importance: evaluation of the subjects treated (or not treated), practitioners cited, references made and techniques mentioned, all provide an implicit picture of the contemporary status of the subject. In so unstable a field (as light measurement was over most of the period covered in this thesis), books also served as powerful tools of persuasion and standardization. The numerous texts on colour, each espousing a radically different system of metrics, are an example of this. In the absence of formal educational programmes, books were also a major source of training for many practitioners.

One of the difficulties of studying a peripheral science such as photometry is that unpublished primary source material is hard to come by. For example, the GEC Hirst Research Centre at Wembley, founded in 1919 and responsible for important developments in industrial photoelectric devices in the following decade, discarded 70 years of internal reports during a recent move[1]. A similar fate has been faced by the records of some of the relevant institutions. The Optical Society of America, in existence as a relatively prosperous and stable entity since 1916, has retained no records from its committees of the inter-war period[2]. The Illuminating Engineering Society of London, a locus for the development of the subject in Britain, eventually merged with a society of building engineers and discarded its early records. As another researcher has noted,

> firms are not in business for the benefit of historians and archivists.... [Firms may destroy their archives] because a new office block has been built, or because they have been taken over by a larger concern, or because they want to make more efficient use of the space available.[3]

Without such primary archival sources, information has necessarily been gleaned from published company histories and by trawling through the publications of relevant journals to cross-reference information.

Biographies, except for brief necrologies, are non-existent for the workers who were important in this subject. Similarly, their notebooks, letters and other unpublished works have not, in general, been archived. The interactions between these individuals have become indirectly apparent through co-citations in articles, papers and book dedications; proceedings of question periods at conferences; and common membership in associations and on commissions.

Clifford Paterson is an exception to most of the personalities mentioned. Knighted and made a member of the Royal Society in later life, he was considerably more distinguished than most workers in light measurement. For the most part, these scientists published relatively few papers owing to the applied character of their work or for reasons of commercial secrecy. For the same reason, most practitioners of the subject were unlikely to have their collected works published or to warrant even biographical sketches from the usual institutions[4].

Historians of science have previously little treated the general subject of light measurement. There are, of course, some relevant secondary sources dealing with particular aspects. Hans Kangro published studies of radiometry in Germany, particularly concerning the experimental work of Heinrich Rubens and collaborators surrounding Planck's radiation law[5]. There have also been a handful of publications dealing with the earliest recorded work in photometry by Bouguer and Lambert. These fall outside the main thrust of this book, and moreover discuss the subjects from an 'internalist' viewpoint. Probably the most thorough general history and bibliography of photometry are contained in a chapter of the 1926 text by John Walsh, himself an important player in the field[6]. This is a positivistic account that treats superficially the then ongoing transition to photoelectric methods—a change that reshaped the subject. The techniques of astronomical photometry, which had a much larger scientific component than other usages, have been summarized historically by practising astronomers[7]. There have been, moreover, a number of retrospectives and capsule histories in journals of optics, physics and electrical engineering[8]. These are, for the most part, unsatisfactory in a historiographical sense. In most cases, such histories take the form of reminiscences or first-hand accounts of a period covering some 10–30 years in one of the numerous branches of the subject. Alternatively, they summarize the field in terms of the progress or inventions of an individual, institution or company. Because of the connection between 'actor' and 'playwright', and because successes are more common subjects than failures, such accounts must be suspected of bias towards a celebratory or eulogising perspective. This work, by contrast, has attempted to uncover and inter-relate the important factors in the development of light measurement, many of which were not explicitly visible to practitioners of the time. No attempt has been made to interpolate judgements of 'success' or 'failure' based on modern interpretations, which are themselves the product of particular cultural circumstances. The coverage also draws connections between subjects that have previously been linked only loosely and which straddle the conventional boundaries of science, technology and industry. Indeed, my assertion that photometry has been a subject moulded by technical fragmentation and by its peripheral role in science does not fit well with the types of history mentioned here.

NOTES

1 S L Cundy [director, GEC Hirst Research Centre] personal communication 24 May 1993.
2 OSA president, personal communication 29 Mar 1994.
3 Cardwell 1972 *The Organisation of Science in England* p 175.
4 The identified unpublished source materials include records at the Commission Internationale de l'Éclairage in Geneva, and files (principally post-1920) at the Illuminating Engineering Society of North America, the successor to the IES of New York. As the CIE session minutes, attendee lists and resolutions were published, there is thought to be little relevant unpublished material on file (J Schanda [executive director of CIE] personal communication 30 June 1993).
5 For example Kangro H 1976 *The Early History of Planck's Radiation Law* (London).
6 Walsh J W T 1926 Photometry (London).
7 The most thorough of these are: Müller G 1897 *Die Photometrie der Gestirne* (Leipzig); Lundmark K 'Luminosities, colours, diameters, densities, masses of the stars', in Hälfte E (ed) 1932 *Handbuch der Astrophysik* (Berlin) Band V vol 1 pp 210–574 and Hearnshaw J B 1996 *The Measurement of Starlight: Two Centuries of Astronomical Photometry* (Cambridge).
8 A number of these, published in *JOSA*, *Appl. Opt.* and *Infr. Phys.*, are listed in the notes.

BIBLIOGRAPHY

Abady J 1902 *Gas Analyst's Manual* (London: Spon)
Abney W de W 1876 *Photography* (London: MacMillan)
——1891 *Colour Measurement and Mixture* (London: Society for the Promotion of Christian Knowledge)
——1895 *Colour Vision* (London: Low, Marston)
——1913 *Researches in Colour Vision and the Trichromatic Theory* (New York: Longmans)
Abney W de W and Festing E R 1888 *Colour Photometry* (London: Harrison)
Agar J 1998 *Science and Spectacle: The Work of Jodrell Bank in Post-War British Culture* (Amsterdam: Harwood Academic)
Airy W (ed) 1896 *Autobiography of Sir George Biddell Airy, KCB* (Cambridge: Cambridge University Press)
Aitken H G J 1985 *The Continuous Wave: Technology and American Radio, 1900–1932* (Princeton, NJ: Princeton University Press)
Alglave E and Boulard J 1882 *La Lumière Électrique: Son Histoire, sa Production et son Emploi* (Paris: Firmin-Didot)
Allen H S 1913 *Photo-Electricity: the Liberation of Electrons by Light* (London: Green)
American National Standard Institute 1986 *Nomenclature and Definitions for Illumination Engineering* (New York: ANSI)
Bachelard G 1933 *Les Intuitions Atomistiques* (Paris: Boivin)
Barnes B and Edge D (eds) 1982 *Readings in the Sociology of Science* (Milton Keynes: Open University Press)
Barrows W E 1938 *Light, Photometry and Illuminating Engineering* (London: McGraw-Hill)
Bell L 1912 *The Art of Illumination* 2nd edn (London: McGraw-Hill)

Benecke T and Quicke A W (eds) 1956 *History of German Guided Missiles Development* (Brunswick, Germany: Verlag)
Bennett S 1979 *A History of Control Engineering 1800–1930* (London: Peregrinus)
——1993 *A History of Control Engineering 1930–1955* (London: Peregrinus)
Bernhard C G, Crawford E and Sörbom P (eds) 1982 *Science, Technology and Society in the Time of Alfred Nobel* (Oxford: Pergamon)
Bijker W E and Pinch T J (eds) 1989 *The Social Construction of Technological Systems* (Cambridge, MA: MIT Press)
Blok A 1914 *The Elementary Principles of Illumination and Artificial Lighting* (London: Scott, Greenwood & Son)
Blondlot C 1905 *'N' Rays* (London: Longmans, Green)
Boast W B 1942 *Illumination Engineering* (New York: McGraw-Hill)
Bohle H 1912 *Electrical Photometry and Illumination* (London: Griffen)
Boring E G 1959 *A History of Experimental Psychology* (Englewood Cliffs, NJ: Prentice-Hall)
Bouasse H 1917 *Vision et Reproduction des Formes et des Couleurs* (Paris: Delagrave)
Bouguer P 1729 *Essai d'Optique sur la Gradation de la Lumière* (Paris: Jombert) reprinted 1921 (Paris: Gauthier-Villars).
——1749 *La Figure de la Terre, déterminée par les observations de Messieurs Bouguer et De la Condamine, de l'Académie Royale des Sciences envoyés par order du Roy au Pérou pour observer aux environs de l'Equateur* (Paris: Jombert)
——1760 *Traité d'Optique sur la Gradation de la Lumière. Ouvrage posthume de M. Bouguer, et publié par M. l'Abbé de la Caille* 2nd edn (Paris: Guerin and Delatour), transl. by Middleton W E K 1961 as *Pierre Bouguer's Optical Treatise on the Gradation of Light* (Toronto: University of Toronto)
Bouma P J 1944 *Physical Aspects of Colour: An Introduction to the Scientific Study of Colour Stimuli and Colour Sensations* (Eindhoven: Philips)
Boutry G A 1934 *Mésure de Densités Photographiques par la Methode Photo-Électrique* (Paris: Gauthier-Villars)
Bower T 1987 *The Paperclip Conspiracy: the Battle for Spoils and Secrets of Nazi Germany* (London: Little, Brown)
Bramson M A 1968 *Infrared Radiation: a Handbook for Applications* (New York: Plenum)
Bridgman P 1927 *The Logic of Modern Physics* (London: MacMillan)
Bright Jr A A 1949 *The Electric-Lamp Industry: Technological Change and Economic Development from 1800 to 1947* (New York: MacMillan)
British Colour Council 1934 *Dictionary of Colour Standards* (London: British Colour Council)
Brown G I 1999 *Scientist, Soldier, Statesman, Spy: Count Rumford* (London: Sutton)
Brown J T 1883 *Photometry and Gas Analysis* (London: Scientific Publishing)
Brush S G 1988 *A Guide to the Second Scientific Revolution, 1800–1950: The History of Modern Science* (Ames, IA: Iowa State University)
Brush S G and Belloni L 1983 *The History of Modern Physics: an International Bibliography* (New York: Garland)
Buchanan P D 1982 *Quantitative Measurement and the Design of the Chemical Balance* (unpublished PhD thesis, University of London).
Buchanan R A 1989 *The Engineers: a History of the Engineering Profession in Britain 1750–1914* (London: Jessica Kingsley)
Buchwald J Z 1994 *The Creation of Scientific Effects: Heinrich Hertz and Electric Waves* (Chicago, IL: University of Chicago)

Buckmaster J C 1875 *The Elements of Acoustics, Light, and Heat* (London: Simpkin, Marshall)

Bud R and Cozzens S E (eds) 1992 *Invisible Connections: Instruments, Institutions and Science* (Bellingham, WA: SPIE Optical Engineering Press)

Bud R and Warner D J W 1998 *Instruments of Science: An Historical Encyclopedia* (New York: Garland)

Burnham R W, Hanes R M and Bartleson C J 1963 *Color: a Guide to Basic Facts and Concepts* (New York: Wiley)

Butler C J and Elliot I (eds) 1993, *Stellar Photometry—Current Techniques and Future Development* (London: Cambridge University Press)

Butterworth H 1968 *The Science and Art Department, 1853–1900* (unpublished PhD thesis, University of Manchester)

Cady F E and Dates H B (eds) 1925 *Illumination Engineering for Students and Engineers* (New York: Wiley)

Cahan D 1989 *An Institute for an Empire: the Physikalisch-Technische Reichsanstalt 1871–1918* (Cambridge: Cambridge University Press)

——(ed) 1993 *Hermann von Helmholtz and the Foundations of Nineteenth Century Science* (Berkeley, CA: University of California Press)

Callon M, Law J and Rip A 1986 *Mapping the Dynamics of Science and Technology: Sociology of Science in the Real World* (London: MacMillan)

Campbell N R 1928 *An Account of the Principles of Measurement and Calculation* (London: Longman)

Campbell N R and Ritchie D 1929 *Photoelectric Cells: Their Properties, Use and Applications* (London: Pitman)

Capron J R 1877 *Photographed Spectra: One Hundred and Thirty-Six Photographs of Metallic, Gaseous and Other Spectra* (London: Spon)

Cardwell D S L 1972 *The Organisation of Science in England* (London: Heinemann)

Carnt P S and Townsend G B 1961 *Colour Television: N.T.S.C. System, Principles and Practice* (London: Iliffe)

Cattermole M J G and Wolfe A F 1987 *Horace Darwin's Shop: a History of the Cambridge Scientific Instrument Company 1878 to 1968* (Bristol: Hilger)

Caullery M 1933 *La Science Française depuis le XVIIe Siècle* (Paris: Armand Colin)

Chevreul M E 1858 *The Laws of Contrast of Colour* (London: Routledge)

Clayton R and Algar J 1991 *A Scientist's War: The War Diary of Sir Clifford Paterson 1939–45* (London: Peter Peregrinus)

——1989 *The GEC Research Laboratories 1919–1984* (London: Peter Peregrinus)

Clerke A M 1893 *History of Astronomy During the Nineteenth Century* (London: Adam & Charles Black)

Clerq P R (ed) 1985 *Nineteenth Century Scientific Instruments and their Makers* (Amsterdam: Rodopi)

Clewell C E 1913 *Factory Lighting* (New York: McGraw-Hill)

Clifton G C 1992 *Professionalism, Patronage & Public Service in Victorian London: the Staff of the Metropolitan Board of Works 1856–1889* (London: Athlone)

Cochrane R C 1966 *Measures for Progress: a History of the National Bureau of Standards* (Washington, DC: US Department of Commerce)

Cole S 1992 *Making Science: Between Nature and Society* (Cambridge, MA: Harvard University Press)

Collins H M (ed) 1982 *Sociology of Scientific Knowledge: a Source Book* (Bath: Bath University Press)

——1985 *Changing Order: Replication and Induction in Scientific Practice* (London: Sage)

Collins H M and Pinch T J 1982 *Frames of Meaning: the Social Construction of Extraordinary Science* (London: Routledge & Kegan Paul)

Commission Internationale de l'Éclairage 1921, 1924, 1928, 1931, 1935, 1939, 1948, 1964, 1968 *Compte Rendu des Séances de la CIE* (Teddington: National Physical Laboratory)

——1981 *Golden Jubilee of Colour in the CIE* (Bradford)

——1990 *History of the CIE 1913–1988* Publication CIE82-1990 (Geneva: CIE)

Conrady A E 1924 *Photography as a Scientific Implement* (London: Blackie & Son)

Coser L A 1975 *The Idea of Social Structure* (New York: Harcourt)

Cox J R 1980 *A Century of Light* (New York: Benjamin)

Crary J 1990 *Techniques of the Observer: On Vision and Modernity in the Nineteenth Century* (Cambridge, MA: MIT Press)

Crawford E 1992 *Nationalism and Internationalism in Science, 1880–1939: Four Studies of the Nobel Population* (Cambridge: Cambridge University Press)

Danziger K 1994 *Constructing the Subject: Historical Origins of Psychological Research* (Cambridge: Cambridge University Press)

Darmois E and Cohu M 1939 *La Photométrie Industrielle* (Paris: Gauthier-Villars)

Daumas M 1953 *Les Instruments Scientifiques aux 17e et 18e Siècles* (Paris: Presses Universitaires de France)

de Candolle A 1885 *Histoire de Sciences et des Savante depuis Deux Siècles* (Geneva: Bale)

de Solla Price D J 1963 *Little Science, Big Science* (New York: Columbia University Press)

Dibdin W J 1889 *Practical Photometry: a Guide to the Study of the Measurement of Light* (London: Walter King)

Dingle H 1950 *Practical Applications of Spectrum Analysis* (London: Chapman and Hall)

Divall C and Johnston S F 2000 *Scaling Up: The Institution of Chemical Engineers and the Rise of a New Profession* (Dordrecht: Kluwer)

Dobson G M B, Griffith I O and Harrison D N 1926 *Photographic Photometry: a Study of Methods of Measuring Radiation by Photographic Means* (Oxford: Clarendon)

Dreyer J L E and Turner H H (eds) 1987 *History of the Royal Astronomical Society Vol I: 1820–1920* (London: Blackwell Scientific)

Eberhard G, Kohlschütter A and Ludendorff H (eds) *Handbuch der Astrophysik 1* (Berlin: Springer)

Edge D O and Mulkay M J 1976 *Astronomy Transformed: the Emergence of Radio Astronomy in Britain* (New York: Wiley)

Elliott B (ed) 1988 *Technology and Social Process* (Edinburgh: Edinburgh University Press)

Englehardt H T and Caplan A L (eds) 1987 *Scientific Controversies: Case Studies in the Resolution and Closure of Disputes in Science and Technology* (Cambridge: Cambridge University Press)

Etzkowitz H and Leydesdorff L A (eds) 1997 *Universities and the Global Knowledge Economy: a Triple Helix of University–Industry–Government Relations* (London: Pinter)

Fabre C 1890 *Traité Encyclopédique de Photographie* (Paris: Gauthier-Villars)

——1928 *Leçons de Photométrie Professées à l'Institut d'Optique Théorique et Appliquée* (Paris: Revue d'Optique Théorique et Instrumentale)

——1938 *Oeuvres Choicies: Section IV: Photométrie* (Paris: Gauthier-Villars)

Fawcett A J 1951 *Electric Eyes: a Concise and Elementary Description of the Photo-Electric Cell, for the Non-Technical Reader; Its Uses in Industry, and Its Uses and Short-Comings in Colorimetry* (Salisbury: Tintometer)
Fleming A J P and Pearce J G 1922 *Research in Industry: the Basis of Economic Progress* (London: Pitman)
Fleming J A 1907 *A Handbook for the Electrical Laboratory & Testing Room* (London: Electrician Printing and Publishing)
Fleury P 1932 *Encyclopédie Photométrique* (Paris: Editions de la Revue d'Optique)
Forsythe W E (ed) 1937 *Measurement of Radiant Energy* (New York: McGraw-Hill)
Foucault L 1878 *Receuil des Traveaux Scientifiques de Léon Foucault* (Paris: Gauthier-Villars)
Fox R, Morrell J B, Cardwell D S L, MacLeod R M and Brock W H 1976 *The Patronage of Science in the Nineteenth Century* (Leyden: Noordhoff)
Fox R, Morrell J B, Cardwell D S L, MacLeod R M, Brock W H and Weisz G (eds) 1980 *The Organisation of Science and Technology in France 1808–1914* (Cambridge: Cambridge University Press)
François-Marie R P 1700 *Nouvelle Découverte sur la Lumière pour la Mésurer et en Compter les Degrés* (Paris: Louis Sevestre)
Frängsmyr T (ed) 1990 *Solomon's House Revisited: the Organisation and Institutionalization of Science* (Canton, MA: Science History Publications)
Frängsmyr T, Heilbron J L and Rider R E (eds) 1990 *The Quantifying Spirit in the 18th Century* (Berkeley, CA: University of California Press)
Franklin A 1986 *The Neglect of Experiment* (Cambridge: Cambridge University Press)
——1990 *Experiment, Right or Wrong* (Cambridge: Cambridge University Press)
Friedman J S 1968 *History of Colour Photography* (London: Focal)
Galison P L 1987 *How Experiments End* (Chicago, IL: University of Chicago Press)
——1997 *Image and Logic: a Material Culture of Microphysics* (Chicago, IL: University of Chicago Press)
Gaster L and Dow J S 1920 *Modern Illuminants and Illuminating Engineering* 2nd edn (London: Pitman)
Gelly A and Tardif H P 1995 *Defence Research Establishment Valcartier 1945–1995: 50 Years of History and Scientific Progress* (Ottawa: Minister of Public Works and Government Services)
Gernsheim H and Gernsheim A 1955 *The History of Photography* (Oxford: Oxford University Press)
Gibb T R P 1942 *Optical Methods of Chemical Analysis* (New York: McGraw-Hill)
Gibson K S 1919 *Photoelectric Spectrophotometry by the Null Method* (Washington, DC: US Government Printing Office)
Gooch F A 1916 *Representative Procedures in Quantitative Chemical Analysis* (New York: Wiley)
Gould S J 1981 *The Mismeasure of Man* (New York: Norton)
Gudden B 1928 *Lichtelektrische Erscheinungen* (Berlin: Springer)
Guillemin A 1868 *Les Phenomènes de la Physique* (Paris: Hachette)
Hacking I 1983 *Representing and Intervening* (Cambridge: Cambridge University Press)
——1990 *The Taming of Chance* (Cambridge: Cambridge University Press)
Hahn R 1980 *A Bibliography of Quantitative Studies on Science and its History* (Berkeley, CA: University of California Press)
Hälfte E (ed) 1932 *Handbuch der Astrophysik* (Berlin: Springer)
Hamelin C 1990 *A Science of Impurity: Water Analysis in Nineteenth Century Britain*

(Berkeley, CA: University of California Press)

Hammond J W 1941 *Men and Volts: the Story of General Electric* (Philadelphia, PA: Lippincott)

Hardy A C 1936 *Handbook of Colorimetry* (Cambridge MA: Technology Press)

Hargreave D 1973 *Thomas Young's Theory of Color Vision: its Roots, Development, and Acceptance by the British Scientific Community* (unpublished PhD thesis, University of Wisconsin)

Harvey E N 1957 *A History of Luminescence from the Earliest Times Until 1900* (Philadelphia, PA: American Philosophical Society)

Hay D R 1846 *A Nomenclature of Colours, Applicable to the Arts and Natural Sciences To Manufactures and Other Purposes of General Utility* (Edinburgh: Blackwood)

Hearnshaw J B 1986 *The Analysis of Starlight: One Hundred and Fifty Years of Astronomical Spectroscopy* (Cambridge: Cambridge University Press)

——1996 *The Measurement of Starlight: Two Centuries of Astronomical Photometry* (Cambridge: Cambridge University Press)

Heerding A 1986 *The History of N.V. Philips' Gloelampenfabrieken. Vol I: the Origin of the Dutch Incandescent Lamp Industry* (Cambridge: Cambridge University Press)

Helmholtz H [transl. J P C Southall] 1924 *Helmholtz's Physiological Optics. Vol. II: The Sensations of Vision* (New York: Optical Society of America)

Hempstead C A 1977 *Semiconductors 1833–1919: an Historical Study of Selenium and Some Related Materials* (unpublished PhD thesis, University of Durham)

Hendry J 1986 *James Clerk Maxwell and the Theory of the Electromagnetic Field* (Bristol: Hilger)

Henden A A and Kaitchuck R H 1982 *Astronomical Photometry* (New York: Willman-Bell)

Herschel J F W 1833 *A Treatise on Astronomy* (London: Longman, Brown & Green)

Hilbert D R 1987 *Color and Color Perception: a Study in Anthropocentric Realism* (Menlo Park, CA: CSLI)

Hoffleit D 1992 *Astronomy at Yale* (New Haven, CT: Connecticut Academy of Arts)

Hoskin M A 1963 *William Herschel and the Construction of the Heavens* (London: Oldbourne)

Hudson R D Jr and Hudson J W (eds) *Infrared Detectors* (Stroudberg: Hutchinson and Ross)

Hughes A L 1914 *Photo-Electricity* (Cambridge: Cambridge University Press)

Hughes T P 1983 *Networks of Power: Electrification of Western Society, 1880–1930* (Baltimore, MD: MIT Press)

——1989 *American Genesis: a Century of Invention and Technological Enthusiasm 1870–1970* (New York: Penguin)

Illuminating Engineering Society (NY) 1917 *Illuminating Engineering Practice: Lectures on Illuminating Engineering Delivered at the University of Pennsylvania, Philadelphia, September 20 to 28, 1916* (New York: IES)

James W 1982 *Psychology* (London: MacMillan)

Jamieson J A, McFee R H, Plass G N, Grube R H and Richards R G 1963 *Infrared Physics and Engineering* (New York: McGraw-Hill)

Joerges B and Shinn T (eds) *2001 Instrumentation: Between Science, State and Industry* (Dordrecht: Kluwer)

Johnston S F 1991 *Fourier Transform Infrared: a Constantly Evolving Technology* (Chichester: Ellis Horwood)

Jones O C and Preston J S 1969 *Photometric Standards and the Unit of Light* (London)

Jones R V 1979 *Most Secret War: British Scientific Intelligence 1939–45* (London: Hodder & Stoughton)
Judd D B 1979 *Contributions to Color Science* (Washington, DC: NBS)
Judt M and Ciesla B (eds) 1996 *Technology Transfer Out of Germany After 1945* (Amsterdam: Harwood Academic)
Kangro H 1976 *Early History of Planck's Radiation Law*, transl. from *Vorgeschichte des Planckschen Strahlungsgesetzes* by R E W Maddison (London: Taylor and Francis)
Katz D 1935 *The World of Colour* (London: Kegan, Paul, Trench, Trubner)
Keating P W 1954 *Lamps for a Brighter America* (New York: McGraw-Hill)
Keitz H A E 1955 *Light Calculations and Measurements: an Introduction to the System of Quantities and Units in Light-Technology and to Photometry* (Eindhoven: Philips)
Kelly K L 1974 *Colorimetry & Spectrophotometry: a Bibliography of NBS Publications January 1906 through January 1973*, NBS Special Publication 393 (Washington, DC: NBS)
Kevles D J 1978 *The Physicists: A History of a Scientific Community in Modern America* (New York: Alfred A Knopf)
Krisciunas K 1988 *Astronomical Centers of the World* (Cambridge: Cambridge University Press)
Kruger L, Daston J and Heidelberger M (eds) 1987 *The Probabilistic Revolution: The Objectification of Observation, Measurement and Statistical Methods in the Nineteenth Century* (Cambridge, MA: MIT Press)
Kruse P W, McGlauchlin L D and McQuistan R B 1962 *Elements of Infrared Technology: Generation, Transmission and Detection* (New York: John Wiley)
Kuhn T S 1970 *The Structure of Scientific Revolutions* 2nd edn (Chicago, IL: University of Chicago Press)
——1977 *The Essential Tension* (Chicago, IL: University of Chicago Press)
——1978 *Black-Body Theory and the Quantum Discontinuity* (Oxford: Oxford University Press)
Lambert J H 1760 *Photometry, or the Measurement and Classification of Light, Colour and Shadow* [abridged German translation in *Ostwald's Klassiker der exakten Wissenschaften* Nos 31, 32 and 33, 1892]
Langley S P 1891 *Energy and Vision* (Washington, DC: National Academy of Sciences)
Lardner D 1866 *Handbook of Natural Philosophy: Optics* (London: Walton and Maberly)
Latour B 1987 *Science in Action* (Milton Keynes: Open University Press)
——1988 *The Pasteurization of France* (Cambridge, MA: Harvard University Press)
——1993 *We Have Never Been Modern* (New York: Harvester Wheatsheaf)
Langley S P and Woolgar S 1979 *Laboratory Life: the Social Construction of Scientific Facts* (Beverley Hills, CA: Sage)
Law J (ed) 1986 *Power, Action and Belief: a New Sociology of Knowledge?* (London: Routledge and Kegan Paul)
——1991 (ed) *A Sociology of Monsters: Essays on Power, Technology and Domination* (London: Routledge)
Lecomte J 1928 *Le Spectre Infrarouge* (Paris: Albert Blanchard)
Legrand Y 1968 *Light, Colour and Vision* (New York: Chapman and Hall)
Lemaire G, McLeod R, Mulkay M and Weingart P (eds) 1976 *Perspectives on the Emergence of Scientific Disciplines* (The Hague: Aldine)
Leslie S W 1993 *The Cold War and American Science: the Military–Industrial–Academic Complex at MIT and Stanford* (New York: Columbia University Press)
Liebenau J (ed) 1988 *The Challenge of New Technology: Innovation in British Business*

Since 1850 (Aldershot: Gower)

Lindqvist S 1993 *Center on the Periphery: Historical Aspects of 20th-Century Physics* (Canton, MA: Science History)

Lobel L and Dubois M 1967 *Basic Sensitometry: the Technique of Measuring Photographic Materials* (London: Focal)

Lockyer J N 1873 *The Spectroscope and its Applications* (London: MacMillan)

Lovibond J W 1897 *Measurement of Light and Colour Sensations* (London: Gill)

——1915 *Light and Colour Theories* (London: Spon)

Luckiesh M 1915 *Color and Its Applications* (London: Constable)

MacAdam D L 1970 *Sources of Colour Science* (Cambridge, MA: MIT Press)

Mack J E and Martin M J 1939 *The Photographic Process* (New York: McGraw-Hill)

Margolin I A and Rumyanstev N P 1957 *Fundamentals of Infrared Technology* 2nd edn (Moscow: Voenizdat)

Martin L C 1924 *Optical Measuring Instruments: Their Construction, Theory, and Use* (London: Blackie)

Marvin C 1988 *When Old Technologies Were New: Thinking About Electric Communication in the Late Nineteenth Century* (New York: Oxford University Press)

Matthaei R and Aach H (eds) 1971 *Goethe's Colour Theory* (New York: Van Nostrand Reinhold)

Maunder E W 1900 *The Royal Greenwich Observatory: a Glance at its History and Work* (London: Religious Tract Society)

McDougall W A 1985 *The Heavens and the Earth: A Political History of the Space Age* (Baltimore, MD: Johns Hopkins)

McGucken W 1969 *Nineteenth-Century Spectroscopy: Development of the Understanding of Spectra 1802–1897* (Baltimore, MD: Johns Hopkins)

——1984 *Scientists, Society, and State: the Social Relations of Science Movement in Great Britain 1931–1947* (Columbus, OH: Ohio State University Press)

Melville H 1962 *The Department of Scientific and Industrial Research* (London: Allen and Unwin)

Mendelssohn K S 1973 *The Rise and Fall of German Science: the World of Walther Nernst* (London: MacMillan)

Mendelsohn E, Smith M E and Weingart P (eds) 1988 *Science, Technology and the Military* (Dordrecht: Kluwer)

Merton R K and Gaston J (eds) 1977 *The Sociology of Science in Europe* (Carbondale, IL: Southern Illinois University Press)

Michelson A A 1901 *Light Waves and Their Uses* (Chicago, IL: University of Chicago Press)

Moon P H 1936 *The Scientific Basis of Illuminating Engineering* (New York: McGraw-Hill)

Morgan G T and Pratt D D 1938 *British Chemical Industry: its Rise and Development* (London: Arnold)

Moseley R 1976 *Science, Government and Industrial Research: the Origins and Development of the National Physical Laboratory, 1900–75* (unpublished PhD thesis, University of Sussex)

Müller G 1897 *Die Photometrie der Gestirne* (Leipzig: Engelmann)

Munsell A H 1907 *A Color Notation* (Boston: Munsell)

Murray H D (ed) 1952 *Colour in Theory and Practice* (London: Chapman and Hall)

National Electric Light Association 1909 *National Electric Light Association 32nd Convention* (Atlantic City: Nela)

Newall N F 1910 *The Spectroscope and its Work* (London: Society for the Promotion of Christian Knowledge)

Nickerson D 1946 *Color Measurement and its Application to the Grading of Agricultural Products; a Handbook on the Method of Disk Colorimetry* (Washington, DC: US Government Printing Office)

Noble D F 1979 *America by Design: Science, Technology, and the Rise of Corporate Capitalism* (New York: Knopf)

Olby R C, Cantor G N, Christie J R R and Hodge M J S (eds) 1990 *Companion to the History of Modern Science* (London: Routledge)

Pais A 1982 *'Subtle is the Lord...': The Science and the Life of Albert Einstein* (London: Oxford University Press)

Palaz A, transl. Patterson G W and Patterson M R 1894 *A Treatise on Industrial Photometry, With Special Application to Electric Lighting* (New York: Van Nostrand)

Parkhurst J A 1906 *Researches in Stellar Photometry During the Years 1894 to 1906* (Washington, DC: Carnegie)

Paul H W 1985 *From Knowledge to Power: the Rise of the Science Empire in France, 1860–1939* (Cambridge: Cambridge University Press)

Pestre D 1984 *Physique et Physiciens en France, 1918–1940* (Paris: Editions des Archives Contemporaines)

Pickering A 1984 *Constructing Quarks: a Sociological History of Particle Physics* (Edinburgh: Edinburgh University Press)

Pinch T J and Bijker W E (eds) 1987 *The Social Construction of Technological Systems* (London: MIT Press)

Poole J B and Andrews K (eds) 1972 *The Government of Science in Britain* (London: Weidenfeld and Nicolson)

Popper K 1972 *Conjectures and Refutations* (London: Routledge and Kegan Paul)

Prideaux E B R 1917 *The Theory and Use of Indicators* (London: Constable)

Pyatt E 1983 *The National Physical Laboratory: a History* (Bristol: Hilger)

Rea M S (ed) 1993 *The Illuminating Engineering Society Lighting Handbook* (New York: Illuminating Engineering Society of North America)

Reich L S 1985 *The Making of American Industrial Research: Science and Business at GE and Bell, 1876–1926* (Cambridge: Cambridge University Press)

Reingold N (ed) 1966 *Science in Nineteenth-Century America: a Documentary History* (London: MacMillan)

Renner P 1964 transl. A Nesbitt *Color: Order and Harmony; a Color Theory for Artists and Craftmen* (New York: Reinhold)

Roche J (ed) 1990 *Physicists Look Back* (Bristol: Institute of Physics Publishing)

Ronchi V 1956 *Histoire de la Lumière* (Paris: Armand Colin)

Rood N R 1879 *Modern Chromatics* (New York: Appleton)

Rosa E B 1911 *Photometric Units and Nomenclature* (Washington, DC: US Government Printing Office)

Ross F E 1924 *The Physics of the Developed Image* (New York: Van Nostrand)

Sabra A I 1967 *Theories of Light from Descartes to Newton* (London: Oldbourne)

Salet G 1888 *Traité Elémentaire de Spectroscopie* (Paris: Masson)

Sapolsky H M 1990 *Science and the Navy: the History of the Office of Naval Research* (Princeton, NJ: Princeton University Press)

Sawyer R A 1951 *Experimental Spectroscopy* (New York: Dover)

Schivelbusch W 1986 *Disenchanted Night: the Industrialisation of Light in the Nineteenth Century* (Oxford: Berg)

Schrøder M 1969 *The Argand Burner, its Origin and Development in France and England, 1780–1800* (Odense: Odense University Press)
Searle G F C 1935 *Experimental Optics—a Manual for the Laboratory* (Cambridge: Cambridge University Press)
Sharp C H 1895 *A Bolometric Study of Light Standards* (New York: MacMillan)
Sherman P D 1971 *Problems in the Theory and Perception of Colour, 1800–1860* (unpublished PhD thesis, University of London)
Smith C 1998 *The Science of Energy: a Cultural History of Energy Physics in Victorian Britain* (London: Athlone)
Smith E E 1975 *Radiation Science at the National Physical Laboratory 1912–55* (London: Department of Industry)
Smith R 1738 *A Compleat System of Optiks in Four Books, Viz. a Popular, a Mathematical, a Mechanical, and a Philosophical Treatise. To Which are Added Remarks Upon the Whole* (Cambridge: Crownfield)
Smith R E, Jones F E and Chasmar R P 1957 *Detection and Measurement of Infrared Radiation* (Oxford: Clarendon)
Snell F D 1922 *Colorimetric Analysis* (New York: Van Nostrand)
Society of Dyers and Colourists 1981 *The Golden Jubilee of Colour in the CIE* (Bradford: SDC)
Stebbins J 1929 *Photoelectric Photometry of Stars* (Madison: Washburn Observatory)
Steiner B 1974 *The Impact of Radiometry and Photometry and the Role of the NBS* (Washington, DC: NBS)
Stephenson H K and Jones E L, with Harrison G R (ed) 1948, 'OPTICS: a History of Divisions 16 and 17, NDRC', in *Science in World War II: Applied Physics* (Boston: Little, Brown)
Stevenson J 1984 *British Society 1914–1945* (London: Penguin)
Stine W M 1900 *Photometric Measurements and Manual for the Practice of Photometry: With Especial Reference to the Photometry of Arc and Incandescent Lamps* (New York: MacMillan)
Strafford N 1933 *The Detection and Determination of Small Amounts of Inorganic Substances by Colorimetric Methods* (London: Institute of Chemistry)
Stuewer R H 1975 *The Compton Effect: Turning Point in Physics* (New York: Science History)
Sturrock W 1953 *Light: Its Measurement and Control* (Cleveland, OH: General Electric)
Summer W 1957 *Photosensitors* (London: Chapman and Hall)
Swarbrick J 1929 *Easements of Light: the Depreciation in Value of Property Due to High Buildings* (Manchester: Wykham). The 1938 edition, *Easements of Light: a Synopsis of Modern Practice and a Brief Explanation of Simplified Methods of Measuring Daylight and Assessing Compensation*, centred on the employment of photoelectric devices
Sydenham P H 1979 *Measuring Instruments: Tools of Knowledge and Control* (London: Peregrinus)
Thomas D B 1969 *The Science Museum Photography Collection* (London: HMSO)
Thompson J S and Thompson H G 1920 *Silvanus Phillips Thompson: his Life and Letters* (London: Unwin)
Thompson S P 1906 *The Manufacture of Light* (London: MacMillan)
Tobey R C 1971 *The American Ideology of National Science, 1919–1930* (Pittsburgh, PA: University of Pittsburgh Press)
Troland L T 1929 *The Principles of Psychophysiology* (New York: Greenwood)

Trotter A P 1911 *Illumination: Its Distribution and Measurement* (London: MacMillan)
Turner G L'E (ed) 1976 *The Patronage of Science in the Nineteenth Century* (London: Noordhoff)
——1981 *Nineteenth Century Scientific Instruments* (San Francisco, CA: University of California Press)
——1990 *Scientific Instruments and Experimental Philosophy 1550–1850* (London: Variorum)
Turner R S 1994 *In the Eye's Mind: Vision and the Helmholtz–Hering Controversy* (Princeton, NJ: Princeton University Press)
Varcoe I 1974 *Organising for Science in Britain: a Case Study* (Oxford: Oxford University Press)
Walker O J 1932 *Recent Applications of Absorption Spectrophotometry* (London: Hilger)
——1939 *Absorption Spectrophotometry and its Applications: Bibliography and Abstracts 1932 to 1938* (London: Hilger)
Walker R C and Lance T M C 1933 *Photoelectric Cell Applications* (London: Pitman)
Walsh J W T 1923 *The Elementary Principles of Lighting and Photometry* (London: Methuen)
——1926 *Photometry* (London: Constable)
——1947 *Text-Book of Illuminating Engineering: Intermediate Level* (London)
Waterfield R L 1938 *A Hundred Years of Astronomy* (London: Duckworth)
Westphal J 1987 *Colour: a Philosophical Introduction* (London: Blackwell)
Whitley R (ed) 1974 *Social Processes of Scientific Development* (London: Routledge and Kegan Paul)
Whitmell C T 1888 *Colour: an Elementary Treatise* (London: Lewis)
Wickenden W E 1910 *Illumination and Photometry* (New York: McGraw-Hill)
Wilkes J D 1973 *Power and Pedagogy: the National Electric Light Association and Public Education, 1919–1928* (unpublished PhD thesis, University of Tennessee)
Williams M E W 1994 *The Precision Makers: a History of the Instruments Industry in Britain and France, 1870–1939* (London: Routledge)
Williams T I 1983 *A History of the British Gas Industry* (Oxford: Oxford University Press)
Williams W M 1870 *The Fuel of the Sun* (London: Simpkin, Marshall)
Wise G 1985 *Willis R. Whitney, General Electric, and the Origins of US Industrial Research* (New York: Columbia University Press)
Wolfe W L (ed) 1965 *Handbook of Military Infrared Technology* (Washington, DC: Office of Naval Research)
Wolfe W L and Zissis G J (eds) 1978 *The Infrared Handbook* (Washington, DC: Office of Naval Research)
Woolf H (ed) 1962 *Quantification: a History of the Meaning of Measurement in the Natural and Social Sciences* (Indianapolis: Bobbs-Merrill)
Wright W D 1944 *The Measurement of Colour* (London: Hilger and Watts)
——1949 *Photometry and the Eye* (London: Hatton)
Yearley S 1988 *Science, Technology, and Social Change* (London: Unwin Hyman)
Zachary G P 1999 *Endless Frontier: Vannevar Bush, Engineer of the American Century* (Chicago, IL: MIT Press)
Zajonc A 1993 *Catching the Light: the Entwined History of Light and Mind* (Oxford: Oxford University Press)

Index